Theory of Electromagnetics

THEORY OF ELECTROMAGNETICS

Dr. Yudhishthir Pandey
Assistant Professor
Rajkiya Engineering College
Ambedkar Nagar
UP, India

Manakin PRESS

CRC Press
Taylor & Francis Group
Boca Raton London New York

CRC Press is an imprint of the
Taylor & Francis Group, an **informa** business

First published 2025
by CRC Press
4 Park Square, Milton Park, Abingdon, Oxon, OX14 4RN

and by CRC Press
2385 NW Executive Center Drive, Suite 320, Boca Raton FL 33431

CRC Press is an imprint of Informa UK Limited

British Library Cataloguing-in-Publication Data
A catalogue record for this book is available from the British Library

ISBN: 9781041023180 (hbk)
ISBN: 9781041023197 (pbk)
ISBN: 9781003618713 (ebk)

DOI: 10.1201/9781003618713

Typeset in Times New Roman
by Manakin Press, Delhi

Manakin
PRESS

PREFACE

The book titled "Theory of Electromagnetics" is designed to cover a comprehensive one-semester course on electromagnetics.

Below is an overview of the major topics covered in the book:

1. Fundamentals of Vector Theory
2. Electrostatic Field and Energy
3. Conductors, Dielectrics, and Capacitance
4. Poisson's and Laplace's Equations and their Applications
5. Steady Magnetic Field
6. Time Varying Fields and Maxwell's Equations
7. Uniform Plane Waves
8. Reflection, Refraction, and Dispersion of Plane Waves
9. Transmission Lines

These topics collectively form a robust framework for understanding the fundamental principles and applications of electromagnetics, making the book a valuable resource for students and professionals alike. Each section delves into the theoretical aspects and practical implications, ensuring a well-rounded grasp of electromagnetics.

Author

CONTENTS

UNIT-I

UNIT-II

UNIT-IV

UNIT-V

UNIT - I

1. Fundamentals of Vector Theory
2. Electrostatic Field and Energy

1

Fundamentals of Vector Theory

1.1 SCALAR AND VECTOR QUANTITIES

Scalar quantities are represented only by magnitude. The examples scalar quantities are mass, time, length, temperature, energy, work, pressure, specific heat, electric charge, current, frequency, potential etc. The scalar quantity can be completely defined by a number and a unit. The statement that length of the rod is 30 cm is complete information.

Vector quantities are represented by magnitude as well as direction. Examples of vector quantities are velocity, acceleration, torque, momentum, impulse, amplitude, wavelength etc. To express a vector quantity completely we must specify the direction along with the magnitude. The statement that the college is situated at a distance of 5 km from my residence is incomplete until the direction is specify. As shown in Fig. (1), a vector is denoted $\overrightarrow{OA}$ *i.e.*, with an arrow over *OA*. The point *O* is known as origin while *A* as the terminus. The magnitude of the vector is given by length *OA*. Vectors are denoted by single letter in bold faced as **A** or by putting an arrow on letter as **A**. In the book, we shall use the faced letters to denote the vectors.

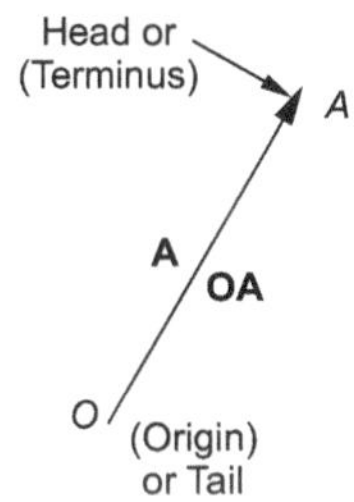

Fig. (1)

1.2 SUMMATION OF VECTORS

Here we shall illustrate the method of addition of two vectors **A** and **B** of magnitudes *A* and *B* (see Fig. 2).

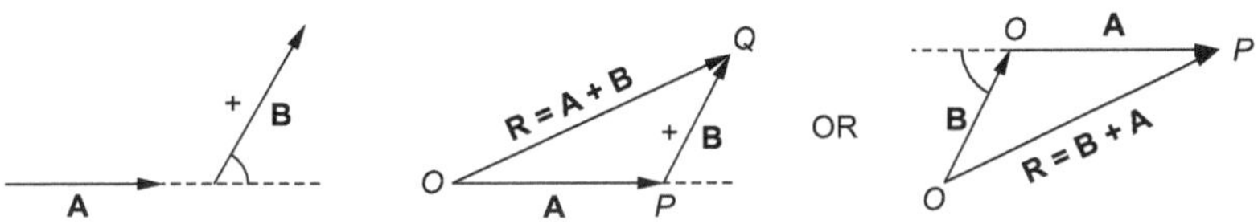

Fig. (2)

First of all we draw a line *OP* of magnitude *A* and put an arrow at *P*. Now **OP** represents vector **A.** From the head of **A** *i.e.*, from *P*, we draw a vector **B**. We join *OQ* *i.e.*, a line is drawn from tail of **A** to head of **B**. The vector **OQ** represents the resultant of two vectors. Symbolically, the addition of two vectors can be represented as

$$\mathbf{R} = \mathbf{A} + \mathbf{B}$$

We can also draw vector **B** first and then from the head of vector **B**, we draw vector **A.** Now by joining the tail of **B** to the head of vector **A**, we get the sum of two vectors. In this case,

$$\mathbf{R} = \mathbf{B} + \mathbf{A}$$

It should be remembered that

The vector sum is commutative, $\mathbf{A} + \mathbf{B} = \mathbf{B} + \mathbf{A}$

Let **A** and **B** be two vectors acting at a point O making an angle θ as shown in Fig. (3).

Now we set the vector **B** at the end of vector **A** and join OQ. Then $\boldsymbol{OQ}$ (vector **R**) represents the sum of two vectors **A** and **B** both in magnitude and direction. From figure

$$|\mathbf{R}| = |\mathbf{A} + \mathbf{B}|$$

$$= \sqrt{[A^2 + B^2 - 2AB\cos(180 - \theta)]}$$

$$= \sqrt{\left[A^2 + B^2 + 2AB\cos\theta\right]} \qquad ...(1)$$

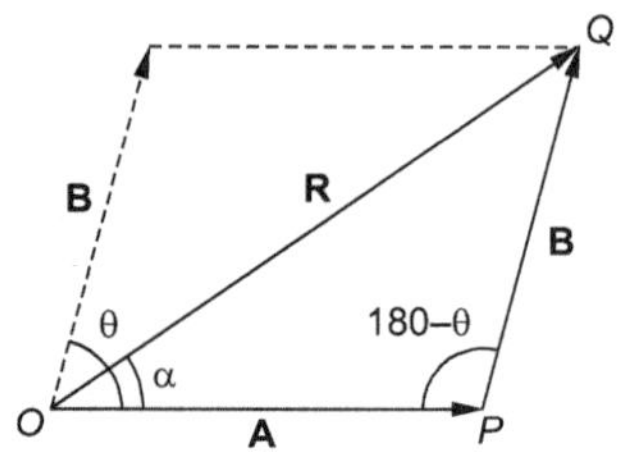

Fig. (3)

Eq. (1) gives the magnitude of vector **R**.

Let α be the angle between **A** and **R**, then

$$\tan\alpha = \frac{A\sin\theta}{A + B\cos\theta} \qquad ...(2)$$

Summation of more than two Vectors

Suppose we want to add three vectors **A, B** and **C**. This may be done with the help of ***Polygon law of vector addition***. According to this law, if a vector polygon be drawn, placing the tail-end of each succeeding vector at the head or arrow-end of the preceding one, (see Fig. 4), their resultant **R** is drawn from tail-end of the first to the head or arrow-end of the last.

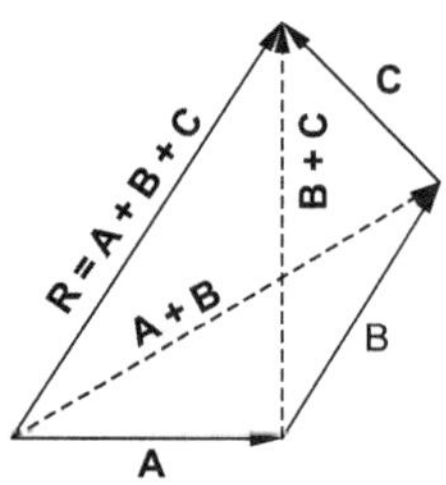

Fig. (4)

Here, first of all we obtain the resultant of two vectors **A** and **B** *i.e.*, (**A** + **B**) by triangle law and then the resultant **R** of (**A** + **B**) and **C** by the same law. Thus

$$\mathbf{R} = (\mathbf{A} + \mathbf{B}) + \mathbf{C}$$

Or

first of all we obtain the resultant vectors **B** and **C** *i.e.*, (**B** + **C**) and then the resultant **R** of (**B** + **C**) and **A** by triangle law. Thus

$$\mathbf{R} = (\mathbf{B} + \mathbf{C}) + \mathbf{A} = \mathbf{A} + (\mathbf{B} + \mathbf{C})$$

$$\therefore \qquad \mathbf{R} = \mathbf{A} + \mathbf{B} + \mathbf{C} = (\mathbf{A} + \mathbf{B}) + \mathbf{C} = \mathbf{A} + (\mathbf{B} + \mathbf{C})$$

Hence, the vector sum is associative.

So, the vector addition has the following two important properties:

(*a*) The vector sum is commutative, *i.e.*,

$$\mathbf{A} + \mathbf{B} = \mathbf{B} + \mathbf{A}$$

This law is commutative of vector addition.

(*b*) The vector sum is association, *i.e.*,

$$\mathbf{A} + \mathbf{B} + \mathbf{C} = (\mathbf{A} + \mathbf{B}) + \mathbf{C} = \mathbf{A} + (\mathbf{B} + \mathbf{C})$$

This law is known as associative law of vector addition.

1.3 SUBTRACTION OF VECTORS

Here we shall subtract a vector **B** from the vector **A.** The subtraction of the vector **B** from **A** is equivalent to the addition of (– **B**) with **A**, *i.e.*,

$$\mathbf{A} - \mathbf{B} = \mathbf{A} + (-\mathbf{B})$$

The negative sign means a vector of equal magnitude but opposite direction. The procedure is illustrated in Fig. (5)

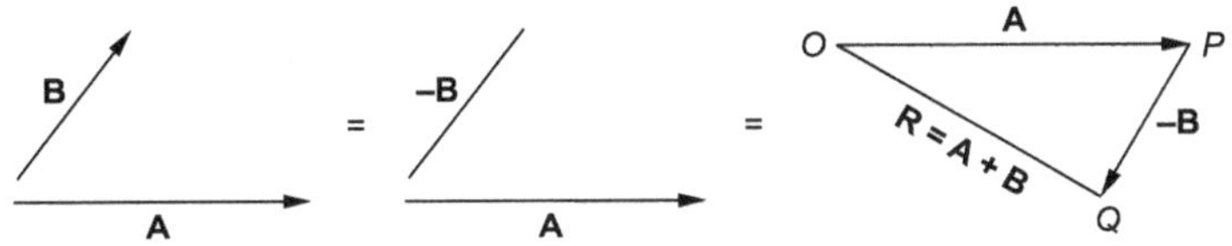

Fig. (5)

Here we draw a line *OP* to represent the vector **A** and then from the head **A** another line *PQ* is drawn to represent vector (– **B**). Further the tail of **A** is joined to head of (– **B**). Now the line *OQ* represents a vector **R** which is difference of two vectors **A** and **B**.

When **A** and **B** are in the same direction, then **A** – **B** is a vector whose length is (**A** – **B**) and the direction is along the longer vector. If the two vectors **A** and **B** have the same length and direction, then difference is a null vector. Further

$$(\mathbf{A} - \mathbf{B}) = -(\mathbf{B} - \mathbf{A})$$

Let **A** and **B** be two vectors acting at a point *O* at an angle θ as shown in Fig. (6).

To obtain the subtraction of two vectors, we set vector **B** at the end of vector **A**. So **PQ** represents vectors – **B**. Then we join *O* to *Q*. Hence **OQ** gives (**A** – **B**) both in magnitude and direction.

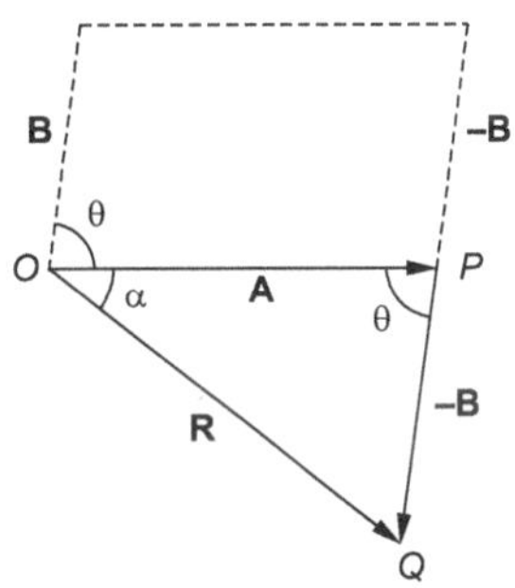

Fig. (6)

$$\therefore \qquad |\mathbf{R}| = \sqrt{[A^2 + B^2 - 2AB\cos\theta]}$$

If **OQ** makes an angle α with **OP** then

$$\tan\alpha = \frac{B\sin\theta}{A - B\cos\theta}$$

Subtraction does not obey commutative or associative law.

1.4 COMPONENT OF VECTORS

Consider the case of vector **A** with its tail at the origin of rectangular coordinate system as shown in Fig. (7). We can resolve this vector **A** in two components. The procedure is as follows:

We draw two perpendiculars from the head of **A** to *X* and *Y* axes respectively. The intercepts on these axes are called the scalar components A_x and A_y. The process is called as resolution of vectors

If resolved vectors of **A** are $\mathbf{A}_x$ and $\mathbf{A}_y$, then $\mathbf{A} = \mathbf{A}_x + \mathbf{A}_y$.

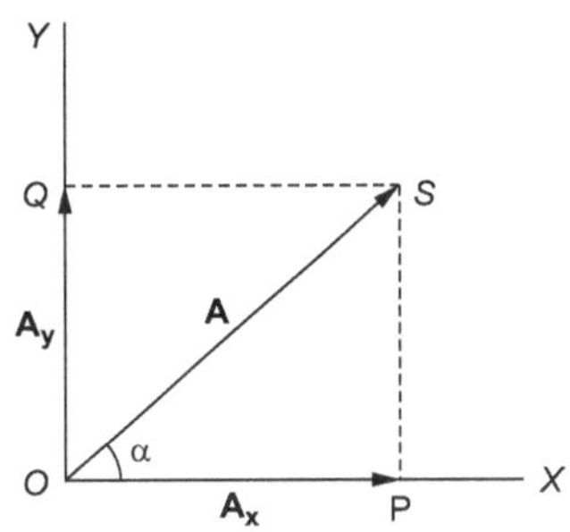

Fig. (7)

Let $\hat{\mathbf{a}}_x$ and $\hat{\mathbf{a}}_y$ be the unit vectors along the axis of X and Y respectively. Then if A_x and A_y be the scalar magnitudes of $\mathbf{A}_x$ and $\mathbf{A}_y$, we write

$$\mathbf{A}_x = \hat{\mathbf{a}}_x A_x \quad \text{and} \quad \mathbf{A}_y = \hat{\mathbf{a}}_y A_y$$

$$\mathbf{A} = \hat{\mathbf{a}}_x A_x + \hat{\mathbf{a}}_y A_y$$

If α be the angle made by $\mathbf{A}$ with positive X-axis, then

$$A_x = A \cos \alpha \quad \text{and} \quad A_y = A \sin \alpha$$

$$\therefore \qquad A = \sqrt{\{A_x^2 + A_y^2\}} = \sqrt{(A^2 \cos^2 \alpha \; A^2 \sin^2 \alpha)}$$

and

$$\tan \alpha = \frac{A_y}{A_x} \quad \text{or} \quad \alpha = \tan^{-1}\left(\frac{A_y}{A_x}\right)$$

In three dimensions $\mathbf{A} = \hat{\mathbf{a}}_x A_x + \hat{\mathbf{a}}_y A_y + \hat{\mathbf{a}}_z A_z$.

Summation of Vectors (Component Method)

Let us add two vectors **A** and **B** by component method. The vectors in terms of components in three dimensions can expressed as

$$\mathbf{A} = \hat{\mathbf{a}}_x A_x + \hat{\mathbf{a}}_y A_y + \hat{\mathbf{a}}_z A_z$$

and

$$\mathbf{B} = \hat{\mathbf{a}}_x B_x + \hat{\mathbf{a}}_y B_y + \hat{\mathbf{a}}_z B_z$$

$$\therefore \qquad \mathbf{A} + \mathbf{B} = \hat{\mathbf{a}}_x (A_x + B_x) + \hat{\mathbf{a}}_y (A_y + B_y) + \hat{\mathbf{a}}_z (A_z + B_z)$$

This method can be applied to any number of vectors.

The magnitude is given by

$$|\mathbf{A} + \mathbf{B}| = [(A_x + B_x)^2 + (A_y + B_y)^2 + (A_z + B_z)^2]^{1/2}$$

Similarly, for subtraction of two vectors, we have

$$\mathbf{A} - \mathbf{B} = \hat{\mathbf{a}}_x (A_x - B_x) + \hat{\mathbf{a}}_y (A_y - B_y) + \hat{\mathbf{a}}_z (A_z - B_z)$$

and

$$|\mathbf{A} - \mathbf{B}| = [(A_x - B_x)^2 + (A_y - B_y)^2 + (A_z - B_z)^2]^{1/2}$$

1.5 MULTIPLICATION OF A VECTOR BY A SCALAR

When a vector **A** is multiplied by scalar or a pure number n, then the resultant vector may be written as

$$\mathbf{R} = n\,\mathbf{A}$$

The magnitude of vector **R** being n times that of vector **A** and the direction is the same as that of **A** when n is positive but opposite when n is negative.

When the scalar n is a physical quantity with a unit, then the unit of **R** will be different from that of **A**. For example, when mass (scalar) is multiplied by velocity (vector) then the product $m\,\mathbf{v}$ represents the momentum, a different vector quantity.

1.6 SCALAR PRODUCT OR DOT PRODUCT OF TWO VECTORS

The scalar product or dot product of two vectors* A *and* B *is defined as the product of the magnitudes of two vectors (i.e.,* A *and* B*) and the cosine of the angle between them.

$$\mathbf{A} \cdot \mathbf{B} = AB \cos \theta.$$

where θ is the angle between two vectors

Note: Angle between two vectors can be determined by joining two vectors either head to head or tail or tail only).

$$\therefore \qquad \cos\theta = \frac{\mathbf{A}\cdot\mathbf{B}}{AB}$$

The product is scalar quantity. The product may be positive or negative depending upon the angle θ. The product is negative when θ is between π/2 and 3π/2.

$$\mathbf{A}\cdot\mathbf{B} = AB\cos\theta = A(B\cos\theta) = (A\cos\theta)B$$

Hence, the scalar product of two vectors will be equal to the product of the length of one vector and the length of the component of the second vector along the first vector direction.

Properties of Scalar Product

(*i*) The scalar product is commutative, *i.e.*,

$$\mathbf{A}\cdot\mathbf{B} = \mathbf{B}\cdot\mathbf{A}$$

(*ii*) The scalar product follows the distributive law, *i.e.*,

$$\mathbf{A}\cdot(\mathbf{B}+\mathbf{C}) = \mathbf{A}\cdot\mathbf{B}+\mathbf{A}\cdot\mathbf{C}$$

(*iii*) The scalar product of a vector by itself is equal to the square of the scalar magnitude, *i.e.*,

$$\mathbf{A}\cdot\mathbf{A} = \mathbf{A}^2$$

(*iv*) The scalar product of two vectors vanishes when the vectors are at right angles, *i.e.*,

$$\mathbf{A}\cdot\mathbf{B} = AB\cos 90^\circ = 0 \qquad (\because \cos 90^\circ = 0)$$

Thus $\mathbf{A}\cdot\mathbf{B} = 0$ means that vectors **A** and **B** are perpendicular.

(*v*) When the vectors are parallel, the scalar product of two vectors is equal to the product of their scalar magnitudes.

$$\mathbf{A}\cdot\mathbf{B} = AB\cos\theta = AB \qquad (\because \cos\theta = 1)$$

(*vi*) The scalar product of unit orthogonal vectors $\hat{\mathbf{a}}_x$, $\hat{\mathbf{a}}_y$ and $\hat{\mathbf{a}}_z$ have the following relations.

$$\hat{\mathbf{a}}_x\cdot\hat{\mathbf{a}}_x = \hat{\mathbf{a}}_y\cdot\hat{\mathbf{a}}_y = \hat{\mathbf{a}}_z\cdot\hat{\mathbf{a}}_z = 1$$

$$\hat{\mathbf{a}}_x\cdot\hat{\mathbf{a}}_y = \hat{\mathbf{a}}_y\cdot\hat{\mathbf{a}}_z = \hat{\mathbf{a}}_z\cdot\hat{\mathbf{a}}_x = 0$$

(*vii*) The scalar product of two vectors is equal to the sum of the products of their corresponding *x*, *y* and *z* components.

If $\qquad \mathbf{A} = \hat{\mathbf{a}}_x A_x + \hat{\mathbf{a}}_y A_y + \hat{\mathbf{a}}_z A_z$

and $\qquad \mathbf{B} = \hat{\mathbf{a}}_x B_x + \hat{\mathbf{a}}_y B_y + \hat{\mathbf{a}}_z B_z$

then $\qquad \mathbf{A}\cdot\mathbf{B} = A_x B_x + A_y B_y + A_z B_z$

Example: (1) If **A** = **B**, then

$$\mathbf{A}\cdot\mathbf{B} = \mathbf{A}\cdot\mathbf{A} = \mathbf{A}^2 = |\mathbf{A}|^2$$

Thus the vector dotted with itself gives the square of its magnitude. This provides a method of finding the magnitude of a vector.

(2) If $\mathbf{A}\cdot\mathbf{B} = 0$ but $\mathbf{A} \neq 0$ and $\mathbf{B} \neq 0$, then the two vectors *i.e.*, **A** and **B** are *perpendicular* to each other or the two vectors are *orthogonal*.

Applications of Scalar Product

(*i*) The **workdone** ***W*** by a force **F** acting on a body is given by

$$W = \mathbf{F}\cdot\mathbf{s}$$

where **s** is displacement produced in the body by the force.

(*ii*) The potential difference V between two point is given by

$$V = \int \mathbf{E} \cdot dl$$

where **E** is electric field strength between two points.

(*iii*) For a conservative field **F**

$$\int \mathbf{F} \cdot dl = 0$$

(*iv*) Divergence of field is a result of dot product of two vectors $\vec{\nabla}$ and field **F**.

$$\text{Divergence} = \vec{\nabla} \cdot \mathbf{F}.$$

1.7 VECTOR PRODUCT OF TWO VECTORS

The vector product or cross product of two vectors is defined as a vector having magnitude equal to the product of the magnitude of two vectors and the sine of the angle between them. The direction being perpendicular to the plane containing the two vectors.

If **A** and **B** are two vectors then their vector product is $\mathbf{A} \times \mathbf{B}$ which can expressed as

$$\mathbf{A} \times \mathbf{B} = AB \sin\theta\, \hat{\mathbf{n}}$$

where A and B are magnitudes of **A** and **B**, θ is the angle between them and $\hat{\mathbf{n}}$ is a unit vector perpendicular to the plane **A** and **B** (see Fig. 8). The direction of $\mathbf{A} \times \mathbf{B}$ is given by the right hand rule.

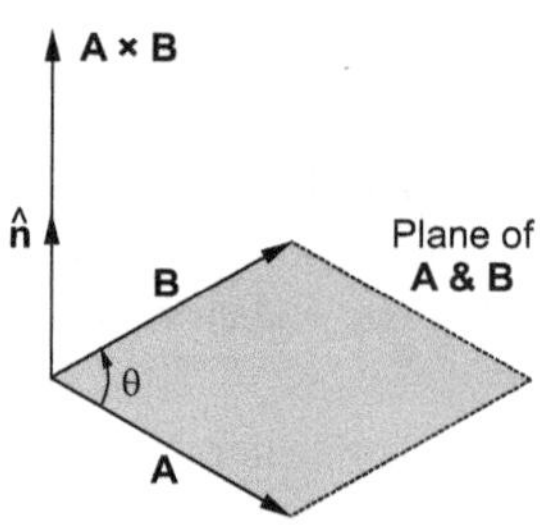

Fig. (8)

According to this rule, if a **right handed** screw is imagined to be placed at the common tail point of the two vectors [whose vector product is to be obtained Fig. 9*a*] and is given a rotation from first vector **A** to second vector **B**, then the advancement of the screw gives the direction of

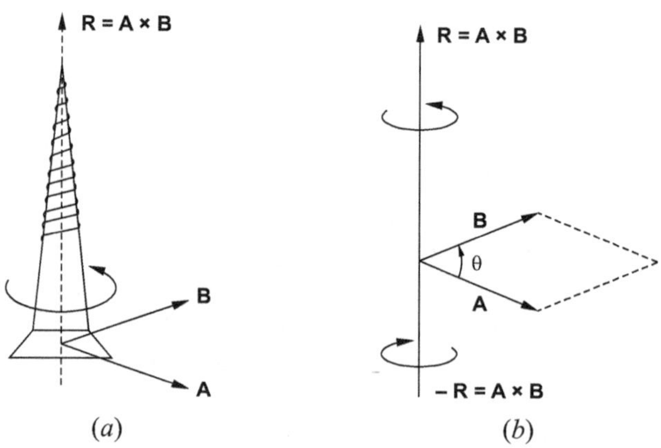

Fig. (9)

resultant vector $\mathbf{R} = \mathbf{A} \times \mathbf{B}$. It is obvious from Fig. (9*b*), that if the screw is rotated from **B** to **A**, then the direction of advancement of the screw will be just opposite. Therefore

$$\mathbf{B} \times \mathbf{A} = \text{vector having magnitude of } \mathbf{R} \text{ but opposite in direction} = -\mathbf{R}.$$

Thus $$\mathbf{A} \times \mathbf{B} \neq \mathbf{B} \times \mathbf{A}$$

Properties of Vector Product

(*i*) The vector product is not commutative

$$\mathbf{A} \times \mathbf{B} \neq \mathbf{B} \times \mathbf{A}$$

(*ii*) The vector product is distributive

$$\mathbf{A} \times (\mathbf{B} + \mathbf{C}) = \mathbf{A} \times \mathbf{B} + \mathbf{A} \times \mathbf{C}$$

(*iii*) The vector product of two parallel vectors is a null vector *i.e.*,

$$\mathbf{A} \times \mathbf{B} = AB \sin\theta\, \hat{\mathbf{n}} = AB \sin 0\, \hat{\mathbf{n}} = 0$$

(*iv*) The vector product of a vector by itself is a null vector (zero)

$$\mathbf{A} \times \mathbf{A} = AA \sin\theta\, \hat{\mathbf{n}} = AA \sin 0\, \hat{\mathbf{n}} = 0$$

(*v*) The magnitude of the vector product of two vectors mutually at right angle is equal to the product of the magnitude of the vectors.

$$\mathbf{A} \times \mathbf{B} = AB \sin\theta\, \hat{\mathbf{n}} = A\,B \sin 90^\circ\, \hat{\mathbf{n}} = A\,B\, \hat{\mathbf{n}}$$

(*vi*) The vector product of unit orthogonal vectors, $\hat{\mathbf{a}}_x, \hat{\mathbf{a}}_y$ and $\hat{\mathbf{a}}_z$ have the following relations:

$$\hat{\mathbf{a}}_x \times \hat{\mathbf{a}}_x = \hat{\mathbf{a}}_y \times \hat{\mathbf{a}}_y = \hat{\mathbf{a}}_z \times \hat{\mathbf{a}}_z = 0$$

$$\hat{\mathbf{a}}_x \times \hat{\mathbf{a}}_y = -\hat{\mathbf{a}}_y \times \hat{\mathbf{a}}_x = \hat{\mathbf{a}}_z,$$

$$\hat{\mathbf{a}}_y \times \hat{\mathbf{a}}_z = -\hat{\mathbf{a}}_z \times \hat{\mathbf{a}}_y = \hat{\mathbf{a}}_x,$$

and

$$\hat{\mathbf{a}}_z \times \hat{\mathbf{a}}_x = -\hat{\mathbf{a}}_x \times \hat{\mathbf{a}}_z = \hat{\mathbf{a}}_y$$

(*vii*) The vector product of two vectors in terms of their *x*, *y* and *z* components can be expressed in the form of determinant. **(Important Method)**

If

$$\mathbf{A} = \hat{\mathbf{a}}_x A_x + \hat{\mathbf{a}}_y A_y + \hat{\mathbf{a}}_z A_z$$

and

$$\mathbf{B} = \hat{\mathbf{a}}_x B_x + \hat{\mathbf{a}}_y B_y + \hat{\mathbf{a}}_z B_z$$

then

$$\mathbf{A} \times \mathbf{B} = \begin{vmatrix} \hat{\mathbf{a}}_x & \hat{\mathbf{a}}_y & \hat{\mathbf{a}}_z \\ A_x & A_y & A_z \\ B_x & B_y & B_z \end{vmatrix}$$

(*viii*) The vector product of two vectors **A** and **B** gives the area of the parallelogram formed by the vectors. Let two vectors **A** and **B** are inclined at an angle θ as shown in Fig. (10). *OPMQ* is the parallelogram formed by the vectors **A** and **B**. *QN* is the perpendicular drawn from *Q* on *OP*.

The area of parallelogram

$$= OP \times \text{perpendicular distance } QN$$

$$= OP \times OQ \sin\theta$$

$$= AB \sin\theta$$

$$= \mathbf{A} \times \mathbf{B}.$$

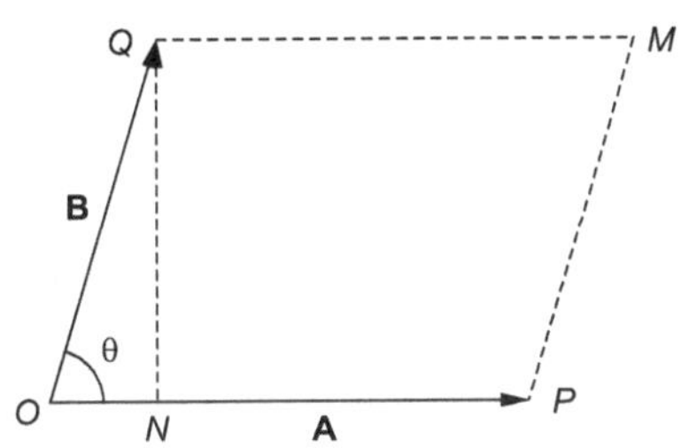

Fig. (10)

Applications of Vector Product

(*i*) Let a body is moving along a circular path with angular velocity ω. If **v** be the linear velocity of the body, then the relation between angular velocity ω and linear velocity *v* is given by

$$\mathbf{v} = \mathbf{r} \times \omega$$

where **r** is the radius of circular path.

(*ii*) In a rotatory motion, the torque acting on the body is given by

$$\tau = \mathbf{r} \times \mathbf{F}$$

where **F** is the force acting on a point in the body and **r** is the radius vector of the point with respect to rotation axis.

(*iii*) The angular momentum **L** and linear momentum **p** are related as

$$\mathbf{L} = \mathbf{r} \times \mathbf{p}$$

where **r** is the radius vector

(*iv*) When a charged particle with charge q moves in a magnetic field of induction **B** then the force **F** acting on the particle is given by

$$\mathbf{F} = q\,(\mathbf{v} \times \mathbf{B})$$

where v is the velocity of charged particle in magnetic field.

(*v*) The Lorentz force **F** acting on a charged particle q moving in electromagnetic field of magnetic induction **B** and electric field **E** is given by

$$\mathbf{F} = q\,[\mathbf{E} + \mathbf{v} \times \mathbf{B}]$$

1.8 SCALAR TRIPLE PRODUCT

The scalar product of a vector **A** with vector product of two other vectors **B** and **C** is termed as scalar triple product. This is written as

$$\mathbf{A} \cdot (\mathbf{B} \times \mathbf{C})$$

The scalar product **A** · (**B** × **C**) gives a scalar. This gives the volume of the parallelopiped formed by vectors **A**, **B** and **C**.

Expressing vectors in their components forms, we have

$$\mathbf{A} = \hat{\mathbf{a}}_x A_x + \hat{\mathbf{a}}_y A_y + \hat{\mathbf{a}}_z A_z$$

$$\mathbf{B} = \hat{\mathbf{a}}_x B_x + \hat{\mathbf{a}}_y B_y + \hat{\mathbf{a}}_z B_z$$

and

$$\mathbf{C} = \hat{\mathbf{a}}_x + C_x + \hat{\mathbf{a}}_y C_y + \hat{\mathbf{a}}_z C_z$$

$$\therefore \quad \mathbf{A} \cdot (\mathbf{B} \times \mathbf{C}) = A_x (B_y C_z - B_z C_y) + A_y (B_z C_x - B_x C_z) + A_z (B_x C_y - B_y C_x)$$

In compact form, the above expression can expressed as

$$\mathbf{A} \cdot (\mathbf{B} \times \mathbf{C}) = \begin{vmatrix} A_x & A_y & A_z \\ B_x & B_y & B_z \\ C_x & C_y & C_z \end{vmatrix}$$

Properties of Scalar Triple Product

(*i*) If the scalar triple product is zero, then the three vectors are coplanar.

(*ii*) The following cyclic relation hold good in scalar triple product

$$\mathbf{A} \cdot (\mathbf{B} \times \mathbf{C}) = \mathbf{B} \cdot (\mathbf{C} \times \mathbf{A}) = \mathbf{C} \cdot (\mathbf{A} \times \mathbf{B})$$

(*iii*) The dot and cross can be interchanged in a scalar triple product

$$\mathbf{A} \cdot (\mathbf{B} \times \mathbf{C}) = (\mathbf{A} \times \mathbf{B}) \cdot \mathrm{C}$$

(*iv*) In scalar triple product

$$\mathbf{A} \cdot (\mathbf{B} \times \mathbf{C}) = -\mathbf{A} \cdot (\mathbf{C} \times \mathbf{B})$$

1.9 VECTOR TRIPLE PRODUCT

The vector product of a vector **A** with the vector product of two other vectors **B** and **C** is termed as vector triple product. This is written as

$$\mathbf{A} \times (\mathbf{B} \times \mathbf{C})$$

By the property of vectors product, **A** × (**B** × **C**) is a vector perpendicular to **A** and also to (**B** × **C**). Further (**B** × **C**) is perpendicular to plane of **B** and **C.** Hence **A** × (**B** × **C**) must be in the plane of **B** and **C**.

$$\mathbf{A} \times (\mathbf{B} \times \mathbf{C}) = \begin{vmatrix} \hat{\mathbf{a}}_x & \hat{\mathbf{a}}_y & \hat{\mathbf{a}}_z \\ A_x & A_y & A_z \\ (B_y C_z - B_z C_y) & (B_z C_x - B_x C_z) & (B_x C_y - B_y C_x) \end{vmatrix}$$

Properties of Vector Triple Product

(*i*) $\mathbf{A} \times (\mathbf{B} \times \mathbf{C}) \neq (\mathbf{A} \times \mathbf{B}) \times \mathbf{C}$

The associative law does not apply to vector products.

(*ii*) $\mathbf{A} \times (\mathbf{B} \times \mathbf{C}) = \mathbf{B}\,(\mathbf{A} \cdot \mathrm{C}) - \mathrm{C}\,(\mathbf{A} \cdot \mathbf{B})$

(*iii*) $\mathbf{A} \times (\mathbf{B} \times \mathbf{C}) = -\,(\mathbf{B} \times \mathbf{C}) \times \mathrm{A}$

(*iv*) $\mathbf{A} \times (\mathbf{B} \times \mathbf{C}) + \mathbf{B} \times (\mathbf{C} \times \mathbf{A}) + \mathbf{C} \times (\mathbf{A} \times \mathbf{C}) = \mathbf{0}$

1.10 DIFFERENT COORDINATE SYSTEMS

To describe a vector accurately, a number of coordinate systems are used. In each coordinate system, the vector can be specified by its length, directions, angles and components. The coordinate systems may be *orthogonal* or *non-orthogonal*. The orthogonal systems are those systems in which the coordinates are mutually perpendicular to each other. It is easier to work with orthogonal systems. Therefore, orthogonal systems are generally used. Following are some of commonly used coordinate systems:

1. Cartesian or rectangular coordinate system.
2. Cylindrical coordinate system
3. Spherical coordinate system.

We shall discuss these coordinate systems one by one.

1.11 CARTESIAN OR RECTANGULAR COORDINATE SYSTEM

In cartesian coordinate system, there are three axes which are mutually perpendicular to each other. The axes are denoted as *X*, *Y* and *Z* axes. The three axes meet at a point known as origin. It is customary to work either with a right handed system or a left handed system. These systems are shown in Fig. (11*a*) and Fig. (11*b*) respectively.

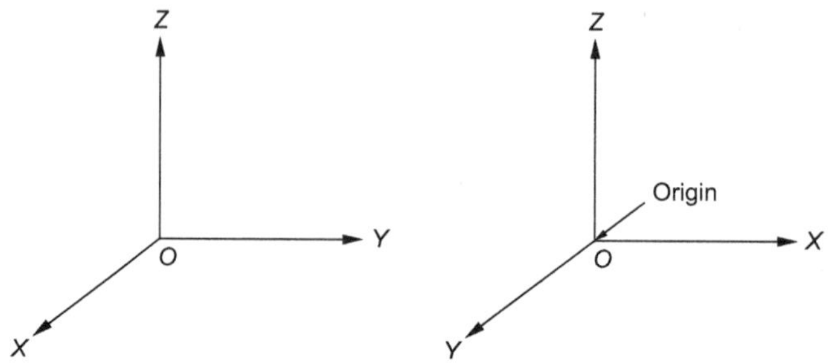

Fig. 11 (a) Right handed system (b) Left handed system

In a right handed coordinate system, the rotation of *X*-axis into *Y*-axis would cause a right handed screw to progress in the direction of *Z*-axis. To understand a right handed system, we stretch the thumb and fingers of right hand in mutually perpendicular directions. Now, the thumb, forefinger and middle finger will be identical to *X*, *Y* and *Z* axes. This system is mostly used.

In cartesian coordinate system, there are three planes as shown in Fig. (12). These planes are mutually perpendicular to each other.

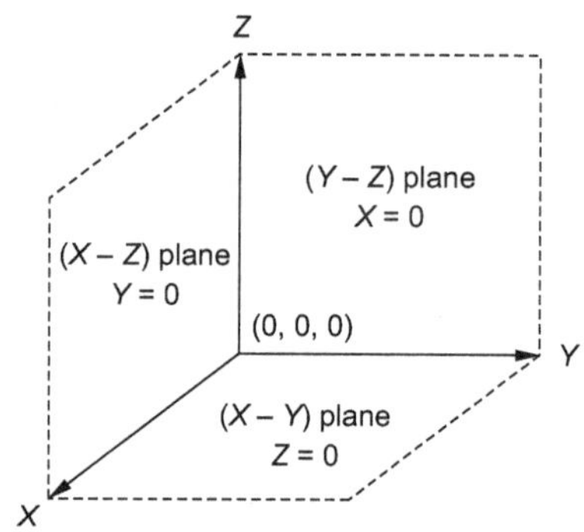

Fig. (12) Showing different planes in Cartesian coordinate

The three planes are:

(*i*) (*Y*-*Z*) plane for which $X = 0$

(*ii*) (*X*-*Z*) plane for which $Y = 0$

(*iii*) (*X*-*Y*) plane for which $Z = 0$

If the above planes have some constant value instead of zero, then the planes are called $X =$ constant plane, $Y =$ constant plane and $Z =$ constant plane as shown in Fig. (13).

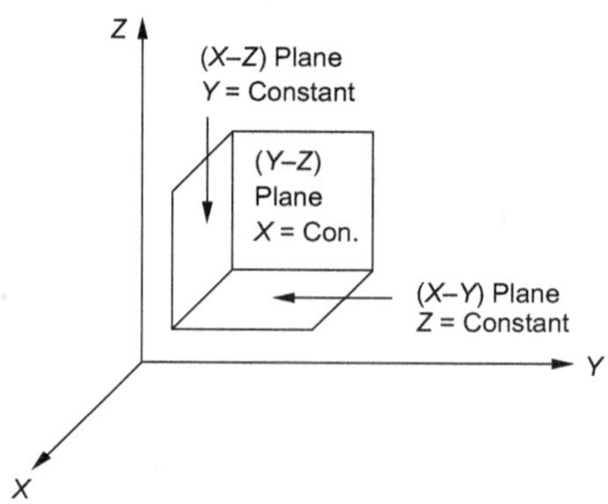

Fig. (13) Showing different planes with X, Y and Z constants

Let us define the position if a point P in cartesian coordinate system. Suppose the position P of the point in this coordinate system be represented by coordinate (x, y, z) along the three rectangular X, Y and Z axis as shown in Fig. (14). The position vector $\mathbf{r}$ of point $P(x, y, z)$ can be represented as shown in Fig. (15).

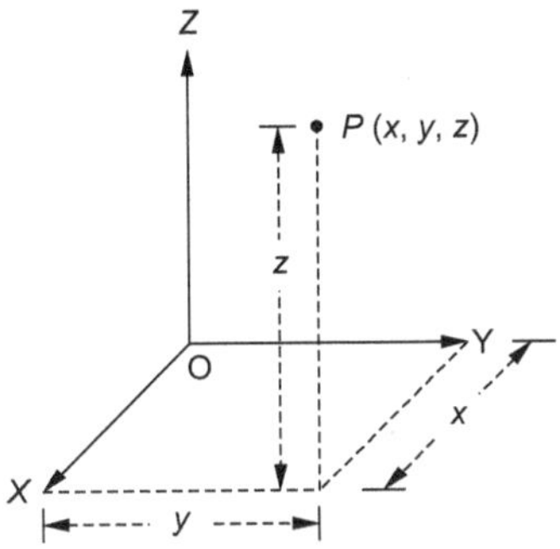

Fig. (14) Showing the position of a point P

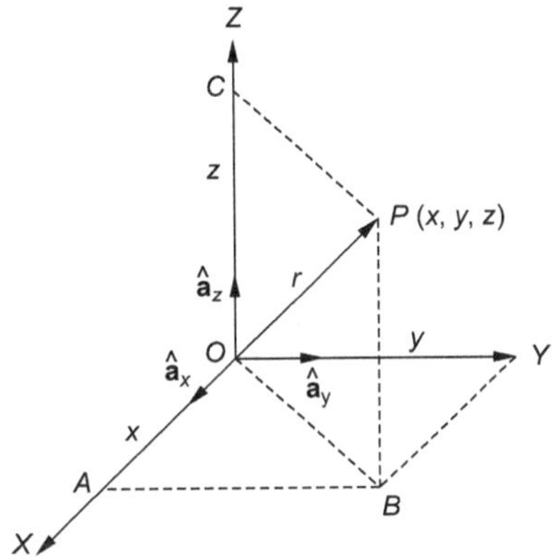

Fig. (15) Showing position vector $\mathbf{r}$ of a point $P(x, y, z)$

If $\hat{\mathbf{a}}_x$, $\hat{\mathbf{a}}_y$ and $\hat{\mathbf{a}}_z$ be the unit vectors along the three axes respectively, then

$$\mathbf{r} = \hat{\mathbf{a}}_z x + \hat{\mathbf{a}}_y y + \hat{\mathbf{a}}_z z \qquad \ldots(1)$$

where $OA = x$, $OB = y$ and $OC = z$.

Differential Length, Differential Area and Differential Volume

In order to understand the idea of differential length, differential area and differential volume, consider a point $P(x, y, z)$ in cartesian coordinate system as shown in Fig. (16). Let the coordinate of point P have been increased by small differential lengths dx, dy and dz. Now a new point P' is obtained with coordinates $(x + dx, y + dy, z + dz)$. Again we can consider three planes related with $P(x, y, z)$ and three planes with $P'(x + dx, y + dy, z + dz)$, *i.e.*, total six planes. The intersection of these six planes gives rise to a small cube having sides dx, dy and dz as shown in Fig. (16).

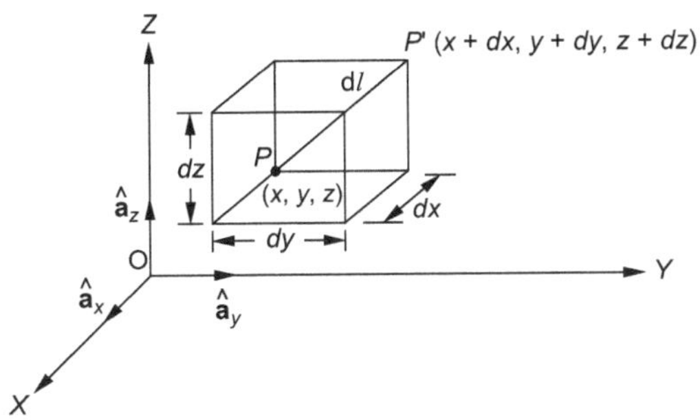

Fig. (16) Showing differential lengths of points P and P'

The differential length in vector form is given by

$$d\mathbf{l} = \hat{\mathbf{a}}_x\, dx + \hat{\mathbf{a}}_y\, dy + \hat{\mathbf{a}}_z\, dz \qquad \ldots(2)$$

Note: Length and Surface area are vector quantity. dl and ds are expressed as vectors while volume is a scalar quantity.

The magnitude of differential is given by

$$dl = \sqrt{[(dx)^2 + (dy)^2 + (dz)^2]} \quad ...(3)$$

Taking area as a vector quantity, three areas *dx dy*, *dy dz* and *dx dz* will be positive while three areas *dx dy*, *dy dz* and *dx dz* will be negative. In vector form the area, the magnitude is equal to area and whose direction is that of normal to the area. Therefore, the differential area *d***s** can be expressed as

$$d\mathbf{s} = [\pm \hat{\mathbf{a}}_x \, dy \, dz \pm \hat{\mathbf{a}}_y \, dx \, dz \pm \hat{\mathbf{a}}_z \, dx \, dz] \quad ...(4)$$

The differential volume is given by

$$dv = dx \, dy \, dz \quad ...(5)$$

1.12 CYLINDRICAL COORDINATE SYSTEM

In electromagnetic field theory, there are some specific problems of cylindrical shapes. These problems can easily be solved by using cylindrical coordinate system rather than cartesian coordinate system. The objective of the section is have understanding of cylindrical coordinate system. The cylindrical coordinates are expressed as (ρ, ϕ, z).

We have studied the cartesian coordinate system where the three surfaces are plane surfaces. The cylinder coordinate system involves a set of three mutually orthogonal surfaces where, of course, we have three surfaces. Out of the three surfaces, one is cylindrical while the other two are plane surfaces as shown in Fig. (17). Any point *P* in this system is considered lying at the intersection of three mutually perpendicular surfaces.

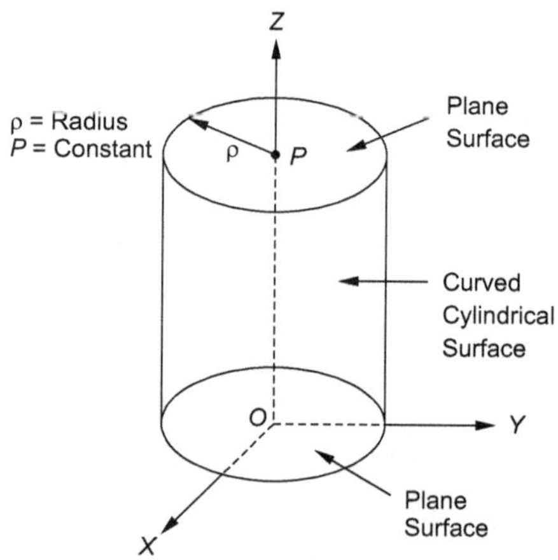

Fig. (17) Cylindrical surface along Z-axis with radius ρ

The three variables in this system are ρ, ϕ and z. Let us define these variables.

(*i*) $\rho \rightarrow$ This is the radial distance of the point *P* from *Z*-axis in fig (18). It may also be considered as the radius of the imaginary cylinder passing through point *P*. The value of ρ may vary from zero to infinity, *i.e.*, $0 < \rho < \infty$.

(*ii*) $\phi \rightarrow$ To defined ϕ, consider an imaginary plane extending out from Z-axis as shown in Fig. (18). The angle made by this plane with respect to X-axis is defined as angle ϕ.

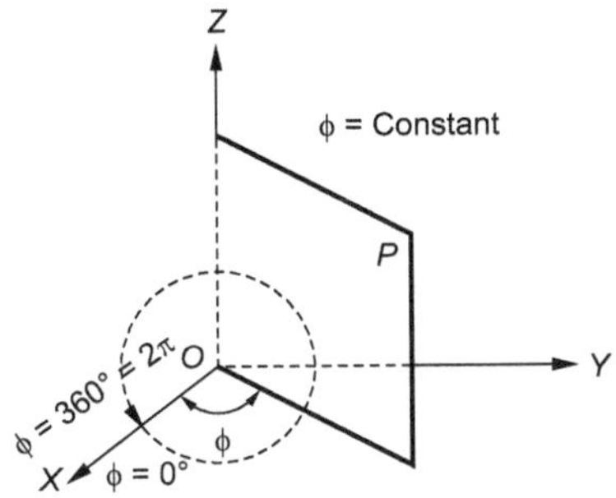

Fig. (18)

This angle can make full rotation along Z-axis, *i.e.*, the value of ϕ varies from 0 to 360°. Therefore,

$$0° \leq \phi \leq 2\pi$$

(*iii*) $z \rightarrow$ This variable is similar to z-variable is cartesian coordinate system. The value of z indicates the vertical height of any point from ground level. A plane which is parallel to XY plane at a distance z is called as z = constant plane as shown in Fig. (19).

The value of z varies from $-\infty$ to $+\infty$, *i.e.*, $-\infty < z < +\infty$.

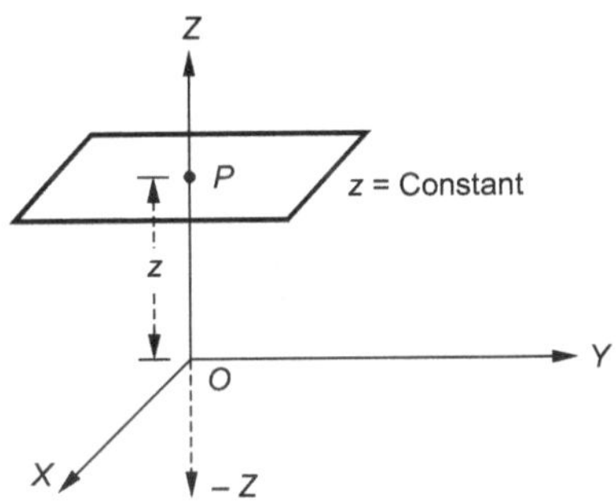

Fig. (19) Showing z = constant plane

Let us define a point $P(\rho, \phi, z)$. This is shown in Fig. (20). As shown,

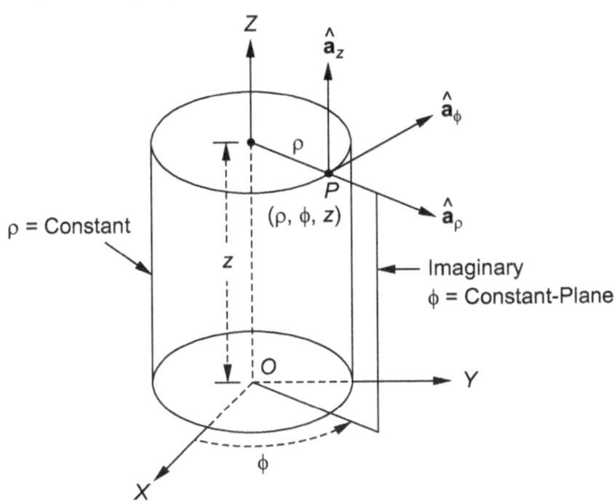

Fig. (20) Showing the position of a point $P(\rho, \phi, z)$

Therefore, P lies at the intersection of cylindrical surface having radius ρ, a plane that makes an angle ϕ with X-axis and a plane for which z is constant.

The three unit vectors are $\hat{\mathbf{a}}_\rho, \hat{\mathbf{a}}_\phi$ and $\hat{\mathbf{a}}_z$ as shown in the figure (20). These are mutually perpendicular. The description is as follow

$\hat{\mathbf{a}}_\rho$ → This unit vector is radially outward, normal to cylindrical surface at the point P under consideration.

$\hat{\mathbf{a}}_\phi$ → This unit vector is perpendicular to ϕ = constant plane. This is directed outward in the advance ϕ direction.

$\hat{\mathbf{a}}_z$ → This is perpendicular to (ρ-ϕ plane) = constant plane and is like cartesian coordinate system.

The unit vectors are mutually perpendicular and hence, follows the following relations:

$$\begin{aligned} \hat{\mathbf{a}}_\rho \cdot \hat{\mathbf{a}}_\rho &= 1, \quad & \hat{\mathbf{a}}_\phi \cdot \hat{\mathbf{a}}_\phi &= 1 \quad \text{and} \quad & \hat{\mathbf{a}}_z \cdot \hat{\mathbf{a}}_z &= 1 \\ \hat{\mathbf{a}}_\rho \cdot \hat{\mathbf{a}}_\phi &= 0, \quad & \hat{\mathbf{a}}_\phi \cdot \hat{\mathbf{a}}_z &= 0 \quad \text{and} \quad & \hat{\mathbf{a}}_z \cdot \hat{\mathbf{a}}_\rho &= 0 \\ \hat{\mathbf{a}}_\rho \times \hat{\mathbf{a}}_\phi &= \hat{\mathbf{a}}_z, \quad & \hat{\mathbf{a}}_\phi \times \hat{\mathbf{a}}_z &= \hat{\mathbf{a}}_\rho \quad \text{and} \quad & \hat{\mathbf{a}}_z \times \hat{\mathbf{a}}_\rho &= \hat{\mathbf{a}}_\phi \end{aligned} \quad \text{...(1)}$$

Any vector in cylindrical coordinate system can be represented as

$$\boxed{\mathbf{A} = \hat{\mathbf{a}}_\rho A_\rho + \hat{\mathbf{a}}_\phi A_\phi + \hat{\mathbf{a}}_z A_z} \quad \text{...(2)}$$

Differential Lengths, Differential Area and Differential Volume

Fig. (21) shown two points $P(\rho, \phi, z)$ and $Q(\rho + d\rho, \phi + d\phi, z + dz)$. Here Q is obtained by incrementing infinitesimally each coordinate from its value at P. These increments make a small box of differential length elements $d\rho$, $\rho\, d\phi$ and dz as shown in the figure.

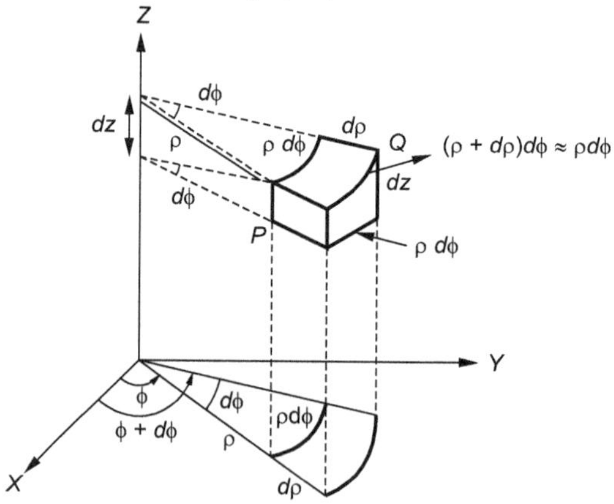

Fig. (21) Differential volume formed by incrementing the coordinates

The vector displacement $d\mathbf{l}$ from P to Q is given by

$$d\mathbf{l} = \hat{\mathbf{a}}_\rho\, d_\rho + \hat{\mathbf{a}}_\phi\, \rho\, d\phi + \hat{\mathbf{a}}_z\, dz \quad \text{...(3)}$$

Note: $(\rho + d\rho)d\phi = \rho d\phi + d\rho d\phi \approx \rho d\phi$

In this equation, $d\rho d\phi$ have been neglected because product of two infinitesmall will become very small. So neglected.

The expression of differential areas are

$$\boxed{\begin{aligned}(d\mathbf{s})_\rho &= \rho\, d\phi\, dz\, \hat{\mathbf{a}}_\rho \\ (d\mathbf{s})_\phi &= d\rho\, dz\, \hat{\mathbf{a}}_\phi \\ (d\mathbf{s})_z &= \rho\, d\rho\, d_\phi\, \hat{\mathbf{a}}_z\end{aligned}} \qquad ...(4)$$

The incremental volume is given by

$$dv = \rho\, d\rho\, d\phi\, dz \qquad ...(5)$$

EXAMPLE 1. *Use cylindrical coordinates to find the area of curved surface of a rigid circular cylinder of radius a and height h. Also obtain the volume.*

Solution. We know that in cylindrical coordinate

$$(ds)_\rho = \rho\, d\phi\, dz$$

where $\rho = a$, $\rho = a$, $\phi = 0$ to 2π and $z = 0$ to h.

$$\therefore \qquad \text{Area} = \iint_S (ds)_\rho = \int_{z=0}^{h}\int_{\phi=0}^{2\pi} \rho d\phi\, dz\,|_{\rho=a}$$

$$= a\int_0^{2\pi} d\phi \int_0^h dz = a\, 2\pi h = 2\pi\, a\, h$$

Now volume,

$$dv = \rho\, d\rho\, d\phi\, dz = \int_0^a \rho d\rho \int_0^{2\pi} d\phi \int_0^h dz$$

$$= \frac{a^2}{2} 2\pi h = \pi\, a^2\, h$$

1.13 SPHERICAL COORDINATE SYSTEM

In electromagnetic field theory, there are some specific problems of spherical shapes and cone shapes. These problems can easily be solved by using spherical coordinate system. The objective of this section is to understand the spherical coordinate system. The spherical coordinates are expressed with parameters such as $r\ \theta\ \phi$. The three mutually orthogonal surfaces are a sphere, a cone and a plane.

Let us define these variables:

$r \rightarrow$ This is radius of an imaginary sphere whose centre is at origin as shown in Fig. (22). The value of r may vary from zero to infinity, *i.e.*, $0 < r < \infty$.

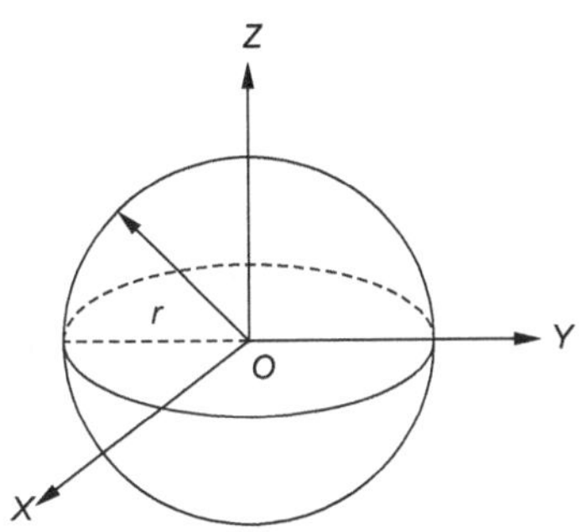

Fig. (22) Showing spherical surface

$\theta \rightarrow$ This the angle made by right circular cone with respect to Z-axis whose apex is at the origin. This is shown in Fig. (23). θ is constant & ϕ varies from 0 to 2π.

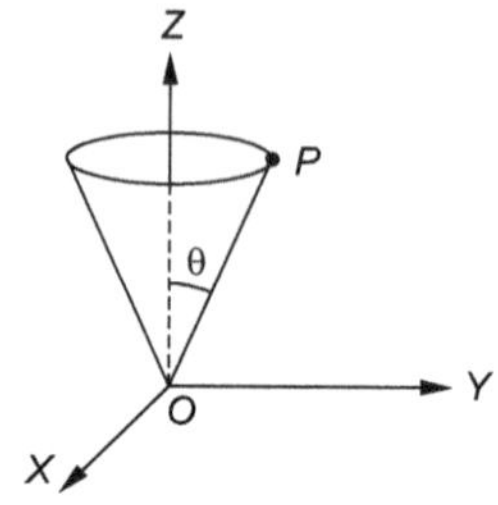

Fig. (23) Showing conical surface

$\phi \rightarrow$ This the angle made by a plane with respect to X-axis. This angle is exactly the same as the angle ϕ of cylindrical coordinates. Thus, ϕ is the angle between X-axis and the projection in the $z - 0$ plane of the line draw from the origin to the point.

Let us consider a point $P(r, \theta, \phi)$ in spherical coordinate system. This is shown in Fig. (24). It is observed from the Fig. (24) that point P lies on the three mutually perpendicular surface.

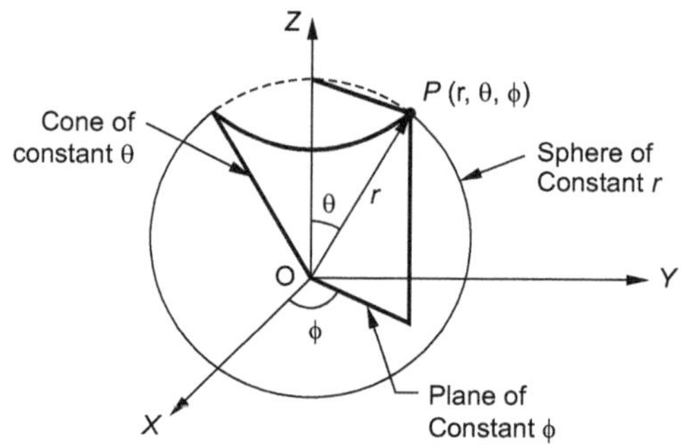

Fig. (24) Showing the position of a point P

These surface are:

(*i*) a sphere of radius r,

(*ii*) a cone with conical surface making an angle θ with Z-axis

(*iii*) a plane making an angle ϕ with X-axis.

The three unit vectors of spherical coordinate system are shown in Fig. (25). The unit vector are $\hat{\mathbf{a}}_r$, $\hat{\mathbf{a}}_\theta$ and $\hat{\mathbf{a}}_\phi$. Each unit vector is perpendicular to one of the three mutually perpendicular

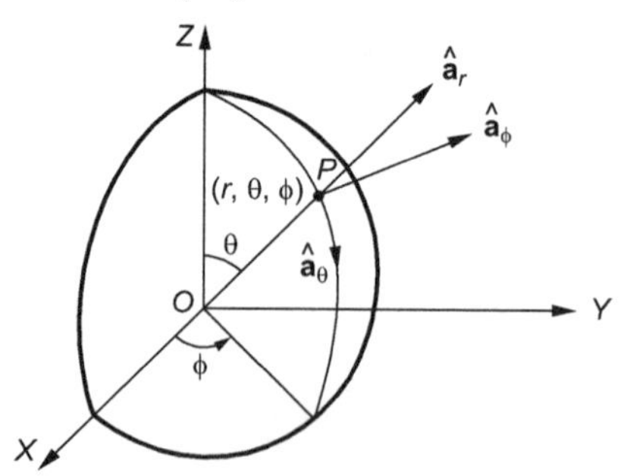

Fig. (25) Unit vectors of spherical coordinates

surfaces and oriented in that direction in which the coordinate increases. The description of these vectors is as follows:

$\hat{\mathbf{a}}_r$ → This unit vector is perpendicular to radius and is directed in the direction of increasing r.

$\hat{\mathbf{a}}_\theta$ → This unit vector is normal to the conical surface. It is directed in the direction of increasing value of θ.

$\hat{\mathbf{a}}_\phi$ → This is unit vector perpendicular to ϕ = constant plane. This is same as in cylindrical coordinate. It is directed in the direction of increasing ϕ.

A vector **A** in spherical coordinate system may by written as

$$\boxed{A = \hat{\mathbf{a}}_r\, A_r + \hat{\mathbf{a}}_\theta\, A_\theta + \hat{\mathbf{a}}_\phi\, A_\phi} \qquad ...(1)$$

Differential Length, Differential Area and Differential Volume

A differential volume element in spherical coordinate system is shown in Fig. (26). This is constructed by increasing r, θ and ϕ by dr, $d\theta$ and $d\phi$ respectively. We observe that:

(*i*) The distance between two spherical surfaces of radius r and $r + dr$ is dr.

(*ii*) The distance between two cones generating angles θ and $\theta + d\theta$ is $r\,d\theta$.

(*iii*) The distance between the two radius planed at angles ϕ and $\phi + d\phi$ is $r \sin\theta\, d\phi$.

The differential lengths are dr, $(r\,d\theta)$ and $(r \sin\theta\, d\phi)$.

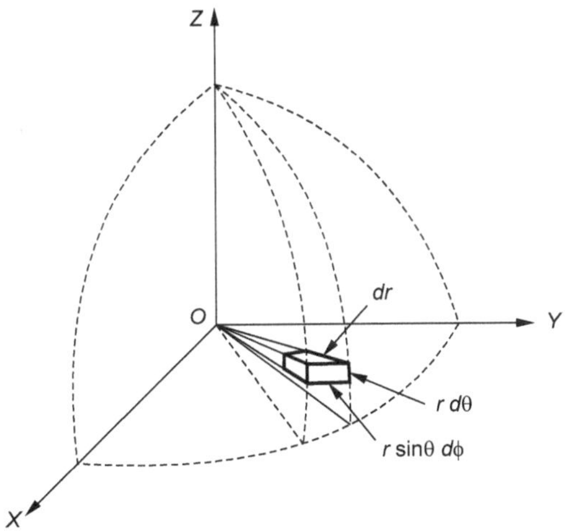

Fig. (26) Differential volume element in spherical coordinate system

The differential lengths $d\mathbf{l}$ in vector form is given by

$$\boxed{d\mathbf{l} = \hat{\mathbf{a}}_r\, dr + \hat{\mathbf{a}}_\theta (r\,d\theta) + \hat{\mathbf{a}}_\phi\, (r \sin\theta\, d\phi)} \qquad ...(2)$$

The expression of differential surface areas are

$$\boxed{d\mathbf{s} = \pm r^2 \sin\theta\, d\theta\, \hat{\mathbf{a}}_r, \pm r \sin\theta\, d\theta\, d\phi\, \hat{\mathbf{a}}_\theta, \pm r\, d\theta\, dr\, \hat{\mathbf{a}}_\phi} \qquad ...(3)$$

The differential volume is

$$dv = (dr)\,(r\,d\theta)\,(r \sin\theta\, d\phi)$$

or

$$\boxed{dv = r^2 \sin\theta\, dr\, d\theta\, d\phi} \qquad ...(4)$$

EXAMPLE 1. *Use spherical coordinates, to find of the region $0 \leq \phi \leq \alpha$ on the spherical shell of radius a.*

Solution. In spherical coordinates, $ds_r = r^2 \sin\theta \, d\theta \, d\phi$

Here $r = a$, $\theta = 0$ to π and $\phi = 0$ to α

$$\therefore \quad \text{Surface area } S = a^2 \int_0^{\pi} \sin\theta \, d\theta \int_0^{\alpha} d\phi = a^2 \left[-\cos\theta\right]_0^{\pi} \left[\phi\right]_0^{\alpha}$$

$$= a^2 [-(-1-1)]\,\alpha = 2a^2\alpha$$

1.14 TRANSFORMATION BETWEEN CARTESIAN AND CYLINDRICAL COORDINATE SYSTEM

We need to convert or realize point in one coordinate system into another. In this section, we will learn to transformation from cartesian to cylindrical or vice-versa. It is required to transform a point or a vector from one coordinate system to another coordinate system. Let us first consider the transformation between cartesian and cylindrical coordinate systems.

Transformation of the Coordinates of a Point

A point P in cartesian coordinate system specified by $P(x, v, z)$ while in case of cylindrical coordinate system it is specified by $P(\rho, \phi, z)$. This is shown in Fig. (27).

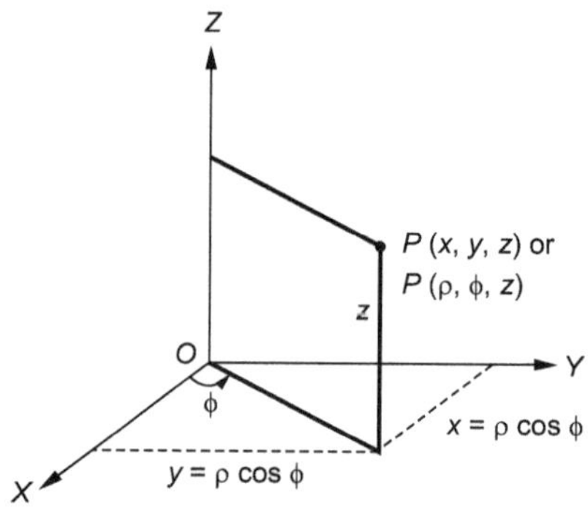

Fig. (27) Showing the relationship between two coordinate systems

By drawing projections of the point P on X-Y plane, we have the following equations:

$$\left.\begin{aligned} x &= \rho\cos\phi \\ y &= \rho\sin\phi \\ z &= z \end{aligned}\right\} \qquad \text{...(1)}$$

From eq. (1), we get.

$$\rho = \sqrt{(x^2 + y^2)} \quad \text{and} \quad \phi = \tan^{-1}\left(\frac{y}{x}\right) \qquad \text{...(2)}$$

Therefore, the transformation of a point from cylindrical to cartesian system is given by

Cylindrical	Cartesian
ρ	$x = \rho\cos\phi$
ϕ	$y = \rho\sin\phi$
z	$z = z$

The transformation of a point from cartesian to cylindrical coordinate system is expressed as

Cartesian	Cylindrical
x	$\rho = \sqrt{(x^2 + y^2)}$
y	$\phi = \tan^{-1}(y/x)$
z	$z = z$

Transformation of a Vector from One System to Other System

Let us consider the transformation of a vector **A** from cartesian coordinate system to cylindrical coordinate system. The vector **A** in cartesian coordinate system is expressed as

$$\mathbf{A} = \hat{\mathbf{a}}_x A_x + \hat{\mathbf{a}}_y A_y + \hat{\mathbf{a}}_z A_z \qquad \text{...(3)}$$

where $\hat{\mathbf{a}}_x$, $\hat{\mathbf{a}}_y$ and $\hat{\mathbf{a}}_z$ are the unit vector along the X, Y and Z-axes respectively.

The same vector **A** in cylindrical coordinate system can be expressed as

$$\mathbf{A} = \hat{\mathbf{a}}_\rho A_\rho + \hat{\mathbf{a}}_\phi A_\phi + \hat{\mathbf{a}}_z A_z \qquad \text{...(4)}$$

where $\hat{\mathbf{a}}_\rho$, $\hat{\mathbf{a}}_\phi$ and $\hat{\mathbf{a}}_z$ are unit vectors along the three axes respectively.

The component of a vector in a desired direction can be obtained by dot multiplication of the vector and unit vector in that direction, *i.e.*,

$$A_\rho = (\text{given vector}) \cdot (\text{unit vector in that direction})$$

or

$$A_\rho = \mathbf{A} \cdot \hat{\mathbf{a}}_\rho \qquad \text{...(5)}$$

Similarly,

$$A_\phi = \mathbf{A} \cdot \hat{\mathbf{a}}_\phi \qquad \text{...(6)}$$

Let us consider the dot product of eq. (5), we get

$$A_\rho = (\hat{\mathbf{a}}_x A_x + \hat{\mathbf{a}}_y A_y + \hat{\mathbf{a}}_z A_z) \cdot \hat{\mathbf{a}}_\rho$$

or

$$A_\rho = \hat{\mathbf{a}}_x \cdot \hat{\mathbf{a}}_\rho A_x + \hat{\mathbf{a}}_y \cdot \hat{\mathbf{a}}_\rho A_y + \hat{\mathbf{a}}_z \cdot \hat{\mathbf{a}}_\rho A_z \qquad \text{...(7)}$$

Similarly,

$$A_\phi = \hat{\mathbf{a}}_x \cdot \hat{\mathbf{a}}_\phi A_x + \hat{\mathbf{a}}_y \cdot \hat{\mathbf{a}}_\phi A_y + \hat{\mathbf{a}}_z \cdot \hat{\mathbf{a}}_\phi A_z \qquad \text{...(8)}$$

and

$$A_z = \hat{\mathbf{a}}_x \cdot \hat{\mathbf{a}}_z A_x + \hat{\mathbf{a}}_y \cdot \hat{\mathbf{a}}_z A_y + \hat{\mathbf{a}}_z \cdot \hat{\mathbf{a}}_z A_z \qquad \text{...(9)}$$

It is clear from eqs. (7), (8) and (9) that the transformation of a vector from one system to other system requires the dot products of unit vectors. The following table can be used to find the dot products:

Table for dot products of unit vectors

• Dot product	$\hat{\mathbf{a}}_\rho$	$\hat{\mathbf{a}}_\phi$	$\hat{\mathbf{a}}_z$
$\hat{\mathbf{a}}_x$	$\cos\phi$	$-\sin\phi$	0
$\hat{\mathbf{a}}_y$	$\sin\phi$	$\cos\phi$	0
$\hat{\mathbf{a}}_z$	0	0	1

Thus $\hat{\mathbf{a}}_x \cdot \hat{\mathbf{a}}_\rho = \cos\phi$, $\hat{\mathbf{a}}_y \cdot \hat{\mathbf{a}}_\rho = \sin\phi$ and $\hat{\mathbf{a}}_z \cdot \hat{\mathbf{a}}_\rho = 0$

Similarly, we can obtain other dot products.

To obtain the values of few dot products see example 2 after this article.

Substituting the values of dot products in eqs. (7), (8) and (9), we get

$$A_\rho = A_x \cos\phi + A_y \sin\phi \quad ...(10)$$

$$A_\phi = A_x \sin\phi + A_y \cos\phi \quad ...(11)$$

and $$A_z = A_x \quad ...(12)$$

Above equation can be expressed in the matrix form as

$$\begin{bmatrix} A_\rho \\ A_\phi \\ A_z \end{bmatrix} = \begin{bmatrix} \cos\phi & \sin\phi & 0 \\ -\sin\phi & \cos\phi & 0 \\ 0 & 0 & 1 \end{bmatrix} \begin{bmatrix} A_x \\ A_y \\ A_z \end{bmatrix}$$

EXAMPLE 1. *Transform a vector from cylindrical coordinate system to cartesian coordinate system.*

Solution. In cylindrical coordinate system, a vector **A** is expressed as

$$\mathbf{A} = \hat{\mathbf{a}}_\rho A_\rho + \hat{\mathbf{a}}_\phi A_\phi + \hat{\mathbf{a}}_z A_z$$

In cartesian coordinate system, a vector **A** is expressed as

$$\mathbf{A} = \hat{\mathbf{a}}_x A_x + \hat{\mathbf{a}}_y A_y + \hat{\mathbf{a}}_z A_z$$

We know that $$A_x = \hat{\mathbf{a}}_x \cdot A$$

or $$A_x = (\hat{\mathbf{a}}_\rho A_\rho + \hat{\mathbf{a}}_\phi A_\phi + \hat{\mathbf{a}}_z A_z) \cdot \hat{\mathbf{a}}_x$$

Similarly, $$A_y = (\hat{\mathbf{a}}_\rho A_\rho + \hat{\mathbf{a}}_\phi A_\phi + \hat{\mathbf{a}}_z A_z) \cdot \hat{\mathbf{a}}_y$$

and $$A_z = (\hat{\mathbf{a}}_\rho A_\rho + \hat{\mathbf{a}}_\phi A_\phi + \hat{\mathbf{a}}_z A_z) \cdot \hat{\mathbf{a}}_z$$

Considering these dot products, we have

$$A_x = \hat{\mathbf{a}}_\rho \cdot \hat{\mathbf{a}}_x A_\rho + \hat{\mathbf{a}}_\phi \cdot \hat{\mathbf{a}}_x A_\phi + \hat{\mathbf{a}}_z \cdot \hat{\mathbf{a}}_x A_z$$

$$A_y = \hat{\mathbf{a}}_\rho \cdot \hat{\mathbf{a}}_y A_\rho + \hat{\mathbf{a}}_\phi \cdot \hat{\mathbf{a}}_y A_\phi + \hat{\mathbf{a}}_z \cdot \hat{\mathbf{a}}_y A_z$$

$$A_z = \hat{\mathbf{a}}_\rho \cdot \hat{\mathbf{a}}_z A_\rho + \hat{\mathbf{a}}_\phi \cdot \hat{\mathbf{a}}_z A_\phi + \hat{\mathbf{a}}_z \cdot \hat{\mathbf{a}}_z A_z$$

Substituting the value of these dot products, we get

$$A_x = \cos\phi\, A_\rho - \sin\phi\, A_\rho$$

$$A_y = \sin\phi\, A_\rho + \cos\phi\, A_\phi$$

$$A_z = A_z$$

$\therefore$ $$\mathbf{A} = \hat{\mathbf{a}}_x (\cos\phi\, A_\rho - \sin\phi\, A_\phi) + \hat{\mathbf{a}}_y (\cos\phi\, A_\rho + \cos\phi\, A_\phi) + \hat{\mathbf{a}}_z A_z$$

In matrix form

$$\begin{bmatrix} A_x \\ A_y \\ A_z \end{bmatrix} = \begin{bmatrix} \cos\phi & \sin\phi & 0 \\ -\sin\phi & \cos\phi & 0 \\ 0 & 0 & 1 \end{bmatrix} \begin{bmatrix} A_\rho \\ A_\phi \\ A_z \end{bmatrix}$$

EXAMPLE 2. *Obtain the dot products* $\hat{\mathbf{a}}_x \cdot \hat{\mathbf{a}}_\rho$ *and* $\hat{\mathbf{a}}_y \cdot \hat{\mathbf{a}}_\rho$.

Solution. To obtain the values of these dot products, let us consider a point $P(\rho, \phi, z)$ in cylindrical coordinate system with unit vectors as shown in Fig. (28).

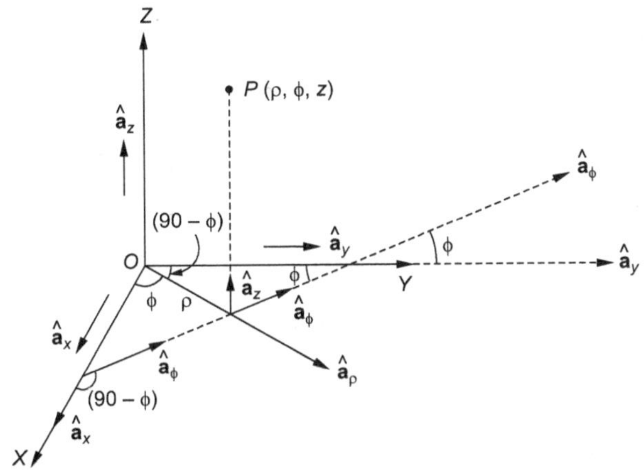

Fig. (28) A point P with its unit vectors in cylindrical coordinate system.

From figure, $\hat{\mathbf{a}}_x \cdot \hat{\mathbf{a}}_\rho = |\hat{\mathbf{a}}_x||\hat{\mathbf{a}}_\rho| \times \cos\phi = 1 \cdot 1 \cdot \cos\phi = \cos\phi$

Similarly, $\hat{\mathbf{a}}_y \cdot \hat{\mathbf{a}}_\rho = |\hat{\mathbf{a}}_x||\hat{\mathbf{a}}_\rho| \cos(90 - \phi) = 1 \cdot 1 \cdot \sin\phi = \sin\phi$

1.15 TRANSFORMATION BETWEEN CARTESIAN AND SPHERICAL COORDINATE SYSTEM

The coordinate of a point or a vector specified in cartesian coordinate system can be transformed into spherical coordinate system and vice-versa. Here, we shall consider both the transformations:

Transformation of Coordinates of a Point

The transformation from cartesian to spherical coordinate system can easily be made by using Fig. (29).

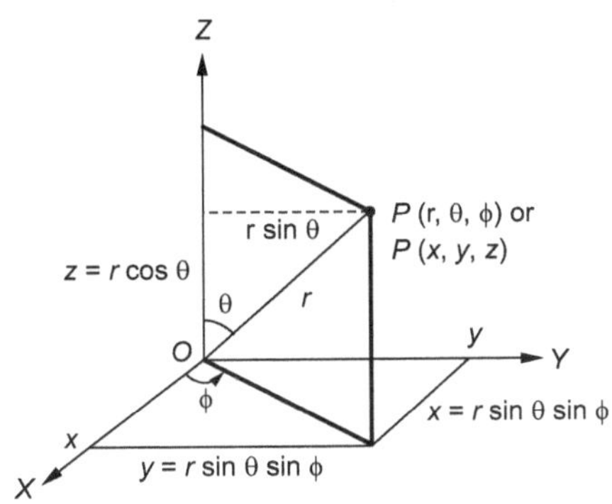

Fig. (29) Showing the transformation diagram

We have the following transformation equations

$$\left.\begin{aligned} x &= r\sin\theta\cos\phi \\ y &= r\sin\theta\sin\phi \\ z &= r\cos\theta \end{aligned}\right\} \quad \text{...(1)}$$

The transformation in the reverse direction is achieved by the following equations

$$\left.\begin{aligned} r &= \sqrt{(x^2 + y^2 + z^2)} \\ \theta &= \cos^{-1}\left[\frac{z}{\sqrt{x^2 + y^2 + z^2}}\right] \\ \phi &= \tan^{-1}\left(\frac{y}{x}\right) \end{aligned}\right\} \quad \text{...(2)}$$

Transformation of a Vector from Cartesian to Spherical Coordinates

In cartesian coordinate system, a vector **A** can be expressed as

$$\mathbf{A} = \hat{\mathbf{a}}_x A_x + \hat{\mathbf{a}}_y A_y + \hat{\mathbf{a}}_z A_z \quad \text{...(3)}$$

The same vector in spherical coordinates can be expressed as

$$\mathbf{A} = \hat{\mathbf{a}}_r A_r + \hat{\mathbf{a}}_\theta A_\theta + \hat{\mathbf{a}}_\phi A_\phi \quad \text{...(4)}$$

We know that

$$A_r = (\text{given vector}) \cdot (\hat{\mathbf{a}}_r) = \mathbf{A} \cdot \hat{\mathbf{a}}_r$$

or $\quad A_r = (\hat{\mathbf{a}}_x A_x + \hat{\mathbf{a}}_y A_y + \hat{\mathbf{a}}_z A_z) \cdot \hat{\mathbf{a}}_r$

Similarly $\quad A_\theta = (\hat{\mathbf{a}}_x A_x + \hat{\mathbf{a}}_y A_y + \hat{\mathbf{a}}_z A_z) \cdot \hat{\mathbf{a}}_\theta$ ...(5)

and $\quad A_\phi = (\hat{\mathbf{a}}_x A_x + \hat{\mathbf{a}}_y A_y + \hat{\mathbf{a}}_z A_z) \cdot \hat{\mathbf{a}}_\phi$

Let us consider the value of A_r. We have

$$A_r = \hat{\mathbf{a}}_x \cdot \hat{\mathbf{a}}_r A_x + \hat{\mathbf{a}}_y \cdot \hat{\mathbf{a}}_r A_y + \hat{\mathbf{a}}_z \cdot \hat{\mathbf{a}}_r A_z \quad \text{...(6)}$$

The dot products can be obtained with the help of following table:

Table for dot-products of unit vectors

• Dot product	$\hat{\mathbf{a}}_r$	$\hat{\mathbf{a}}_\theta$	$\hat{\mathbf{a}}_\phi$
$\hat{\mathbf{a}}_x$	$\sin\theta\cos\phi$	$\cos\theta\cos\phi$	$-\sin\phi$
$\hat{\mathbf{a}}_y$	$\sin\theta\sin\phi$	$\cos\theta\sin\phi$	$\cos\phi$
$\hat{\mathbf{a}}_z$	$\cos\theta$	$-\sin\theta$	0

Hence, $\quad \hat{\mathbf{a}}_x \cdot \hat{\mathbf{a}}_r = \sin\theta\cos\phi, \quad \hat{\mathbf{a}}_y \cdot \hat{\mathbf{a}}_r = \sin\theta\sin\phi$

and $\quad \hat{\mathbf{a}}_z \cdot \hat{\mathbf{a}}_r = \cos\theta$

Substituting these values in eq. (6), we get

$$A_r = \sin\theta\cos\phi\, A_x + \sin\theta\sin\phi\, A_y + \cos\theta\, A_z \qquad ...(7)$$

Similarly, we can obtain

$$A_\theta = \cos\theta\cos\phi\, A_x + \cos\theta\sin\phi\, A_y - \sin\theta\, A_z \qquad ...(8)$$

and

$$A_\phi = -\sin\phi\, A_x + \cos\phi\, A_y \qquad ...(9)$$

In matrix form, the transformation is expressed as

$$\begin{bmatrix} A_r \\ A_\theta \\ A_\phi \end{bmatrix} = \begin{bmatrix} \sin\theta\cos\phi & \sin\theta\sin\phi & \cos\theta \\ \cos\theta\cos\phi & \cos\theta\sin\phi & -\sin\theta \\ -\sin\phi & \cos\phi & 0 \end{bmatrix} \begin{bmatrix} A_x \\ A_y \\ A_z \end{bmatrix} \qquad ...(10)$$

EXAMPLE 1. *Transform a vector A from spherical coordinate system to cartesian coordinate system.*

Solution. We know that

$$\begin{aligned} A_x &= (\text{given vector}) \cdot \hat{\mathbf{a}}_x \\ &= (\hat{\mathbf{a}}_r A_r + \hat{\mathbf{a}}_\theta A_\theta + \hat{\mathbf{a}}_\phi A_\phi) \cdot \hat{\mathbf{a}}_x \\ &= \hat{\mathbf{a}}_r \cdot \hat{\mathbf{a}}_x A_r + \hat{\mathbf{a}}_\theta\, \hat{\mathbf{a}}_x A_\theta + \hat{\mathbf{a}}_\phi \cdot \hat{\mathbf{a}}_x A_\phi \\ &= \sin\theta\cos\phi\, A_r + \cos\theta\cos\phi\, A_\theta - \sin\phi\, A_\phi \end{aligned} \qquad ..(11)$$

Using eq. (1), we get

$$\sin\theta\cos\phi = \frac{x}{r} = \frac{x}{\sqrt{(x^2+y^2+z^2)}}$$

$$\cos\theta\cos\phi = \frac{zx}{\sqrt{(x^2+y^2)}\sqrt{(x^2+y^2+z^2)}}$$

$$\sin\phi = \frac{y}{\sqrt{(x^2+y^2)}}$$

Substituting these values in eq. (11), we get

$$A_x = \frac{x}{\sqrt{(x^2+y^2+z^2)}} A_r + \frac{xz}{\sqrt{(x^2+y^2)}\sqrt{(x^2+y^2+z^2)}} A_\theta - \frac{y}{\sqrt{(x^2+y^2)}} A_\phi \qquad ...(12)$$

Similarly, we can write

$$A_y = \frac{z}{\sqrt{(x^2+y^2+z^2)}} A_r + \frac{yz}{\sqrt{(x^2+y^2)}\sqrt{(x^2+y^2+z^2)}} A_\theta - \frac{x}{\sqrt{(x^2+y^2)}} A_\phi$$

and

$$A_z = \frac{z}{\sqrt{(x^2+y^2+z^2)}} A_r - \frac{\sqrt{x^2+y^2}}{\sqrt{(x^2+y^2+z^2)}} A_\theta$$

In matrix form, the transformation of a vector **A** from spherical coordinate to cartesian coordinate system can be expressed as:

$$\begin{bmatrix} A_x \\ A_y \\ A_z \end{bmatrix} = \begin{bmatrix} \sin\theta\cos\phi & \cos\theta\cos\phi & -\sin\phi \\ \sin\theta\sin\phi & \cos\theta\sin\phi & \cos\phi \\ \cos\theta & -\sin\theta & 0 \end{bmatrix} \begin{bmatrix} A_r \\ A_\theta \\ A_\phi \end{bmatrix}$$

Few *Worked* EXAMPLES

EXAMPLE 1. *A point P is specified in the cartesian coordinate system as P(2, 1, 3). Express the point in cylindrical coordinate system.*

Solution. The given point in cartesian coordinate system is $P(2, 1, 3)$, *i.e.*, $x = 2, y = 1$ and $z = 3$. We have to express this point in cylindrical coordinate system. *i.e.*, $P(\rho, \phi, z)$.

The conversion formulae are

$$\rho = \sqrt{(x^2 + y^2)} = \sqrt{[(2)^2 + (1)^2]} = \sqrt{5} = \mathbf{2.24}$$

$$\phi = \tan^{-1}\left(\frac{y}{x}\right) = \tan^{-1}\left(\frac{1}{2}\right) = \mathbf{26.56°}$$

$$z = \mathbf{z = 3}$$

∴ Point in cylindrical coordinate system is expressed as $P(2.24, 26.56°, z = 3)$.

EXAMPLE 2. *Transform the point P(3, 45°, 60°) in (i) Cartesian coordinates and (ii) Cylindrical coordinates.*

Solution. The given point is in spherical coordinate system, *i.e.*, $r = 3$, $\theta = 45°$ and $\phi = 60°$

(*i*) Transformation of the point in cartesian coordinate system. Here

$$x = r \sin\theta \cos\phi = 3 \sin 45° \cos 60° = \mathbf{1.06}$$

$$y = r \sin\theta \sin\phi = 3 \sin 45° \sin 60° = \mathbf{1.5}$$

$$z = r \cos\theta = 3 \cos 45° = \mathbf{2.12}$$

∴ Point in cartesian coordinate can be expressed as $P(1.06, 1.5, 2.12)$

(*ii*) Transformation of the point in cylindrical coordinate system. Here

$$\rho = r \sin\theta = 3 \sin 45° = 3/\sqrt{2} = \mathbf{2.12}$$

$$\phi = \phi = 60°$$

$$z = r \cos\theta = 3 \cos 45° = 3/\sqrt{2} = \mathbf{2.12}$$

∴ Point in cylindrical coordinates can be expressed as $P(2.12, 60°, 2.12)$

EXAMPLE 3. *A point P is expressed as P(x = 2, y = 3 and z = – 1). Find the coordinate of this point in spherical coordinate system.*

Solution. In spherical coordinate system

$$r = \sqrt{(x^2 + y^2 + z^2)} = \sqrt{[(2)^2 + (3)^2 + (-1)^2]} = \sqrt{(14)} = \mathbf{3.74}$$

$$\theta = \cos^{-1}\left[\frac{z}{\sqrt{(x^2 + y^2 + z^2)}}\right] = \cos^{-1}\left[\frac{-1}{3.74}\right] = \mathbf{105.5°}$$

$$\phi = \tan^{-1}\left(\frac{y}{x}\right) = \tan^{-1}\left(\frac{3}{2}\right) = \mathbf{56.31°}$$

The spherical coordinate of point are P (3.74, 105.5°, 56.31)

EXAMPLE 4. *A point P in cylindrical coordinate system is expressed as P(3, 45°, 2). Find the cartesian coordinates.*

Solution. In cartesian coordinate system

$$x = \rho \cos \phi = 3 \cos 45° = 3/\sqrt{2} = \mathbf{2.12}$$

$$y = \rho \sin \phi = 3 \sin 45° = 3/\sqrt{2} = \mathbf{2.12}$$

$$z = \mathbf{2}$$

∴ Point in cartesian coordinate system is P**(2.12, 2, 12, 2)**

EXAMPLE 5. *Find the location of point P(2, – 1, 3) in cylindrical and spherical coordinates.*

Solution.

1. In cylindrical coordinate system

$$\rho = \sqrt{(x^2 + y^2)} = \sqrt{[(2)^2 + (-1)^2]} = \sqrt{\mathbf{5}}$$

$$\phi = \tan^{-1}\left(\frac{y}{x}\right) = \tan^{-1}\left(\frac{-1}{2}\right) = \mathbf{-26.6°}$$

or $$\phi = 2\pi - 26.6° = 360° - 26.6° = \mathbf{333.4°}$$

$$z = \mathbf{3}$$

∴ $$P(\rho, \phi\ z) = P(\sqrt{\mathbf{5}}, \mathbf{33.4°, 3})$$

2. In spherical coordinate system

$$r = \sqrt{(x^2 + y^2 + z^2)} = \sqrt{[(2)^2 + (-1)^2] + (3)^2} = \sqrt{14} = \mathbf{3.742}$$

$$\theta = \cos^{-1}\left[\frac{z}{\sqrt{(x^2 + y^2 + z^2)}}\right] = \cos^{-1}\left[\frac{3}{3.724}\right] = \cos^{-1}[0.8] = \mathbf{36.7°}$$

$$\phi = \tan^{-1}\left(\frac{y}{x}\right) = \tan^{-1}\left(\frac{-1}{2}\right) = \mathbf{-26.6°}$$

or $$\phi = 2\pi - 26.6° = \mathbf{333.4°}$$

∴ $$P(r, \theta, \phi) = P\mathbf{(3.742, 36.7°, 333.4)}$$

EXAMPLE 6. *The two points P and Q are specified as*

P(1, – 3, 4) and Q(2, 60°, 1)

Find the distance between these two points.

Solution. Let us first expressed the point Q in cartesian coordinate system.

From the given point $Q(\rho, \phi\ z)$, we have $\rho = 2$, $\phi = 60°$ and $z = 1$.

Now, $$x = \rho \cos \phi = 2 \cos 60° = 2 \times \left(\frac{1}{2}\right) = 1$$

$$y = \rho \sin \phi = 2 \sin 60° = 2 \times \left(\frac{\sqrt{3}}{2}\right) = 1.73$$

$$z = 1$$

$\therefore \quad Q(x, y, z) = Q\,\mathbf{(1, 1.73, 1)}$

The distance between points P and Q is given by

$$d = |\,\mathbf{PQ}\,| = \sqrt{[(1-1)^2 + \{1.73-(-3)^2 + (1-4)^2\}]} = \mathbf{6.88}$$

EXAMPLE 7. *Transform the following vector to spherical coordinates. The vector is* $\mathbf{A} = 5\,\hat{\mathbf{a}}_x$ *and point* $P = (r = 4, \theta = 25°, \phi = 120°)$.

Solution. The given vector $\mathbf{A} = 5\,\hat{\mathbf{a}}_x$, *i.e.*, $A_x = 5$, $A_y = 0$ and $A_z = 0$

From the given point, $r = 4$, $\theta = 25°$ and $\phi = 120°$

Here $\quad A_r = (\text{vector}) \cdot \hat{\mathbf{a}}_r = 5\,\hat{\mathbf{a}}_x \cdot \hat{\mathbf{a}}_r = 5 \sin\theta \cos\phi \qquad (\because \hat{\mathbf{a}}_x \cdot \hat{\mathbf{a}}_r = \sin\theta\cos\phi)$

$\quad = 5 \sin 25° \cos 120° = -1.057$

$A_\theta = 5\,\hat{\mathbf{a}}_x \cdot \hat{\mathbf{a}}_\theta = 5 \cos\theta \cos\phi \qquad (\because \hat{\mathbf{a}}_x \cdot \hat{\mathbf{a}}_r = \cos\theta\cos\phi)$

$\quad = 5 \cos 25° \cos 120° = -2.27$

$A_\phi = 5\,\hat{\mathbf{a}}_x \cdot \hat{\mathbf{a}}_\phi = 5(-\sin\phi) = -5 \sin 120° = -4.33$

$\therefore \quad A = -1.057\,\hat{\mathbf{a}}_r - 2.27\,\hat{\mathbf{a}}_\theta - 4.33\,\hat{\mathbf{a}}_\phi$

EXAMPLE 8. *Transform a vector* $\mathbf{A} = y\hat{\mathbf{a}}_x - x\hat{\mathbf{a}}_y + z\hat{\mathbf{a}}_z$ *into cylindrical coordinates.*

Solution. Let us consider the cylindrical coordinate of the given vector A_r, A_ϕ and A_z.

These are given as follow:

$$A_r = \mathbf{A} \cdot \hat{\mathbf{a}}_r = (y\hat{\mathbf{a}}_x - x\hat{\mathbf{a}}_y + z\hat{\mathbf{a}}_z) \cdot \hat{\mathbf{a}}_r = y\hat{\mathbf{a}}_x \cdot \hat{\mathbf{a}}_r - x\hat{\mathbf{a}}_y \cdot \hat{\mathbf{a}}_r + z\hat{\mathbf{a}}_z \cdot \hat{\mathbf{a}}_r$$

$$= y\cos\phi - x\sin\phi + 0 = -\rho\sin\phi\cos\phi - \rho\cos\phi\sin\phi + 0 = 0$$

$$A_\phi = (y\hat{\mathbf{a}}_x - x\hat{\mathbf{a}}_y + z\hat{\mathbf{a}}_z) \cdot \hat{\mathbf{a}}_\phi = y\hat{\mathbf{a}}_x \cdot \hat{\mathbf{a}}_\phi - x\hat{\mathbf{a}}_y \cdot \hat{\mathbf{a}}_\phi + z\hat{\mathbf{a}}_z \cdot \hat{\mathbf{a}}_\phi$$

$$= -y\sin\phi - x\cos\phi + 0 = -\rho\sin\phi\sin\phi - \rho\cos\phi\cos\phi$$

$$= -\rho\sin^2\phi - \rho\cos^2\phi = -\rho(\sin^2\phi + \cos^2\phi) = -\rho$$

$$A_z = (y\hat{\mathbf{a}}_x - x\hat{\mathbf{a}}_y + z\hat{\mathbf{a}}_z) \cdot \hat{\mathbf{a}}_z = z$$

Thus, $\quad \mathbf{A} = A_r\,\hat{\mathbf{a}}_\rho + A_\phi\,\hat{\mathbf{a}}_\phi + A_z\,\hat{\mathbf{a}}_z = 0 + (-\rho)\,\hat{\mathbf{a}}_\phi + z\hat{\mathbf{a}}_z$

$\therefore \quad \mathbf{A} = -\rho\hat{\mathbf{a}}_\phi + z\hat{\mathbf{a}}_z$

EXAMPLE 9. *A vector* **A** *in cylindrical coordinate is expressed as*

$$\mathbf{A} = 2\cos\phi\,\hat{\mathbf{a}}_\rho + 3\rho\,\hat{\mathbf{a}}_\phi - 8\,\hat{\mathbf{a}}_z$$

Transform this vector into cartesian coordinate system.

Solution. The components of this vector in cartesian coordinate system are A_x, A_y and A_z respectively. Let us find these components.

$$A_x = (\mathbf{A}) \cdot \hat{\mathbf{a}}_x = (2\cos\phi\,\hat{\mathbf{a}}_\rho + 3\rho\,\hat{\mathbf{a}}_\phi - 8\,\hat{\mathbf{a}}_z)\cdot\hat{\mathbf{a}}_x$$

$$= 2\cos\phi\;\hat{\mathbf{a}}_\rho\cdot\hat{\mathbf{a}}_x + 3\rho\,\hat{\mathbf{a}}_\phi\cdot\hat{\mathbf{a}}_x - 8\,\hat{\mathbf{a}}_z\,.\,\hat{\mathbf{a}}_x$$

$$= 2\cos\phi\,(\cos\phi) + 3\rho\,(-\sin\phi) - 8\cdot(0) = 2\cos^2\phi - 3\,\rho\sin\phi$$

$$= 2\left[\frac{x}{\sqrt{(x^2+y^2)}}\right]^2 - 3\sqrt{(x^2+y^2)}\left[\frac{y}{\sqrt{x^2+y^2}}\right] = \frac{2x^2}{\sqrt{(x^2+y^2)}} - 3y$$

Now, $$A_y = (2\cos\phi\,\hat{\mathbf{a}}_\rho + 3\rho\,\hat{\mathbf{a}}_\phi - 8\,\hat{\mathbf{a}}_z)\cdot\hat{\mathbf{a}}_y = 2\cos\phi\,\hat{\mathbf{a}}_\rho\cdot\hat{\mathbf{a}}_y + 3\rho\,\hat{\mathbf{a}}_\phi\cdot\hat{\mathbf{a}}y - 8\,\hat{\mathbf{a}}_z\cdot\hat{\mathbf{a}}_y$$

$$= 2\cos\phi\,(\text{sins}\,\phi) + 3\rho\cos\phi - 8\,(0)$$

$$= 2\cdot\frac{x}{\sqrt{(x^2+y^2)}}\cdot\frac{y}{\sqrt{(x^2+y^2)}} + 3\sqrt{(x^2+y^2)}\left[\frac{x}{\sqrt{x^2+y^2}}\right]$$

$$= \frac{2xy}{\sqrt{(x^2+y^2)}} + 3x$$

and $$A_z = A_z = -8\,\hat{\mathbf{a}}_z$$

Therefore the given vector in cartesian coordinate can be expressed as

$$\mathbf{A} = \left[\frac{2x^2}{(x^2+y^2)} - 3y\right]\hat{\mathbf{a}}_x + \left[\frac{2xy}{(x^2+y^2)} - 3x\right]\hat{\mathbf{a}}_y - 8\,\hat{\mathbf{a}}_z$$

EXAMPLE 10. *Transform a cartesian vector*

$$A = 3\hat{\mathbf{a}}_x - 2\hat{\mathbf{a}}_y - 3\hat{\mathbf{a}}_z \;\; \textit{at point } P(-2, -3, 4)$$

into spherical coordinate system.

Solution. Let us consider the spherical coordinate of the vector A_r, A_θ and A_ϕ. These are given as

$$A_r = (3\hat{\mathbf{a}}_x - 2\hat{\mathbf{a}}_y - 3\hat{\mathbf{a}}_z)\cdot\hat{\mathbf{a}}_r = 3\hat{\mathbf{a}}_x\cdot\hat{\mathbf{a}}_r - 2\hat{\mathbf{a}}_y\cdot\hat{\mathbf{a}}_r - 3\hat{\mathbf{a}}_z\cdot\hat{\mathbf{a}}_r$$

$$= 3(\sin\theta\cos\phi) - 2(\sin\theta\sin\phi) - 3(\cos\theta)$$

Now, $$A_\theta = (3\hat{\mathbf{a}}_x - 2\hat{\mathbf{a}}_y - 3\hat{\mathbf{a}}_z)\cdot\hat{\mathbf{a}}_\theta = 3\hat{\mathbf{a}}_x\cdot\hat{\mathbf{a}}_\theta - 2\hat{\mathbf{a}}_y\hat{\mathbf{a}}_\theta - 3\hat{\mathbf{a}}_z\cdot\hat{\mathbf{a}}_\theta$$

$$= 3(\cos\theta\cos\phi) - 2\,(\cos\theta\sin\phi) - 3\,(-\sin\theta)$$

$$= 3(\cos\theta\cos\phi) - 2\cos\theta\sin\phi + 3\sin\phi$$

and $$A_\phi = (3\hat{\mathbf{a}}_x - 2\hat{\mathbf{a}}_y - 3\hat{\mathbf{a}}_z)\cdot\hat{\mathbf{a}}_\phi = 3\hat{\mathbf{a}}_x\cdot\hat{\mathbf{a}}_\phi - 2\hat{\mathbf{a}}_y\cdot\hat{\mathbf{a}}_\phi - 3\hat{\mathbf{a}}_z\cdot\hat{\mathbf{a}}_\phi$$

$$= 3(-\sin\phi) - 2(\cos\phi) - 3(0) = -3\sin\phi - 2\cos\phi$$

Let us convert the cartesian coordinate (– 2, – 3, 4) of the given point in spherical coordinate (r, θ, ϕ). We know that

$$r = \sqrt{(x^2+y^2+z^2)} = \sqrt{(-2)^2+(-3)^2+(4)^2} = \sqrt{29} = 5.38$$

$$\theta = \cos^{-1}\left[\frac{z}{\sqrt{(x^2+y^2+z^2)}}\right] = \cos^{-1}\left[\frac{4}{5.38}\right] = 41.97°$$

$$\boxed{\phi = \tan^{-1}\left(\frac{y}{x}\right) = \tan^{-1}\left[\frac{(-3)}{(-2)}\right] = \tan^{-1}\left(\frac{3}{2}\right) = 56.3°}$$

Substituting these values in the expression of A_r, A_θ and A_ϕ, we get

$$A_r = 3\,(\sin 41.97°)\,(\cos 56.3°) - 2(41.97°)\,\sin(56.3°) - 3\cos 41.97°$$

Solving, we get $A_r = 2.23$

Now, $A_\theta = 3(\cos 41.97°)\,(\cos 56.3°) - 2\cos(41.97°)\,(\sin 56.3°) + 3(\sin 41.97°)$

Solving we get $A_\theta = 2$

and $A_\phi = -3\,(\sin 56.3°) - 2\,(\cos 56.3°) = -3.6$

Putting these value, we get

$$A = 2.23\,\hat{\mathbf{a}}_r + 2\,\hat{\mathbf{a}}_\theta - 3.6\,\hat{\mathbf{a}}_\phi$$

EXAMPLE 11. *Express the vector*

$$\mathbf{A} = x\,y^2\,z\,\hat{\mathbf{a}}_x + x^2\,yz\,\hat{\mathbf{a}}_y + xyz^2\,\hat{\mathbf{a}}_z$$

into cylindrical and spherical polar coordinate system.

Solution.

1. Transformation of a vector from cartesian coordinate system to cylindrical coordinate system can be expressed by the following matrix

$$\begin{bmatrix} A_\rho \\ A_\phi \\ A_z \end{bmatrix} = \begin{bmatrix} \cos\phi & \sin\phi & 0 \\ -\sin\phi & \cos\phi & 0 \\ 0 & 0 & 1 \end{bmatrix} \begin{bmatrix} A_x \\ A_y \\ A_z \end{bmatrix}$$

The given vector, $\mathbf{A} = x\,y^2\,z\,\hat{\mathbf{a}}_x + x^2\,yz\,\hat{\mathbf{a}}_y + x\,y\,z^2\,\hat{\mathbf{a}}_z$

i.e., $A_x = x\,y^2\,z,\ A_y = x^2\,y\,z$ and $A_z = x\,y\,z^2$

Substituting these values, we get the following matrix

$$\begin{bmatrix} A_\rho \\ A_\phi \\ A_z \end{bmatrix} = \begin{bmatrix} \cos\phi & \sin\phi & 0 \\ -\sin\phi & \cos\phi & 0 \\ 0 & 0 & 1 \end{bmatrix} \begin{bmatrix} x\,y^2\,z \\ x^2\,y\,z \\ x\,y\,z^2 \end{bmatrix}$$

$\therefore$

$$A_\rho = x\,y^2\,z\cos\phi + x^2\,y\,z\sin\phi$$
$$A_\phi = -x\,y^2\,z\sin\phi + x^2\,y\,z\cos\phi$$
$$A_z = x\,y\,z^2$$

We know that $x = \rho\cos\phi,\quad y = \rho\sin\phi\quad$ and $\quad z = z$

Therefore,

$$A_\rho = \rho^3\,z\cos^2\phi\,\sin^2\phi + \rho^3\,z\sin^2\phi\,\cos^2\phi$$
$$= 2\,\rho^3\,z\cos^2\phi\,\sin^2\phi$$
$$A_\phi = -\rho^3\,z\cos\phi\,\sin^3\phi + \rho^3\,z\cos^3\phi\,\sin\phi$$
$$A_z = \rho^2\,z^2\sin\phi\cos\phi$$

2. Transformation of a vector from cartesian coordinate system to spherical coordinate system can be expressed by the following matrix

$$\begin{bmatrix} A_r \\ A_\theta \\ A_\phi \end{bmatrix} = \begin{bmatrix} \sin\theta\cos\phi & \sin\theta\sin\phi & \cos\theta \\ -\cos\theta\cos\phi & \cos\theta\sin\phi & -\sin\theta \\ -\sin\phi & \cos\phi & 0 \end{bmatrix} \begin{bmatrix} x\,y^2\,z \\ x^2\,y\,z \\ x\,y\,z^2 \end{bmatrix}$$

Here $x = r \sin\theta \cos\phi$

$y = r \sin\theta \sin\phi$

$z = r \cos\theta$

Now,
$$A_r = x y^2 z \sin\theta \cos\phi + x^2 y z \sin\phi \sin\theta + x y z^2 \cos\theta$$
$$= (r \sin\theta \cos\phi)(r \sin\theta \cos\phi)^2 (r \cos\theta) \sin\theta \cos\phi$$
$$+ (r \sin\theta \cos\phi)^2 (r \sin\theta \sin\phi)(r \cos\phi) \sin\theta \sin\theta$$
$$+ (r \sin\theta \cos\phi)(r \sin\theta \sin\phi)(r \cos\theta)^2 \cos\theta$$

$\therefore$
$$A_r = r^4 \sin^4\theta \cos^2\phi \sin^2\phi \cos\theta + r^4 \sin^4\theta \cos^2\phi \cos^2\theta$$
$$+ r^4 \sin^2\theta \cos^2\theta \cos\theta \sin\phi \cos\phi$$

Similarly, we get
$$A_\theta = -r^4 \sin^3\theta \cos^2\phi \sin^3\phi \cos^2\theta + r^4 \cos^2\theta \sin^3\phi \sin^2\theta \cos^2\phi$$
$$- r^4 \cos^2\theta \sin^3\phi \cos\phi$$

and
$$A_\phi = -r^4 \sin^3\theta \sin^3\phi \cos\theta \cos\phi + r^4 \sin^3\theta \cos^3\phi \cos\theta \sin\phi$$

1.16 PARTIAL DERIVATIVES AND DEL OPERATOR ($\vec{\nabla}$)

When a vector is a function of more than one independent scalar variable, then we can extend the general properties of vector differentiation to partial derivatives also. Let **A** be a vector depending on more than one scalar variable such as cartesian coordinates x, y, z of a point in space. So **A** is a function of coordinate (x, y, z). The function can be differentiated with respect to one of the variable, keeping the other variables constant, *i.e.*,

$$\frac{\partial \mathbf{A}}{\partial x} \text{ (with } y \text{ and } z \text{ remaining constant)}$$

$$\frac{\partial \mathbf{A}}{\partial y} \text{ (with } x \text{ and } z \text{ constant)}$$

$$\frac{\partial \mathbf{A}}{\partial z} \text{ (with } x \text{ and } y \text{ constant)}$$

So, $\dfrac{\partial \mathbf{A}}{\partial x}, \dfrac{\partial \mathbf{A}}{\partial y}, \dfrac{\partial \mathbf{A}}{\partial z}$ are the partial derivatives

If now x, y, z change simultaneously by small increments dx, dy, dz then the total change or total derivative of **A** will be

$$d\mathbf{A} = \frac{\partial \mathbf{A}}{\partial x} dx + \frac{\partial \mathbf{A}}{\partial y} dy + \frac{\partial \mathbf{A}}{\partial z} dz = \left(\frac{\partial}{\partial x} dx + \frac{\partial}{\partial y} dy + \frac{\partial}{\partial z} dz\right)\mathbf{A}$$
$$= \left(\frac{\partial}{\partial x}\hat{\mathbf{a}}_x + \frac{\partial}{\partial y}\hat{\mathbf{a}}_y + \frac{\partial}{\partial z}\hat{\mathbf{a}}_z\right)\cdot(dx\,\hat{\mathbf{a}}_x + dy\,\hat{\mathbf{a}}_y + dz\,\hat{\mathbf{a}}_z)\mathbf{A}$$

where $\hat{\mathbf{a}}_x, \hat{\mathbf{a}}_y$ and $\hat{\mathbf{a}}_z$ are the unit vectors along X, Y and Z-axes respectively.

Now, $\left(\dfrac{\partial}{\partial x}\hat{\mathbf{a}}_x + \dfrac{\partial}{\partial y}\hat{\mathbf{a}}_y + \dfrac{\partial}{\partial z}\hat{\mathbf{a}}_z\right)$ is defined as vector differential operator.

This is expressed by $\vec{\nabla}$ (read as 'del')

$$\therefore \quad \boxed{\vec{\nabla} = \frac{\partial}{\partial x}\hat{\mathbf{a}}_x + \frac{\partial}{\partial y}\hat{\mathbf{a}}_y + \frac{\partial}{\partial z}\hat{\mathbf{a}}_y} \quad ...(1)$$

In cylindrical coordinate system, the operator 'del' can be expressed as

$$\boxed{\vec{\nabla} = \frac{\partial}{\partial \rho}\hat{\mathbf{a}}_\rho + \frac{1}{\rho}\frac{\partial}{\partial \phi}\hat{\mathbf{a}}_\phi + \frac{\partial}{\partial z}\hat{\mathbf{a}}_z} \quad ...(2)$$

Similarly, in spherical coordinate system, the operator 'del' can be expressed as

$$\boxed{\vec{\nabla} = \frac{\partial}{\partial r}\hat{\mathbf{a}}_r + \frac{\partial}{r\partial \theta}\hat{\mathbf{a}}_\theta + \frac{1}{r\sin\theta}\frac{\partial}{\partial \phi}\hat{\mathbf{a}}_\phi} \quad ...(3)$$

1.17 GRADIENT AND IT'S PHYSICAL SIGNIFICANCE

Definition of Gradient

The gradient is a dot product of a scalar and del operator $\vec{\nabla}$.

Let V be a scalar function then

$$\text{gradient } V = \vec{\nabla} \cdot V$$

In order to consider the gradient of a scalar, let $V(x, y, z)$ be a scalar function of position of a scalar point of coordinates $(x, y, z.)$ The partial derivatives of V along the three coordinates axes are

$$\frac{\partial V}{\partial x}, \frac{\partial V}{\partial y} \text{ and } \frac{\partial V}{\partial z} \text{ respectively.}$$

The vector differential operator $\vec{\nabla}$ (del) is defined as

$$\vec{\nabla} = \frac{\partial}{\partial x}\hat{\mathbf{a}}_x + \frac{\partial}{\partial y}\hat{\mathbf{a}}_y + \frac{\partial}{\partial z}\hat{\mathbf{a}}_z$$

Therefore,

$$\vec{\nabla} V = \frac{\partial V}{\partial x}\hat{\mathbf{a}}_x + \frac{\partial V}{\partial y}\hat{\mathbf{a}}_y + \frac{\partial V}{\partial z}\hat{\mathbf{a}}_z$$

Gradient of *V* in Different Coordinate Systems

In cartesian coordinate system

$$\boxed{\vec{\nabla} V = \frac{\partial V}{\partial x}\hat{\mathbf{a}}_x + \frac{\partial V}{\partial y}\hat{\mathbf{a}}_y + \frac{\partial V}{\partial z}\hat{\mathbf{a}}_z} \quad ...(1)$$

In cylindrical coordinate system

$$\boxed{\vec{\nabla} V = \frac{\partial V}{\partial \rho}\hat{\mathbf{a}}_\rho + \frac{1}{\rho}\frac{\partial V}{\partial \phi}\hat{\mathbf{a}}_\phi + \frac{\partial V}{\partial z}\hat{\mathbf{a}}_z} \quad ...(2)$$

In spherical coordinate system

$$\boxed{\vec{\nabla} V = \frac{\partial V}{\partial r}\hat{\mathbf{a}}_r + \frac{1}{r}\frac{\partial V}{\partial \theta}\hat{\mathbf{a}}_\theta + \frac{1}{r\sin\theta}\frac{\partial V}{\partial \phi}\hat{\mathbf{a}}_\phi} \quad ...(3)$$

Physical Significance of Gradient

1. *The gradient is differential operator by means of which we can associate a vector field with a scalar field.* For example, the intensity of electric field **E** (a vector quantity) is the gradient of potential V (scalar quantity) with a negative sign *i.e.*,

$$\mathbf{E} = -\text{grad } V.$$

The negative sign indicates that the direction of field intensity is opposite to the direction of increase of potential.

2. If the scalar function P represents the pressure, then $\vec{\nabla} P$ will represent the pressure gradient, *i.e.*, the rate of change of pressure with distance. So, $\vec{\nabla} P$ is a vector quantity having magnitude and direction both. The direction of $\vec{\nabla} P$ will be the direction of maximum rate of change of pressure.

Therefore, *the gradient of a scalar field* (say *P*), *is a vector quantity that represents both the magnitude and the direction of maximum space rate of increase of P.*

1.18 DIVERGENCE AND IT'S PHYSICAL SIGNIFICANCE

Definition of Divergence

The operator $\vec{\nabla}$ can also be involved in the multiplication with a vector. The dot product of operator $\vec{\nabla}$ with a vector **A** (*i.e.*, $\vec{\nabla} \cdot \mathbf{A}$) is called as divergence. The divergence is a scalar.

Let **A** be a vector function at each point (x, y, z) in a region of space. The divergence of A is given by

$$\vec{\nabla} \cdot \mathbf{A} = \left(\hat{\mathbf{a}}_x \frac{\partial}{\partial x} + \hat{\mathbf{a}}_y \frac{\partial}{\partial y} + \hat{\mathbf{a}}_z \frac{\partial}{\partial z}\right) \cdot (\hat{\mathbf{a}}_x A_x + \hat{\mathbf{a}}_y A_y + \hat{\mathbf{a}}_z A_z) = \frac{\partial A_x}{\partial x} + \frac{\partial A_y}{\partial y} + \frac{\partial A_z}{\partial z}$$

$$\therefore \qquad \text{div } \mathbf{A} = \vec{\nabla} \cdot \mathbf{A} = \frac{\partial A_x}{\partial x} + \frac{\partial A_y}{\partial y} + \frac{\partial A_z}{\partial z}$$

The divergence of a vector field at any point is defined as the amount of flux per unit volume diverging from that points.

$$div\, \vec{A} = \lim_{\Delta V \to 0} \frac{\phi \vec{A} . d\vec{s}}{\Delta V}$$

Divergence of a Vector A in Different Coordinate Systems

In cartesian coordinate system

$$\boxed{\text{div } \mathbf{A} = \frac{\partial A_x}{\partial x} + \frac{\partial A_y}{\partial y} + \frac{\partial A_z}{\partial z}} \qquad \text{...(4)}$$

In cylindrical coordinate system

$$\boxed{\text{div } \mathbf{A} = \frac{1}{\rho}\frac{\partial}{\partial \rho}(\rho A_\rho) + \frac{1}{\rho}\frac{\partial}{\partial \theta} A_\phi + \frac{\partial A_z}{\partial z}} \qquad \text{...(5)}$$

In spherical coordinate system

$$\boxed{\text{div } \mathbf{A} = \frac{1}{r^2}\frac{\partial}{\partial r}(r^2 A_r) + \frac{1}{r \sin\theta}\frac{\partial}{\partial \theta}(\sin\theta\, A_\theta) + \frac{1}{r \sin\theta}\frac{\partial A_\phi}{\partial \phi}} \qquad \text{...(6)}$$

Outward flow means divergence as shown in Fig. 29(*a*)

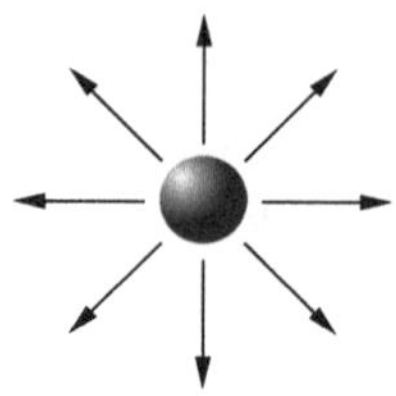

Outward flow

Fig. 29(a)

Inward flow means convergence (negative divergence)

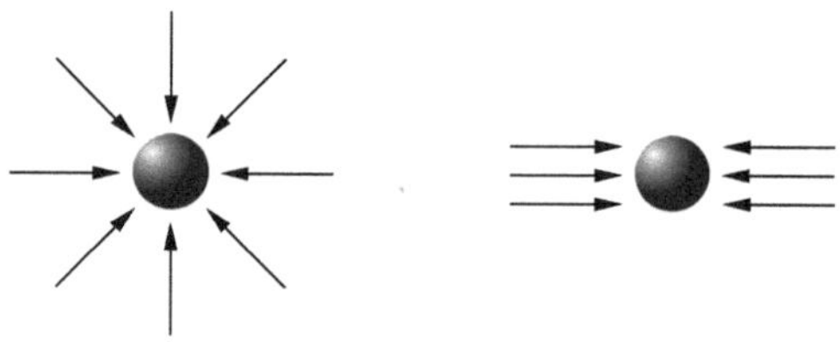

Inward flow (Zero divergence)

Fig. 29(b)

Physical Significance of Divergence

The divergence of a vector function at each point gives the rate per unit volume at which the physical entity is issuing from that point.

If **A** *represents the velocity of a fluid (liquid or gas), then div.* **A** *gives the rate of flow of the fluid at that point per unit volume.* If div **A** is positive at that point, then either the fluid is undergoing expansion (with its density falling proportionately) or the point is a source of fluid (a balloon filled with air is punctured by a pin). Similarly, a negative value of div. **A** means that either the fluid is undergoing, contraction (density is rising at that point) or the point serves as a sink. If However, div. $\mathbf{A} = 0$, then the fluid entering and leaving the element is the same, *i.e.*, there is no change in the density of the fluid (or fluid is incompressible). The vector satisfying this condition is called solenoidal.

For example, in electrostatic, the electric field intensity diverges from a positive charge while converges at a negative charge as shown in Fig. (30 *a* and *b*) respectively.

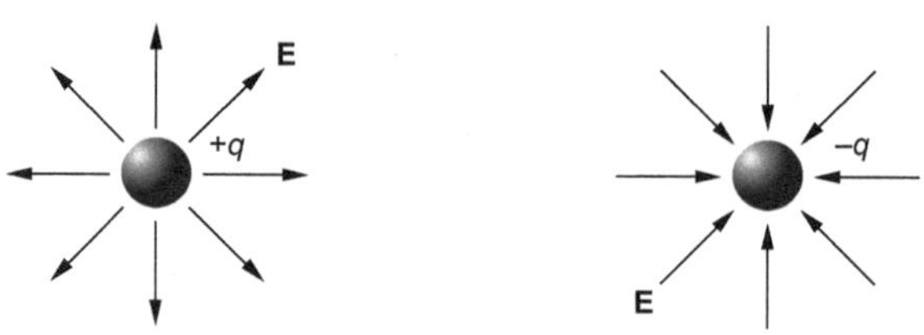

(*a*) Positive divergence (*b*) Negative divergence

Fig. (30)

1.19 CURL AND IT'S PHYSICAL SIGNIFICANCE

Definition of Curl

The cross product of del operator ($\vec{\nabla}$) with any vector (say **A**) is called as curl or rotation of that vector. Thus,

$$\text{curl } \mathbf{A} = \text{rot. } \mathbf{A} = \vec{\nabla} \times \mathbf{A} = \lim_{\Delta S \to 0} \frac{\oint_L \vec{A}.d\vec{l}}{\Delta S}$$

The curl of a vector at any point in a vector field is defined as a vector whose magnitude is equal to the line integral* per unit area along the boundary of an infinitesimal test area drawn around that point and whose direction is along the normal to this area.

$$\begin{aligned}
\text{curl } \mathbf{A} &= \vec{\nabla} \times \mathbf{A} \\
&= \left(\frac{\partial}{\partial x}\hat{\mathbf{a}}_x + \frac{\partial}{\partial y}\hat{\mathbf{a}}_y + \frac{\partial}{\partial z}\hat{\mathbf{a}}_z\right) \times \left[A_x\hat{\mathbf{a}}_x + A_y\hat{\mathbf{a}}_y + A_z\hat{\mathbf{a}}_z\right] \\
&= \frac{\partial A_x}{\partial x}(\hat{\mathbf{a}}_x \times \hat{\mathbf{a}}_x) + \frac{\partial A_x}{\partial y}\left(\hat{\mathbf{a}}_y \times \hat{\mathbf{a}}_x\right) + \frac{\partial A_x}{\partial z}(\hat{\mathbf{a}}_z \times \hat{\mathbf{a}}_x) \\
&\quad + \frac{\partial A_y}{\partial x}(\hat{\mathbf{a}}_x \times \hat{\mathbf{a}}_y) + \frac{\partial A_x}{\partial y}(\hat{\mathbf{a}}_y \times \hat{\mathbf{a}}_y) + \frac{\partial A_y}{\partial z}(\hat{\mathbf{a}}_z \times \hat{\mathbf{a}}_y) \\
&\quad + \frac{\partial A_z}{\partial x}(\hat{\mathbf{a}}_x \times \hat{\mathbf{a}}_z) + \frac{\partial A_z}{\partial y}(\hat{\mathbf{a}}_y \times \hat{\mathbf{a}}_z) + \frac{\partial A_z}{\partial z}(\hat{\mathbf{a}}_z \times \hat{\mathbf{a}}_z) \\
&= 0 - \frac{\partial A_x}{\partial y}\hat{\mathbf{a}}_z + \frac{\partial A_x}{\partial z}\hat{\mathbf{a}}_y + \frac{\partial A_y}{\partial x}\hat{\mathbf{a}}_z + 0 - \frac{\partial A_y}{\partial z}\hat{\mathbf{a}}_x - \frac{\partial A_z}{\partial y}\hat{\mathbf{a}}_y + \frac{\partial A_z}{\partial y}\hat{\mathbf{a}}_x + 0 \\
&= \left[\frac{\partial A_z}{\partial y} - \frac{\partial A_y}{\partial z}\right]\hat{\mathbf{a}}_x + \left[\frac{\partial A_x}{\partial z} - \frac{\partial A_z}{\partial x}\right]\hat{\mathbf{a}}_y + \left[\frac{\partial A_y}{\partial x} - \frac{\partial A_x}{\partial y}\right]\hat{\mathbf{a}}_z
\end{aligned}$$

This can be expressed in the following matrix form.

$$\vec{\nabla} \times \mathbf{A} = \begin{vmatrix} \hat{\mathbf{a}}_x & \hat{\mathbf{a}}_y & \hat{\mathbf{a}}_z \\ \frac{\partial}{\partial x} & \frac{\partial}{\partial y} & \frac{\partial}{\partial z} \\ A_x & A_y & A_z \end{vmatrix}$$

Curl of a Vector A in Different Coordinate systems

In cartesian coordinate system

$$\vec{\nabla} \times \mathbf{A} = \begin{vmatrix} \hat{\mathbf{a}}_x & \hat{\mathbf{a}}_y & \hat{\mathbf{a}}_z \\ \frac{\partial}{\partial x} & \frac{\partial}{\partial y} & \frac{\partial}{\partial z} \\ A_x & A_y & A_z \end{vmatrix} \qquad ...(7)$$

In cylindrical coordinate system

$$\vec{\nabla} \times \mathbf{A} = \frac{1}{\rho}\begin{vmatrix} \hat{\mathbf{a}}_\rho & \rho\,\hat{\mathbf{a}}_\phi & \hat{\mathbf{a}}_z \\ \frac{\partial}{\partial \rho} & \frac{\partial}{\partial \phi} & \frac{\partial}{\partial z} \\ A_\rho & \rho\,A_\phi & A_z \end{vmatrix} \qquad ...(8)$$

* The integration of a vector along a curve is called as line integral.

In spherical coordinate system

$$\vec{\nabla} \times \mathbf{A} = \frac{1}{r^2 \sin\theta} \begin{vmatrix} \hat{\mathbf{a}}_x & r\,\hat{\mathbf{a}}_y & r\sin\theta\,\hat{\mathbf{a}}_z \\ \dfrac{\partial}{\partial r} & \dfrac{\partial}{\partial \theta} & \dfrac{\partial}{\partial \phi} \\ A_r & r\,A_\theta & r\sin\theta\,A_\phi \end{vmatrix} \qquad ...(9)$$

Physical Significance of Curl

The general meaning of curl is rotation. When curl **A** is zero, it means that no rotation is attached with vector **A** whereas if curl **A** is non-zero, it means that rotation is attached with vector **A**. To make it more clear, consider the flow of a liquid. Let a frictionless paddle is placed in the path of liquid flow. In a hypothetical case, if we assume that all the liquid layers are moving with the same velocity, the paddle will not rotate as shown in Fig. (31*a*). This shows that the curl of velocity vector is zero. If we consider that the liquid layers are moving with different velocities (as the actual case is), then the paddle will rotate as shown in Fig. (31*b*). The rotation is due to non-zero value of the curl of velocity vector.

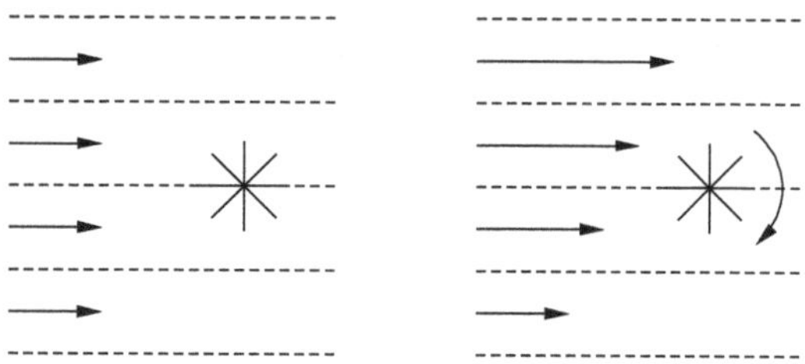

(*a*) (*b*)

Fig. (31)

Examples. Here we consider the following examples:

(*i*) When a rigid body is in motion, then the curl of its linear velocity at any point gives twice its angular velocity in magnitude and direction

$$\text{curl } \mathbf{v} = 2\omega$$

(*ii*) When a current is passed through a conductor, then magnetic field is developed around it. At any nearby point, the curl of the magnetic field represents current per unit area passing through that point. So curl **B** is also known as magneto-motive force.

(*iii*) The curl of an electric field results that rate of change of magnetic field in perpendicular direction.

(*iv*) The curl of magnetic field results the flow of electricity.

Few *Worked* EXAMPLES

Numerical Examples Based on Gradient of Scalar Field

EXAMPLE 1. *If $S = S(x, y, z) = x^2 - x^2 y + x y^2 z^2$, then find the value of grad S at the point* (2, – 1, – 3),

Solution. We know that grad $S = \vec{\nabla} S$ where

$$\vec{\nabla} = \hat{\mathbf{a}}_x \frac{\partial}{\partial x} + \hat{\mathbf{a}}_y \frac{\partial}{\partial y} + \hat{\mathbf{a}}_z \frac{\partial}{\partial z}$$

$$\therefore \qquad \text{grad } S = \left(\hat{\mathbf{a}}_x \frac{\partial}{\partial x} + \hat{\mathbf{a}}_y \frac{\partial}{\partial y} + \hat{\mathbf{a}}_z \frac{\partial}{\partial z}\right)(x^2 - x^2 y + x y^2 z^2)$$

$$= \hat{\mathbf{a}}_x \frac{\partial}{\partial x}(x^2 - x^2 y + x y^2 z^2) + \hat{\mathbf{a}}_z \frac{\partial}{\partial x}(x^2 - x^2 y + x y^2 z^2)$$

$$+ \hat{\mathbf{a}}_z \frac{\partial}{\partial x}(x^2 - x^2 y + xy^2 z^2)$$

$$= \hat{\mathbf{a}}_x (2x - 2xy + y^2 z^2) + \hat{\mathbf{a}}_y (-x^2 + 2xyz^2) + \hat{\mathbf{a}}_z (2zxy^2)$$

Substituting the given values, we get

$$\text{grad } S = \hat{\mathbf{a}}_x \{2 \times 2 - 2 \times 2(-1) + (-1)^2 (-3)^2\}$$

$$+ \hat{\mathbf{a}}_y \{-(2)^2 + 2 \times 2(-1)(-3)^2\} + \hat{\mathbf{a}}_z \{(-3)(2)(-1)^2\}$$

$$= \hat{\mathbf{a}}_x (17) - \hat{\mathbf{a}}_y (40) - \hat{\mathbf{a}}_z (12)$$

EXAMPLE 2. *The electric potential in a region of space is given by*

$$V = 5x - 7x^2y + 8y^2 + 16yz - 5z \text{ volt}$$

where distance are measured in metre. Deduce an expression for the electric field intensity **E**. *Calculate the y-component of the field at point* (2, 4, – 3).

Solution. We know that **E** = – grad V

$$\therefore \qquad E = -\hat{\mathbf{a}}_x \frac{\partial V}{\partial x} - \hat{\mathbf{a}}_y \frac{\partial V}{\partial y} - \hat{\mathbf{a}}_z \frac{\partial V}{\partial z} \qquad ...(1)$$

Given that $V = 5x - 7x^2y + 8y^2 + 16yz - 5z$

$$\therefore \qquad \frac{\partial V}{\partial x} = 5 - 14xy, \quad \frac{\partial V}{\partial y} = -7x^2 + 16y + 16z$$

and

$$\frac{\partial V}{\partial z} = 16y - 5$$

Putting these values in eq. (1), we get

$$\mathbf{E} = -\hat{\mathbf{a}}_x (5 - 14xy) - \hat{\mathbf{a}}_y (-7x^2 + 16y + 16z) - \hat{\mathbf{a}}_z (16y - 5)$$

or

$$\mathbf{E} = \hat{\mathbf{a}}_x (14xy - 5) + \hat{\mathbf{a}}_y (7x^2 - 16y - 16z) - \hat{\mathbf{a}}_z (5 - 16y) \qquad ...(2)$$

The y-component of field is

$$E_y = 7x^2 - 16y - 16z$$

At point (2, 4 – 3), its value is given by

$$E_y = 7(2)^2 - 16(4) - 16(-3) = 28 - 64 + 48 = \mathbf{12 \text{ volt/metre}}.$$

EXAMPLE 3. *If* **r** *is position vector of a point then evaluate* (*a*) *grad r and* (*b*) *grad* (1/*r*)

Solution.

(*a*) We know that $\mathbf{r} = \mathbf{i}x + \mathbf{j}y + \mathbf{k}z$

$$r = (x^2 + y^2 + z^2)^{1/2}$$

$$\text{Now grad } r = \left(\hat{\mathbf{a}}_x \frac{\partial}{\partial x} + \hat{\mathbf{a}}_y \frac{\partial}{\partial y} + \hat{\mathbf{a}}_z \frac{\partial}{\partial z}\right)(x^2 + y^2 + z^2)^{1/2}$$

$$= \hat{\mathbf{a}}_x \frac{\partial}{\partial x}(x^2 + y^2 + z^2)^{1/2} + \hat{\mathbf{a}}_y \frac{\partial}{\partial y}(x^2 + y^2 + z^2)^{1/2}$$

$$= \hat{\mathbf{a}}_z \frac{\partial}{\partial z}(x^2 + y^2 + z^2)^{1/2}$$

$$= \hat{\mathbf{a}}_x \frac{\frac{1}{2} \times 2x}{(x^2 + y^2 + z^2)^{1/2}} + \hat{\mathbf{a}}_y \frac{\frac{1}{2} \times 2y}{(x^2 + y^2 + z^2)^{1/2}} + \hat{\mathbf{a}}_z \frac{\frac{1}{2} \times 2z}{(x^2 + y^2 + z^2)^{1/2}}$$

$$= \frac{\hat{\mathbf{a}}_x x + \hat{\mathbf{a}}_y y + \hat{\mathbf{a}}_z z}{(x^2 + y^2 + z^2)^{1/2}} = \frac{\mathbf{r}}{r}$$

Let $\hat{\mathbf{a}}_r$ be a unit vector along r, then $r = \hat{\mathbf{a}}_r$ or $\mathbf{r}/r = \hat{\mathbf{a}}_r$

$$\therefore \qquad \text{grad } r = \hat{\mathbf{a}}_r$$

(*b*) $$\text{grad}\left(\frac{1}{r}\right) = \left(\hat{\mathbf{a}}_x \frac{\partial}{\partial x} + \hat{\mathbf{a}}_y \frac{\partial}{\partial y} + \hat{\mathbf{a}}_z \frac{\partial}{\partial z}\right)(x^2 + y^2 + z^2)^{-1/2}$$

$$= \hat{\mathbf{a}}_x \frac{\partial}{\partial x}(x^2 + y^2 + z^2)^{-1/2} + \hat{\mathbf{a}}_y \frac{\partial}{\partial y}(x^2 + y^2 + z^2)^{-1/2} + \hat{\mathbf{a}}_z \frac{\partial}{\partial z}(x^2 + y^2 + z^2)^{-1/2}$$

$$= \hat{\mathbf{a}}_x \frac{-\frac{1}{2} \times 2x}{(x^2 + y^2 + z^2)^{3/2}} + \hat{\mathbf{a}}_y \frac{-\frac{1}{2} \times 2y}{(x^2 + y^2 + z^2)^{3/2}} + \hat{\mathbf{a}}_x \frac{-\frac{1}{2} \times 2z}{(x^2 + y^2 + z^2)^{3/2}}$$

$$= \frac{\hat{\mathbf{a}}_x x + \hat{\mathbf{a}}_y y \, \hat{\mathbf{a}}_z z}{(x^2 + y^2 + z^2)^{3/2}} = -\frac{\mathbf{r}}{r^3} = -\frac{\hat{\mathbf{a}}_r}{r^2}$$

EXAMPLE 4. *If* **r** *is the positive vector, such that* $\phi = \log \mathbf{r}$, *find grad* ϕ.

Solution. The positive vector $\mathbf{r} = \hat{\mathbf{a}}_x x + \hat{\mathbf{a}}_y y + \hat{\mathbf{a}}_z z$

and $$r^2 = x^2 + y^2 + z^2 \quad \text{or} \quad r = (x^2 + y^2 + z^2)^{1/2}$$

$$\therefore \qquad \log r = \log(x^2 + y^2 + z^2) = \frac{1}{2}\log(x^2 + y^2 + z^2)$$

Now $$\text{grad } \phi = \text{grad} \log r = \nabla \log \log \mathbf{r} = \frac{1}{2}\vec{\nabla} \log(x^2 + y^2 + z^2)$$

$$= \frac{1}{2}\left[\frac{\partial}{\partial x}\log(x^2 + y^2 + z^2)\,\hat{\mathbf{a}}_x + \frac{\partial}{\partial x}\log(x^2 + y^2 + z^2)\,\hat{\mathbf{a}}_y + \frac{\partial}{\partial x}\log(x^2 + y^2 + z^2)\,\hat{\mathbf{a}}_z\right]$$

$$= \frac{1}{2}\left[\frac{2x}{(x^2 + y^2 + z^2)}\hat{\mathbf{a}}_x + \frac{2y}{(x^2 + y^2 + z^2)}\hat{\mathbf{a}}_y + \frac{2z}{(x^2 + y^2 + z^2)}\hat{\mathbf{a}}_z\right]$$

$$= \frac{x\hat{\mathbf{a}}_x + y\hat{\mathbf{a}}_y + \hat{\mathbf{a}}_z}{(x^2 + y^2 + z^2)} = \frac{\mathbf{r}}{r^2}$$

EXAMPLE 5. *Prove that grad* r^n *or* $\vec{\nabla} r^n = n\, r^n\, \mathbf{r}$ *where is positive vector of any point*

Solution. We know that $r = \hat{\mathbf{a}}_x x + \hat{\mathbf{a}}_y y + \hat{\mathbf{a}}_z z$

So that $r = (x^2 + y^2 + z^2)^{1/2}$

$$\vec{\nabla} r^n = \nabla (x^2 + y^2 + z^2)^{n/2}$$

$$= \hat{\mathbf{a}}_x \frac{\partial}{\partial x}(x^2 + y^2 + z^2)^{n/2} + \hat{\mathbf{a}}_y \frac{\partial}{\partial y}(x^2 + y^2 + z^2)^{n/2} + \hat{\mathbf{a}}_z \frac{\partial}{\partial z}(x^2 + y^2 + z^2)^{n/2}$$

$$= \hat{\mathbf{a}}_x \frac{n}{2}(x^2 + y^2 + z^2)^{\{n/2-1\}} \cdot 2x + \hat{\mathbf{a}}_y \frac{n}{2}(x^2 + y^2 + z^2)^{\{n/2-1\}} \cdot 2y$$

$$+ \hat{\mathbf{a}}_z \frac{n}{2}(x^2 + y^2 + z^2)^{\{n/2-1\}} \cdot 2z$$

$$= n(x^2 + y^2 + z^2)^{\{n/2-1\}} (\hat{\mathbf{a}}_x\, x + \hat{\mathbf{a}}_y\, y + \hat{\mathbf{a}}_z\, z)$$

$$= n(r^2)^{\{n/2-1\}}\, \mathbf{r} = n\, r^{n-2}\, \mathbf{r}$$

Examples Based on Divergence of a Vector Field

EXAMPLE 1. *If* $\mathbf{A} = \hat{\mathbf{a}}_x\, y + \hat{\mathbf{a}}_y\, (x^2 + y^2) + \hat{\mathbf{a}}_z\, (yz + zx)$ *then find div* $\mathbf{A}$ *at point* (1, – 2, 3).

Solution. We know that the div $\mathbf{A} = \nabla \cdot \mathbf{A}$.

where $\vec{\nabla} = \hat{\mathbf{a}}_x \frac{\partial}{\partial x} + \hat{\mathbf{a}}_y \frac{\partial}{\partial y} + \hat{\mathbf{a}}_z \frac{\partial}{\partial z}$

$$\therefore \quad \text{div}\, \mathbf{A} = \left(\hat{\mathbf{a}}_x \frac{\partial}{\partial x} + \hat{\mathbf{a}}_y \frac{\partial}{\partial y} + \hat{\mathbf{a}}_z \frac{\partial}{\partial z}\right) \cdot \{\hat{\mathbf{a}}_x\, y + \hat{\mathbf{a}}_y\, (x^2 + y^2) + \hat{\mathbf{a}}_z\, (yz + zy)\}$$

$$= \frac{\partial}{\partial x} y + \frac{\partial}{\partial y}(x^2 + y^2) + \frac{\partial}{\partial z}(yz + zx) = 0 + 2y + (y + x) = (3\,y + x)$$

At point (1, – 2, 3)

$$\text{div}\, \mathbf{A} = 3 \times (-2) + 1 = \mathbf{-5}.$$

EXAMPLE 2. *Evaluate div.* $\mathbf{F}$*, where*

$$\mathbf{F} = 2x^3\, z\,\hat{\mathbf{a}}_x - xy^2\, z\,\hat{\mathbf{a}}_y + 3\,y^2 x\,\hat{\mathbf{a}}_z$$

Solution. $$\nabla \cdot \mathbf{F} = \left(\frac{\partial}{\partial x}\hat{\mathbf{a}}_x + \frac{\partial}{\partial y}\hat{\mathbf{a}}_y + \frac{\partial}{\partial z}\hat{\mathbf{a}}_z\right) \cdot (2x^3\, z\,\hat{\mathbf{a}}_x - x\,y^2\, z\hat{\mathbf{a}}_y + 3\,y^2\, x\,\hat{\mathbf{a}}_z)$$

$$\therefore \quad \nabla \cdot \mathbf{F} = \frac{\partial}{\partial x}(2x^3\, z) - \frac{\partial}{\partial y}(x\,y^2\, z) + \frac{\partial}{\partial z}(3\,y^2\, x)$$

$$= 6x^2\, z - 2x\,y\,z + 0 = 6x^2\, z - 2x\,y\,z.$$

EXAMPLE 3. *If* $\mathbf{A} = 2\rho\cos^2\phi\; \hat{\mathbf{a}}_\rho + 3\rho^2 \sin z\; \hat{\mathbf{a}}_\phi + 4\,z\sin^2\phi\; \hat{\mathbf{a}}_z$ *then find* $\vec{\nabla} \cdot \mathbf{A}$.

Solution. We know that in cylindrical coordinate system

$$\vec{\nabla} \cdot \mathbf{A} = \frac{1}{\rho}\frac{\partial}{\partial \rho}(\rho A_\rho) + \frac{1}{\rho}\frac{\partial A_\phi}{\partial \phi} + \frac{\partial A_z}{\partial z}$$

Here, $A_\rho = 2\rho\cos^2\phi$, $A\phi = 3\rho^2 \sin z$ and $A_z = 4\,z\sin^2\phi$

Substituting these values, we get

$$\vec{\nabla} \cdot \mathbf{A} = \frac{1}{\rho}\frac{\partial}{\partial \rho}[2\rho^2 \cos^2\phi] + \frac{1}{\rho}\frac{\partial}{\partial \phi}[3\rho^2 \sin z] + \frac{\partial}{\partial z}[4z\sin^2\phi]$$

$$= \frac{1}{\rho}[4\rho \cos^2 \phi] + \frac{1}{\rho}[0] + 4 \sin^2 \phi$$

$$= 4 \cos^2 \phi + 4 \sin^2 \phi = 4 [\cos^2 \phi + \sin^2 \phi] = \mathbf{4}$$

EXAMPLE 4. *If* $A = 10\ \hat{\mathbf{a}}_r + 5 \sin \theta\ \hat{\mathbf{a}}_\theta$ *then find* $\vec{\nabla} \cdot \mathbf{A}$.

Solution. We know that in spherical coordinate system

$$\vec{\nabla} \cdot \mathbf{A} = \frac{1}{r^2}\frac{\partial}{\partial r}(r^2 A_r) + \frac{1}{r \sin \theta}\frac{\partial}{\partial \theta}(\sin \theta\, A_\theta) + \frac{1}{r \sin \theta}\frac{\partial A\phi}{\partial \phi}$$

Here, $A_r = 10, A_\theta = 5 \sin \theta$ and $A_\phi = 0$

Substituting these value, we get

$$\vec{\nabla} \cdot \mathbf{A} = \frac{1}{r^2}\frac{\partial}{\partial r}(r^2 \times 10) + \frac{1}{r \sin \theta}\frac{\partial}{\partial \theta}(5 \sin\theta \times \sin\theta) + \frac{1}{r \sin\theta}\frac{\partial}{\partial \phi}(0)$$

$$= \frac{1}{r^2}(20r) + \frac{1}{r \sin \theta}(10 \sin \theta \cos\theta) + 0$$

$$= \frac{20}{r} + \frac{10}{r} \cos \theta = \frac{10}{r}[2 + \cos\theta]$$

EXAMPLE 5. *Determine the divergence of the vector field*

$$\mathbf{A} = \rho z \sin \phi\ \hat{\mathbf{a}}_\rho + \mathbf{3}\, \rho z^2 \cos \phi\ \hat{\mathbf{a}}_\phi$$

at point (5, π/2, 1).

Solution. In cylindrical coordinate system

$$\vec{\nabla} \cdot \mathbf{A} = \frac{1}{\rho}\frac{\partial}{\partial \rho}(\rho A_\rho) + \frac{1}{\rho}\frac{\partial A_\phi}{\partial \phi} + \frac{\partial A_z}{\partial z}$$

Given that $\mathbf{A} = \rho z \sin \phi\ \hat{\mathbf{a}}_\rho + 3\rho z^2 \cos \phi\ \hat{\mathbf{a}}_\phi$

Therefore, $A_\rho = \rho z \sin \phi$, $A_\phi = 3\rho z^2 \cos \phi$ and $A_z = 0$

Substitution these values, we get

$$\vec{\nabla} \cdot \mathbf{A} = \frac{1}{\rho}\frac{1}{\partial \rho}(\rho \times \rho z \sin \phi) + \frac{1}{\rho}\frac{\partial}{\partial \phi}(3 \rho z^2 \cos \phi) + \frac{\partial}{\partial z}(0)$$

$$= \frac{1}{\rho}[2 \rho z \sin \phi] + \frac{1}{\rho}[-3 \rho z^2 \sin \phi]$$

$$= 2 z \sin \phi - 3 z^2 \sin \phi = z \sin \phi\, [2 - 3 z]$$

The value of $\vec{\nabla} \cdot \mathbf{A}$ at point (5, π/2, 1) is given by

$$\vec{\nabla} \cdot \mathbf{A}\,\big|_{(5,\, \pi/2,\, 1)} = 1.\sin \pi/2.\ (2 - 3 \times 1) = 1.1\ (2 - 3) = 1.1.\ (-1) = \mathbf{-1}$$

EXAMPLE 6. *Determine the divergence of vector field*

$$\mathbf{A} = 2\, r \cos \theta \cos \phi\ \hat{\mathbf{a}}_r + r^{1/2}\ \hat{\mathbf{a}}_\phi$$

at point (1, π/6, π/3)

Solution. In spherical coordinate system

$$\vec{\nabla} \cdot \mathbf{A} = \frac{1}{r^2}\frac{\partial}{\partial r}(r^2 A_r) + \frac{1}{r \sin \theta}\frac{\partial}{\partial \theta}(\sin \theta\, A_\theta) + \frac{1}{r \sin \theta}\frac{\partial A_\phi}{\partial \phi}$$

The given vector field is

$$A = 2\,r\cos\theta\cos\phi\;\hat{\mathbf{a}}_r + r^{1/2}\,\hat{\mathbf{a}}_\phi$$

Therefore $A_r = 2r\cos\theta\cos\phi$, $A_\theta = 0$ and $A_\phi = r^{1/2}$

Substituting these values, we get

$$\vec{\nabla}\cdot\mathbf{A} = \frac{1}{r^2}\frac{\partial}{\partial r}(r^2\,2r\cos\theta\cos\phi) + \frac{1}{r\sin\theta}\frac{\partial}{\partial\theta}(\sin\theta\times 0) + \frac{1}{r\sin\theta}\frac{\partial}{\partial\phi}(r^{1/2})$$

$$= \frac{1}{r^2}\frac{\partial}{\partial r}(2r^3\cos\theta\cos\phi) + \frac{1}{r\sin\theta}\frac{\partial}{\partial\theta}(r^{1/2})$$

$$= \frac{1}{r^2}\,(6\,r^2\cos\theta\cos\phi) + 0 = 6\cos\theta\cos\phi$$

The value of $\vec{\nabla}\cdot\mathbf{A}$ at point $(1, \pi/6, \pi/3)$ is given by

$$\vec{\nabla}\cdot\mathbf{A}\Big|_{(1,\,\pi/6,\,\pi/3)} = 6\cos\frac{\pi}{6}\cos\frac{\pi}{3} = 6\cos 30\cos 60$$

$$= 6\left(\frac{\sqrt{3}}{2}\right)\left(\frac{1}{2}\right) = \frac{(3\sqrt{3})}{2} = \mathbf{2.598}.$$

EXAMPLE 7. *If* **r** *is the position vector of a point, then show that*
(a) div $\mathbf{r} = 3$ *and (b)* $\nabla(\mathbf{r}\cdot\mathbf{A}) = \mathbf{A}$

Solution.

(*a*) The position vector **r** is written as

$$\mathbf{r} = \hat{\mathbf{a}}_x\,x + \hat{\mathbf{a}}_y\,y + \hat{\mathbf{a}}_z\,z$$

So that $r^2 = x^2 + y^2 + z^2$

Now

$$\text{div}\,\mathbf{r} = \vec{\nabla}\cdot\mathbf{r} = \left(\hat{\mathbf{a}}_x\frac{\partial}{\partial x} + \hat{\mathbf{a}}_y\frac{\partial}{\partial y} + \hat{\mathbf{a}}_z\frac{\partial}{\partial z}\right)\cdot(\hat{\mathbf{a}}_x\,x + \hat{\mathbf{a}}_y\,y + \hat{\mathbf{a}}_z\,z)$$

$$= \frac{\partial x}{\partial x} + \frac{\partial y}{\partial y} + \frac{\partial z}{\partial z} = 1 + 1 + 1 = 3$$

(*b*) $\mathbf{r}\,.\,\mathbf{A} = (\hat{\mathbf{a}}_x\,x + \hat{\mathbf{a}}_y\,y + \hat{\mathbf{a}}_z\,z)(\hat{\mathbf{a}}_x\,A_x + \hat{\mathbf{a}}_y\,A_y + \hat{\mathbf{a}}_z\,A_z) = xA_x + yA_y + zA_z$

$$\therefore\quad \nabla\cdot(\mathbf{r}\cdot\mathbf{A}) = \left(\hat{\mathbf{a}}_x\frac{\partial}{\partial x} + \hat{\mathbf{a}}_y\frac{\partial}{\partial y} + \hat{\mathbf{a}}_z\frac{\partial}{\partial z}\right)\cdot(xA_x + yA_y + zA_z)$$

$$= \hat{\mathbf{a}}_x\,A_x + \hat{\mathbf{a}}_y\,A_y + \hat{\mathbf{a}}_z\,A_z = \mathbf{A}.$$

EXAMPLE 8. *If* **r** *is the position vector of a point then show that*
div $(r^n\,\mathbf{r}) = (3+n)\,r^n$

Solution.

$$\nabla\cdot(r^n\,\mathbf{r}) = \left(\hat{\mathbf{a}}_x\frac{\partial}{\partial x} + \hat{\mathbf{a}}_y\frac{\partial}{\partial y} + \hat{\mathbf{a}}_z\frac{\partial}{\partial z}\right)\cdot r^n\,(\hat{\mathbf{a}}_x\,x + \hat{\mathbf{a}}_y\,y + \hat{\mathbf{a}}_z\,z)$$

$$= \frac{\partial}{\partial x}(r^n\,x) + \frac{\partial}{\partial y}(r^n\,y) + \frac{\partial}{\partial z}(r^n\,z)$$

We know that $\mathbf{r} = \hat{\mathbf{a}}_x\,x + \hat{\mathbf{a}}_y\,y + \hat{\mathbf{a}}_z\,z$

and $|\,\mathbf{r}\,| = r = (x^2 + y^2 + z^2)^{1/2}$

$$\therefore \quad \nabla \cdot (r^n \mathbf{r}) = \frac{\partial}{\partial x}\{(x^2+y^2+z^2)^{n/2}x\} + \frac{\partial}{\partial y}\{(x^2+y^2+z^2)^{n/2}y\} + \frac{\partial}{\partial z}\{(x^2+y^2+z^2)^{n/2}z\}$$

Now we consider the x component, hence

$$\frac{\partial}{\partial x}\{(x^2+y^2+z^2)^{n/2}x\} = (x^2+y^2+z^2)^{n/2}\cdot 1 + x\,\frac{n}{2}\,(x^2+y^2+z^2)^{(n/2)-1}\cdot 2x$$
$$= (r^2)^{n/2} + nx^2\,(r^2)^{(n/2)-1} = r^n + nx^2\,r^{n-2}$$

Similarly,

$$\frac{\partial}{\partial y}\{(x^2+y^2+z^2)^{n/2}y\} = r^n + n\,y^2\,r^{n-2}$$

and
$$\frac{\partial}{\partial z}\{(x^2+y^2+z^2)^{n/2}z\} = r^n + n\,y^2\,r^{n-2}$$

$$\therefore \quad \nabla \cdot (r^n r) = \{r^n + nx^2 r^{n-2}\} + \{r^n + ny^2 r^{n-2}\} + \{r^n + nz^2 r^{n-2}\}$$
$$= 3\,r^n + nr^{n-2}\,(x^2+y^2+z^2) = 3\,r^n + r^{n-2}\,r^2$$
$$= 3\,r^n + n\,r^n = r^n\,(3+n)$$

EXAMPLE 9. *If the gravitational potential at any point is $(-GM/r)$, where r is the position vector on the point, find the intensity of gravitational field at the point*

Solution. Here, $\mathbf{r} = \hat{\mathbf{a}}_x\,x + \hat{\mathbf{a}}_y\,y + \hat{\mathbf{a}}_z\,z$ and $r = (x^2+y^2+z^2)^{1/2}$

$$\vec{\nabla}\left(\frac{-GM}{r}\right) = GM\,\vec{\nabla}\left(\frac{1}{r}\right) = -GM\,\vec{\nabla}\,[x^2+y^2+z^2]^{-1/2}$$

$$= GM\left\{\hat{\mathbf{a}}_x\frac{\partial}{\partial x}(x^2+y^2+z^2)^{-1/2} + \hat{\mathbf{a}}_y\frac{\partial}{\partial y}(x^2+y^2+z^2)^{-1/2} + \hat{\mathbf{a}}_z\frac{\partial}{\partial z}(x^2+y^2+z^2)^{-1/2}\right\}$$

$$= -GM\left\{\hat{\mathbf{a}}_x\left(-\frac{1}{2}\right)(x^2+y^2+z^2)^{-3/2}\cdot 2x + \hat{\mathbf{a}}_y\left(-\frac{1}{2}\right)(x^2+y^2+z^2)^{-3/2}\cdot 2y + \hat{\mathbf{a}}_z\left(-\frac{1}{2}\right)(x^2+y^2+z^2)^{-3/2}\cdot 2z\right\}$$

$$= \frac{\hat{\mathbf{a}}_x\,x + \hat{\mathbf{a}}_y\,y + \hat{\mathbf{a}}_z\,z}{(x^2+y^2+z^2)^{3/2}} = GM\left(\frac{r}{r^3}\right)$$

EXAMPLE 10. *Find the value of constant c for which the vector*

$\mathbf{A} = \hat{\mathbf{a}}_x\,(x+3y) + \hat{\mathbf{a}}_y\,(y-2z) + \hat{\mathbf{a}}_z\,(x+cz)$ *is solenoidal.*

Solution. The vector **A** is solenoidal if the divergence is zero. We know that

$$\text{div}\,\mathbf{A} = \frac{\partial A_x}{\partial x} + \frac{\partial A_y}{\partial y} + \frac{\partial A_z}{\partial z} = \frac{\partial}{\partial x}(x+3y) + \frac{\partial}{\partial y}(y-2z) + \frac{\partial}{\partial z}(x+cz)$$
$$= 1 + 1 + c = 2 + c$$
$$2 + c = 0 \quad \text{or} \quad c = -\mathbf{2}.$$

EXAMPLE 11. *Find the nature of the given vector field by determining the divergence of the field*

$$\mathbf{A} = 30\,\hat{\mathbf{a}}_x + 2xy\,\hat{\mathbf{a}}_y + 5xz^2\,\hat{\mathbf{a}}_z$$

Solution. The divergence of the vector field in cartesian coordinate is given by

$$\vec{\nabla}\cdot\mathbf{A} = \frac{\partial}{\partial x}A_x + \frac{\partial}{\partial y}A_y + \frac{\partial}{\partial z}A_z$$

The given vector field is

$$\mathbf{A} = 30\,\hat{\mathbf{a}}_x + 2x\,y\hat{\mathbf{a}}_y + 5x\,z^2\,\hat{\mathbf{a}}_z$$

Therefore, $A_x = 30, A_y = 2xy$ and $A_z = 5\,xz^2$

Substituting these values, we get

$$\vec{\nabla}\cdot\mathbf{A} = \frac{\partial}{\partial x}(30) + \frac{\partial}{\partial y}(2x\,y) + \frac{\partial}{\partial z}(5xz^2) = 0 + 2x + 10xz = 2x\,(1 + 5z)$$

This shown that there exists the divergence of the field. Hence, the field is **non-solenoidal.**

EXAMPLE 12. *If a is constant vector, field. div. (r × a).*

Solution. The vector $\mathbf{r} = \hat{\mathbf{a}}_x\,x + \hat{\mathbf{a}}_y\,y + \hat{\mathbf{a}}_z\,z$

Let $\mathbf{a} = \hat{\mathbf{a}}_x\,a_1 + \hat{\mathbf{a}}_y\,a_2 + \hat{\mathbf{a}}_z\,a_3$

$$\mathbf{r}\times\mathbf{a} = \begin{vmatrix} \hat{\mathbf{a}}_x & \hat{\mathbf{a}}_y & \hat{\mathbf{a}}_z \\ x & y & z \\ a_1 & a_2 & a_3 \end{vmatrix} = \hat{\mathbf{a}}_x\,(a_3\,y - a_2\,z) + \hat{\mathbf{a}}_y\,(a_1 z - a_3 z) + \hat{\mathbf{a}}_z\,(a_2\,x - a_1 y)$$

Now, $$\text{div}\,(\mathbf{r}\times\mathbf{a}) = \frac{\partial}{\partial x}(a_3\,y - a_2\,z) + \frac{\partial}{\partial y}(a_1\,z - a_3\,x) + \frac{\partial}{\partial z}(a_2\,x - a_1\,y) = 0 + 0 + 0 = \mathbf{0}.$$

Numerical Examples Based on Curl

EXAMPLE 1. *If* $\mathbf{A} = \hat{\mathbf{a}}_x\,y + \hat{\mathbf{a}}_y(x^2 + y^2) + \hat{\mathbf{a}}_z\,(yz + zx)$, *then find curl* **A** *at point* (2, 2, – 2).

Solution. We known that curl $\mathbf{A} = \nabla\times\mathbf{A}$

$$\therefore \quad \text{curl}\,\mathbf{A} = \begin{vmatrix} \hat{\mathbf{a}}_x & \hat{\mathbf{a}}_y & \hat{\mathbf{a}}_z \\ \frac{\partial}{\partial x} & \frac{\partial}{\partial y} & \frac{\partial}{\partial z} \\ A_x & A_y & A_z \end{vmatrix} = \begin{vmatrix} \hat{\mathbf{a}}_x & \hat{\mathbf{a}}_y & \hat{\mathbf{a}}_z \\ \frac{\partial}{\partial x} & \frac{\partial}{\partial y} & \frac{\partial}{\partial z} \\ y & (x^2 + y^2) & (yz + zx) \end{vmatrix}$$

$$= \hat{\mathbf{a}}_x\left[\frac{\partial}{\partial y}(yz + zx) - \frac{\partial}{\partial z}(x^2 + y^2)\right]$$

$$+ \hat{\mathbf{a}}_y\left[\frac{\partial}{\partial z}y - \frac{\partial}{\partial x}(yz + zx)\right] + \hat{\mathbf{a}}_z\left[\frac{\partial}{\partial x}(x^2 + y^2) - \frac{\partial}{\partial y}y\right]$$

$$= \hat{\mathbf{a}}_x\,[z - 0] + \hat{\mathbf{a}}_y\,[0 - z] + \hat{\mathbf{a}}_z\,[2x - 1] = \hat{\mathbf{a}}_x\,z - \hat{\mathbf{a}}_y\,z + \hat{\mathbf{a}}_z\,(2x - 1)$$

Substituting the values, we get

$$\text{curl } \mathbf{A} = -2\,\hat{\mathbf{a}}_x + 2\hat{\mathbf{a}}_y + 3\hat{\mathbf{a}}_z$$

EXAMPLE 2. *If* **r** *is the position vector of point, prove that curl* **r** = 0.

Solution. When **r** is the position vector of (x, y, z) then

$$\mathbf{r} = \hat{\mathbf{a}}_x\, x + \hat{\mathbf{a}}_y\, y + \hat{\mathbf{a}}_z\, z$$

Now

$$\text{curl } \mathbf{r} = \nabla \times \mathbf{r} = \left(\hat{\mathbf{a}}_x \frac{\partial}{\partial x} + \hat{\mathbf{a}}_y \frac{\partial}{\partial y} + \hat{\mathbf{a}}_z \frac{\partial}{\partial z}\right) \times (\hat{\mathbf{a}}_x\, x + \hat{\mathbf{a}}_y\, y + \hat{\mathbf{a}}_z\, z)$$

$$= \begin{vmatrix} \hat{\mathbf{a}}_x & \hat{\mathbf{a}}_y & \hat{\mathbf{a}}_z \\ \dfrac{\partial}{\partial x} & \dfrac{\partial}{\partial y} & \dfrac{\partial}{\partial z} \\ x & y & z \end{vmatrix}$$

$$\therefore \quad \text{curl } \mathbf{r} = \hat{\mathbf{a}}_x \left(\frac{\partial z}{\partial x} - \frac{\partial y}{\partial z}\right) + \hat{\mathbf{a}}_y \left(\frac{\partial x}{\partial z} - \frac{\partial z}{\partial x}\right) + \hat{\mathbf{a}}_z \left(\frac{\partial y}{\partial x} - \frac{\partial x}{\partial y}\right)$$

$$= \hat{\mathbf{a}}_x\,(0-0) + \hat{\mathbf{a}}_y\,(0-0) + \hat{\mathbf{a}}_z\,(0-0) = 0$$

EXAMPLE 3. *Evaluate* $\vec{\nabla} \times \mathbf{A}$, *if*

$$\mathbf{A} = \rho^2 z\, \hat{\mathbf{a}}_\rho + \rho^3\, \hat{\mathbf{a}}_\phi + 3\,\rho\, z^2\, \hat{\mathbf{a}}_z$$

Solution. In cylindrical coordinate system,

$$\vec{\nabla} \times \mathbf{A} = \frac{1}{\rho} \begin{vmatrix} \hat{\mathbf{a}}_\rho & \hat{\mathbf{a}}_\phi & \hat{\mathbf{a}}_z \\ \dfrac{\partial}{\partial \rho} & \dfrac{\partial}{\partial \phi} & \dfrac{\partial}{\partial z} \\ A_\rho & \rho A_\phi & A_z \end{vmatrix}$$

The given vector, $\mathbf{A} = \rho^2 z\, \hat{\mathbf{a}}_\rho + \rho^3\, \hat{\mathbf{a}}_\phi + 3\rho z^2\, \hat{\mathbf{a}}_z$

Therefore, $A_\rho = \rho^2 z$, $A_\phi = \rho^3$ and $A_z = 3\,\rho\, z^2$

$$\therefore \quad \vec{\nabla} \times \mathbf{A} = \frac{1}{\rho} \begin{vmatrix} \hat{\mathbf{a}}_\rho & \hat{\mathbf{a}}_\phi & \hat{\mathbf{a}}_z \\ \dfrac{\partial}{\partial \rho} & \dfrac{\partial}{\partial \phi} & \dfrac{\partial}{\partial z} \\ \rho^2 z & \rho^4 & 3\rho z^2 \end{vmatrix}$$

$$= \frac{1}{\rho}\left[\hat{\mathbf{a}}_\rho \left\{\frac{\partial}{\partial \phi}(3\rho z^2) - \frac{\partial}{\partial z}(\rho^4)\right\} + \rho\, \hat{\mathbf{a}}_\phi \left\{\frac{\partial}{\partial z}(\rho^2 z) - \frac{\partial}{\partial \rho}(3\,\rho\, z^2)\right\} + \hat{\mathbf{a}}_z \left\{\frac{\partial}{\partial z}(\rho^4) - \frac{\partial}{\partial \phi}(\rho^2\, z^2)\right\}\right]$$

$$= \frac{1}{\rho}[\hat{\mathbf{a}}_\rho\,(0) + \rho\, \hat{\mathbf{a}}_\phi\, [\rho^2 - 3z^2] + \hat{\mathbf{a}}_z\,(4\rho^3)]$$

$$= \frac{1}{\rho}[(\rho^3 - 3\rho^3 z^2)\hat{\mathbf{a}}_\phi + 4\rho^3 \hat{\mathbf{a}}_z] = \frac{1}{\rho}[(\rho^2 - 3z^2)\hat{\mathbf{a}}_\phi + (4\rho^2)\hat{\mathbf{a}}_z$$

EXAMPLE 4. *Find the curl of a vector*

$$\mathbf{A} = (e^{-r}/r)\,\hat{\mathbf{a}}_\theta$$

Solution. In spherical coordinate system

$$\vec{\nabla} \times \mathbf{A} = \frac{1}{r^2 \sin\theta}\begin{vmatrix} \hat{\mathbf{a}}_r & r\,\hat{\mathbf{a}}_\theta & r\sin\theta\,\hat{\mathbf{a}}_\phi \\ \dfrac{\partial}{\partial r} & \dfrac{\partial}{\partial \theta} & \dfrac{\partial}{\partial \phi} \\ A_r & rA_\theta & r\sin\theta\,A_\phi \end{vmatrix}$$

Given $\mathbf{A} = (e^{-r}/r)\,\hat{\mathbf{a}}_\theta$, *i.e.*, $A_r = 0$, $A_\theta = (e^{-r}/r)$ and $A_\phi = 0$

Substituting these values, we get

$$\vec{\nabla} \times \mathbf{A} = \frac{1}{r^2 \sin\theta}\begin{vmatrix} \hat{\mathbf{a}}_r & \hat{\mathbf{a}}_\theta & r\sin\theta\,\hat{\mathbf{a}}_\phi \\ \dfrac{\partial}{\partial r} & \dfrac{\partial}{\partial \theta} & \dfrac{\partial}{\partial \phi} \\ 0 & r\dfrac{e^{-r}}{r} & 0 \end{vmatrix}$$

$$\therefore \qquad \vec{\nabla} \times \mathbf{A} = \frac{1}{r^2 \sin\theta}\left[\hat{\mathbf{a}}_r\left\{0 - \frac{\partial}{\partial\phi}\left(r\frac{e^{-r}}{r}\right)\right\} + r\,\hat{\mathbf{a}}_\phi\,(0-0) + r\sin\theta\,\hat{\mathbf{a}}_\phi\,\frac{\partial}{\partial r}\left(r\frac{e^{-r}}{r}\right)\right]$$

$$= \frac{1}{r^2 \sin\theta} \times r\sin\theta\,\hat{\mathbf{a}}_\phi\,\frac{\partial}{\partial r}(e^{-r}) = \frac{\hat{\mathbf{a}}_\phi}{r}\frac{\partial}{\partial r}(e^{-r}) = -\frac{e^{-r}}{r}\,\hat{\mathbf{a}}_\phi$$

EXAMPLE 5. *Calculate the workdone by the force experienced by a charge q moving with a velocity* **v** *in a magnetic field* **B**.

Solution. The force **F** experienced by a charged particle (charge q) in magnetic field **B** is given by

$$F = q(\mathbf{v} \times \mathbf{B})$$

$$\text{work done} = \mathbf{F} \cdot \mathbf{s} = \mathbf{F} \cdot \mathbf{v}t \qquad (\because\ \mathbf{s} = \mathbf{v}t)$$

But $(\mathbf{v} \times \mathbf{B})$ and v are perpendicular and hence

$$(\mathbf{v} \times \mathbf{B}) \cdot \mathbf{v} = 0$$

$$\therefore \qquad \text{workdone} = 0$$

So, the work done by force **F** is zero.

EXAMPLE 6. *Show that* $\mathbf{A} = e^{-x}(-yz\,\hat{\mathbf{a}}_x + z\,\hat{\mathbf{a}}_y + y\,\hat{\mathbf{a}}_z)$ *is irrotational.*

Solution. We known that when $\nabla \times \mathbf{A} = 0$, the function is irrotational. Given that

$$\mathbf{A} = -yz\,e^{-x}\,\hat{\mathbf{a}}_x + z\,e^{-x}\,\hat{\mathbf{a}}_y + ye^{-x}\,\hat{\mathbf{a}}_z$$

$$\therefore \qquad \nabla \times \mathbf{A} = \begin{vmatrix} \hat{\mathbf{a}}_x & \hat{\mathbf{a}}_y & \hat{\mathbf{a}}_z \\ \partial/\partial x & \partial/\partial y & \partial/\partial z \\ -yze^{-x} & ze^{-x} & ye^{-x} \end{vmatrix}$$

$$= \hat{\mathbf{a}}_x \left[\frac{\partial}{\partial y}(ye^{-x}) - \frac{\partial}{\partial z}(ze^{-x})\right] - \hat{\mathbf{a}}_y \left[\frac{\partial}{\partial x}(y-x) - \frac{\partial}{\partial z}(-yze^{-x})\right]$$

$$+ \hat{\mathbf{a}}_z \left[\frac{\partial}{\partial x}(ze^{-x}) - \frac{\partial}{\partial y}(-yze^{-x})\right]$$

$$= \hat{\mathbf{a}}_x (e^{-x} - e^{-x}) - \hat{\mathbf{a}}_y (-ye^{-x} + ye^{-x}) + \hat{\mathbf{a}}_z (-ze^{-x} + ze^{-x})$$

So, the function is **irrotational.**

EXAMPLE 7. *Show that the field*

$$\mathbf{F} = \left(\frac{150}{\rho^2}\right)\hat{\mathbf{a}}_\rho + 10\,\hat{\mathbf{a}}_\phi \ \textit{(cylindrical coordinate)}$$

is rotational.

Solution. For rotational field $\vec{\nabla} \times \mathbf{F} \neq 0$

Let us evaluate the value of $\vec{\nabla} \times \mathbf{F}$.

Given that $\mathbf{F} = \left(\frac{150}{\rho^2}\right)\hat{\mathbf{a}}_\rho + 10\hat{\mathbf{a}}_\phi$, *i.e.*, $F_\rho = \frac{150}{\rho^2}$ and $F_\phi = 10$

$$\text{Now, } \vec{\nabla} \times \mathbf{F} = \frac{1}{\rho}\begin{vmatrix} \hat{\mathbf{a}}_\rho & \rho\,\hat{\mathbf{a}}_\phi & \hat{\mathbf{a}}_z \\ \frac{\partial}{\partial \rho} & \frac{\partial}{\partial \phi} & \frac{\partial}{\partial z} \\ F_\rho & \rho F_\phi & F_z \end{vmatrix} = \frac{1}{\rho}\begin{vmatrix} \hat{\mathbf{a}}_\rho & \rho\,\hat{\mathbf{a}}_\phi & \hat{\mathbf{a}}_z \\ \frac{\partial}{\partial \rho} & \frac{\partial}{\partial \phi} & \frac{\partial}{\partial z} \\ \frac{150}{\rho^2} & 10\rho & 0 \end{vmatrix}$$

$$= \frac{1}{\rho}\left[\hat{\mathbf{a}}_\rho \left\{0 - \frac{\partial}{\partial z}(10\rho)\right\} + \rho\,\hat{\mathbf{a}}_\rho \left\{\frac{\partial}{\partial z}(150/\rho^2) - 0\right\} \hat{\mathbf{a}}_z \left\{\frac{\partial}{\partial z}(10\rho) - \frac{\partial}{\partial \phi}(150/\rho^2)\right\}\right]$$

$$= \frac{1}{\rho}[\hat{\mathbf{a}}_\rho (0) + \rho\,\hat{a}_\phi (0) + \hat{\mathbf{a}}_z (10 - 0)] = \frac{10}{\rho}\hat{\mathbf{a}}_z$$

So, the field is **rotational.**

EXAMPLE 8. *Find constant a, b, c so that the vector* $\mathbf{A} = \hat{\mathbf{a}}_x (x + 2y + az) + \hat{\mathbf{a}}_y (bx - 3y - z) + \hat{\mathbf{a}}_z (4x + cy + 2z)$ *is irrotational.*

Solution. The vector is irrotational when curl $\mathbf{A} = 0$

According to the given problem

$$\text{Curl } \mathbf{A} = \begin{vmatrix} \hat{\mathbf{a}}_x & \hat{\mathbf{a}}_y & \hat{\mathbf{a}}_z \\ \frac{\partial}{\partial x} & \frac{\partial}{\partial y} & \frac{\partial}{\partial z} \\ (x+2y+az) & (bx-3y-z) & (4x+cy+2z) \end{vmatrix}$$

$$= \hat{\mathbf{a}}_x \left\{ \frac{\partial}{\partial x}(4x + cy + 2z) - \frac{\partial}{\partial z}(bx - 3y - z) \right\}$$
$$+ \hat{\mathbf{a}}_y \left\{ \frac{\partial}{\partial z}(x + 2y + az) - \frac{\partial}{\partial x}(4x - cy - 2z) \right\}$$
$$+ \hat{\mathbf{a}}_z \left\{ \frac{\partial}{\partial x}(bx - 3y - z) - \frac{\partial}{\partial y}(x + 2y + az) \right\}$$
$$= \hat{\mathbf{a}}_x \{c + 1\} + \hat{\mathbf{a}}_y \{a - 4\} + \hat{\mathbf{a}}_z \{b - 2\}$$

This is equal to zero, when $c = \mathbf{-1}$, $a = \mathbf{4}$ and $b = \mathbf{2}$.

EXAMPLE 9. *If a rigid body is rotating with angular velocity ω prove that*

$$\omega = \frac{1}{2} \text{ curl } \mathbf{v}$$

where **v** *is the linear velocity.*

Solution. $\text{Curl } \mathbf{v} = \vec{\nabla} \times \mathbf{v}$

Now $\mathbf{v} = (\omega \times \mathbf{r})$

$\therefore$ $\mathbf{v} = \hat{\mathbf{a}}_x (\omega_y z - y\omega_z) + \hat{\mathbf{a}}_y (\omega_z x - z\omega_z) + \hat{\mathbf{a}}_z (\omega_x x - x\omega_z)$

Now $\vec{\nabla} \times \mathbf{v} = \vec{\nabla} \times (\omega \times \mathbf{r})$

$$= \begin{vmatrix} \hat{\mathbf{a}}_x & \hat{\mathbf{a}}_y & \hat{\mathbf{a}}_z \\ \partial/\partial x & \partial/\partial y & \partial/\partial z \\ (\omega_y z - y\omega_z) & (\omega_z x - z\omega_x) & (\omega_x x - x\omega_y) \end{vmatrix}$$
$$= \hat{\mathbf{a}}_x \left[\frac{\partial}{\partial y}(\omega_x y - x\omega_y) - \frac{\partial}{\partial z}(\omega_z x - z\omega_x) \right]$$
$$+ \hat{\mathbf{a}}_y \left[\frac{\partial}{\partial z}(\omega_y z - y\omega_z) - \frac{\partial}{\partial x}(\omega_x y - x\omega_y) \right]$$
$$+ \hat{\mathbf{a}}_z \left[\frac{\partial}{\partial x}(\omega_z x - z\omega_x) - \frac{\partial}{\partial y}(\omega_y z - y\omega_z) \right]$$
$$= \hat{\mathbf{a}}_x(\omega_x + \omega_x) + \hat{\mathbf{a}}_y(\omega_y + \omega_y) + \hat{\mathbf{a}}_z(\omega_z + \omega_z)$$
$$= 2[\hat{\mathbf{a}}_x \omega_x + \hat{\mathbf{a}}_y \omega_y + \hat{\mathbf{a}}_z \omega_z] = 2\omega$$

$\therefore$ $$\omega = \frac{1}{2} \text{ curl } \mathbf{v}$$

EXAMPLE 10. *Find the value of curl* **(a × r)**, *where* **a** *is a constant vector.*

Solution.

$$\mathbf{a} \times \mathbf{r} = \begin{vmatrix} \hat{\mathbf{a}}_x & \hat{\mathbf{a}}_y & \hat{\mathbf{a}}_z \\ a_x & a_y & a_z \\ x & y & z \end{vmatrix} = \hat{\mathbf{a}}_x (a_y z - a_z y) + \hat{\mathbf{a}}_y (a_z x - a_x z) + \hat{\mathbf{a}}_z (a_x y - a_y x)$$

Now, curl ($\mathbf{a} \times \mathbf{r}$) = $\nabla \times (\mathbf{a} \times \mathbf{r})$

$$\therefore \quad \vec{\nabla} \times (\mathbf{a} \times \mathbf{r}) = \begin{vmatrix} \hat{\mathbf{a}}_x & \hat{\mathbf{a}}_y & \hat{\mathbf{a}}_z \\ \partial/\partial x & \partial/\partial y & \partial/\partial z \\ (a_y z - a_z y) & (a_z x - a_x z) & (a_x y - a_y x) \end{vmatrix}$$

$$= \hat{\mathbf{a}}_x \left[\frac{\partial}{\partial y}(a_x y - a_y x) - \frac{\partial}{\partial x}(a_z x - a_x z) \right]$$

$$+ \hat{\mathbf{a}}_y \left[\frac{\partial}{\partial z}(a_y z - a_z y) - \frac{\partial}{\partial x}(a_x y - a_y x) \right]$$

$$+ \hat{\mathbf{a}}_z \left[\frac{\partial}{\partial x}(a_z x - a_x z) - \frac{\partial}{\partial y}(a_y z - a_z y) \right]$$

$$= \hat{\mathbf{a}}_x [a_x + a_x] + \hat{\mathbf{a}}_y [a_y + a_y] + \hat{\mathbf{a}}_z [a_z + a_z]$$

$$= 2[\hat{\mathbf{a}}_x a_x + \hat{\mathbf{a}}_y a_y + \hat{\mathbf{a}}_z a_z] = \mathbf{2\,a}.$$

EXAMPLE 11. *Show that curl r^n **r** is zero any value of n.*

Solution. Curl (r^n **r**) = $\vec{\nabla} \times r^n \mathbf{r}$

Now

$$\vec{\nabla} \times r^n \mathbf{r} = \vec{\nabla} \times (r^n x \hat{\mathbf{a}}_x + r^n y \hat{\mathbf{a}}_y + r^n z \hat{\mathbf{a}}_z) = \begin{vmatrix} \hat{\mathbf{a}}_x & \hat{\mathbf{a}}_y & \hat{\mathbf{a}}_z \\ \dfrac{\partial}{\partial x} & \dfrac{\partial}{\partial y} & \dfrac{\partial}{\partial z} \\ r^n x & r^n y & r^n z \end{vmatrix}$$

$$= \hat{\mathbf{a}}_x \left(\frac{\partial r^n}{\partial y} z - \frac{\partial r^n}{\partial z} y \right) + \hat{\mathbf{a}}_y \left(\frac{\partial r^n}{\partial y} x - \frac{\partial r^n}{\partial z} z \right)$$

$$+ \hat{\mathbf{a}}_z \left(\frac{\partial r^n}{\partial x} y - \frac{\partial r^n}{\partial y} x \right) \qquad ...(1)$$

But,

$$\frac{\partial r^n}{\partial x} = \frac{\partial}{\partial x}(x^2 + y^2 + z^2)^{n/2} = \frac{n}{2}(x^2 + y^2 + z^2)^{n/2}(2x)$$

$$= nx (r^2)^{(n/2)-1}$$

$$= nx\, r^{n-2} \qquad ...(2)$$

Similarly,

$$\frac{\partial r^n}{\partial y} = n\, y\, r^n \text{ and } \frac{\partial r^n}{\partial z} = n\, z\, r^{n-2}$$

Substituting these values in eq. (1), we get

$$\text{Curl } (r^n \mathbf{r}) = \hat{\mathbf{a}}_x (nzyr^{n-2} - nzyr^{n-2}) + \hat{\mathbf{a}}_y (nxyr^{n-2} - nxyr^{n-2})$$

$$+ \hat{\mathbf{a}}_z (nxyr^{n-2} - nxyr^{n-2})$$

$$= 0.$$

This shows the vectors r^n **r** is irrotattional whatever be the value of n.

1.20 LINE INTEGRAL

The integration of a vector along a curve is called its integral. As shown in Fig. (32), let PQ be a curve drawn between two points P and Q in vector field. Let $d\mathbf{l}$ be an element of length along the curve PQ at R. Suppose $\mathbf{A}$ is the vector at R making an angle θ with the direction of $d\mathbf{l}$. Then

$$\mathbf{A} \cdot d\mathbf{l} = A\, dl \cos\theta = dl\,(A \cos\theta) \quad \text{...(1)}$$

Equation (1) shows that the value of $\mathbf{A} \,.\, d\mathbf{l}$ at any point of the curve is equal to the product of small element dl and the component (A cos θ) of $\mathbf{A}$ along the direction of dl. The value of $\mathbf{A} \cdot d\mathbf{l}$ for the complete curve PQ can be obtained by integrating eq. (1). Hence,

$$\int_P^Q \mathbf{A} \cdot d\mathbf{l} = \int_P^Q dl\,(A \cos\theta) \quad \text{...(2)}$$

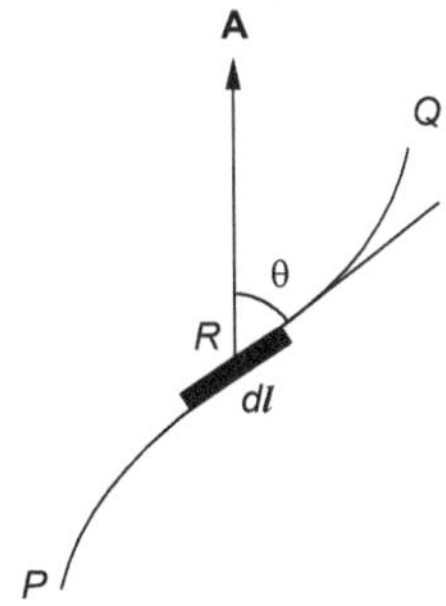

Fig. (32)

Integral $\int_P^Q \mathbf{A} \cdot d\mathbf{l}$ is defined as the line integral of $\mathbf{A}$ along the curve PQ.

Examples.

(1) If $\mathbf{A}$ represents the force acting on a a particle moving along the curve from P to Q, then the line integral $\int_P^Q \mathbf{A} \cdot d\mathbf{l}$ represents the total work done by the force during the motion of the particle over its entire path from P to Q.

(2) If $\mathbf{A}$ denotes the electric field intensity at any point, then the line integral represents the potential difference between P and Q.

1.21 SURFACE INTEGRAL AND VOLUME INTEGRAL

Surface Integral: Consider a simple S in a vector field bounded by a curve as shown in Fig. (33) Let dS be an infinitesimal element of the surface. This surface element of area dS can be represented by area vector $d\mathbf{S}$. If $\hat{\mathbf{n}}$ be a unit positive vector (drawn outward the surface) in the direction of dS, then

$$d\mathbf{S} = \hat{\mathbf{n}}\, dS$$

Let $\mathbf{A}$ be a vector at middle of the element dS in the direction making an angle θ with $\hat{\mathbf{n}}$ Now the scalar product

$$\mathbf{A} \cdot d\mathbf{S} = \mathbf{A} \cdot \mathbf{n}\, dS = A\, dS \cos\theta$$

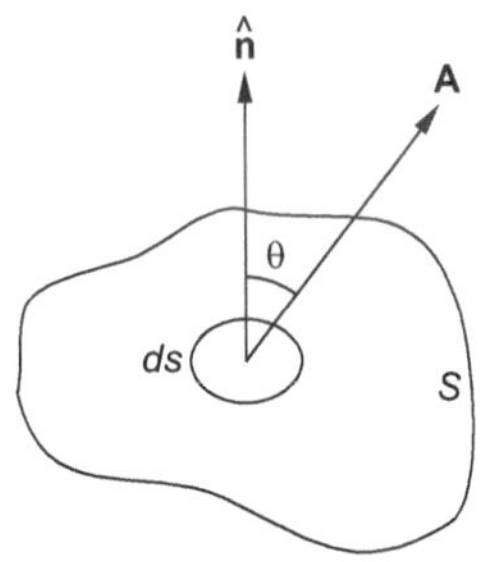

Fig. (33)

is called the flux of vector field $\mathbf{A}$ across the area element dS. The total flux of the vector field across the entire surface area S is given by

$$\iint_S \mathbf{A} \cdot d\mathbf{S} = \iint_S \mathbf{A} \cdot \mathbf{n}\, d\mathbf{S} = \iint_S A \cos\theta\, dS$$

This is defined as the surface integral.

Example. Let A denotes the velocity vector of a moving fluid. Consider that a surface S is drawn in the fluid. Now the surface integral, *i.e.*, represents the amount of the fluid flowing

$$\iint_S \mathbf{v} \cdot d\mathbf{S}$$

per unit time normally to the surface. The surface integral is taken positive when the fluid flows outside the closed surface and negative when the fluid flows into the closed surface.

Volume integral: Consider a closed surface in space enclosing a volume V. If **A** be a vector point function at a point in a small element of volume dV, then the integral is called the volume

$$\iiint_V \mathbf{A}\, dV$$

integral of vector **A** over the surface.

1.20 STOKE'S THEOREM

This theorem states that the line integral of vector field* A *around a curve is equal to the surface integral of the curl of* A *taken over the surface* S *surrounded by the closed curve, i.e.,

$$\oint_C \mathbf{A} \cdot d\mathbf{l} = \iint_S \text{Curl } \mathbf{A} \cdot d\mathbf{S} = \iint_S (\vec{\nabla} \times \mathbf{A})\, d\mathbf{S} \quad ...(1)$$

Proof of Stole's theorem. Consider a surface S enclosed in a vector field **A.** As shown in Fig. (34), the boundary of the surface S is closed PQR. The line integral A around the curve PQR traced counter-clockwise is

$$\oint_C \mathbf{A} \cdot d\mathbf{l}$$

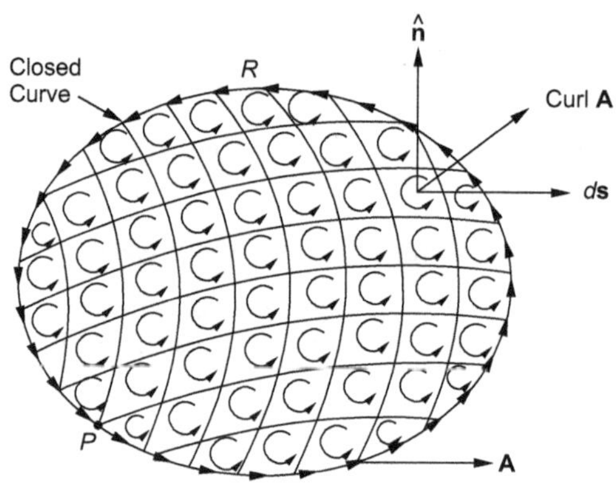

Fig. (34)

Let the entire surface be divided into a large number of square loops. Let the area enclosed by each infinitesimally small loop be dS Suppose $\hat{\mathbf{n}}$ be a unit positive outward normal upon dS. The vector area of the element is

$$\hat{\mathbf{n}}\, dS = d\mathbf{S}$$

We know that the curl of a vector field at any point is the maximum line integral of the vector computed per unit area along the boundary of an infinitesimal area at that point. So the line integral of **A** around the boundary of the area dS is

$$\text{curl } \mathbf{A} \cdot d\mathbf{S}$$

This applies to all surface elements. Hence the sum of the line integrals of **A** around the boundaries all the elements is given by

$$\iint_S \text{curl } \mathbf{A} \cdot d\mathbf{S} \quad ...(2)$$

This is the surface integral of **A.**

It is clear from Fig. (34) that the line integral along the common sides of the continuous elements mutually cancel because they are traversed in opposite directions.

Now the sides of the element which lie in the periphery of the surface (*i.e.*, in the closed curve) contribute to the integral. The sum of the line integrals on the boundary line of the curve is given by eq. (2). This is also given by eq. (1). Hence

$$\boxed{\oint_C \mathbf{A} \cdot d\mathbf{l} = \iint_S \text{curl } \mathbf{A} \cdot d\mathbf{S} = \iint_S (\nabla \times \mathbf{A}) \cdot d\mathbf{S}} \qquad ...(3)$$

This is Stokes' theorem.

It is obvious from eq. (3) that Stokes theorem gives a method to convert a surface integral into a line integral and vice verse. When curl **A** is zero, the line integral of **A** over the closed path is zero and hence the field is conservative.

1.23 GAUSS' THEOREM OF DIVERGENCE

This theorem states that the surface integral of the normal component of vector.* A *taken over a closed surface S is equal to the volume integral of the divergence of vector* A *over the volume V enclosed by the surface S, *i.e.*,

$$\oint_S \vec{A} \cdot d\vec{S} = \iiint_V \text{div } \mathbf{A}\, dV = \iiint_V (\nabla \cdot \mathbf{A})\, dV$$

This theorem provides a method for connecting volume integrals (triple integrals) to surface integral (double integrals).

Proof of Gauss theorem. Consider a closed surface S of any arbitary shape drawn in a vector field **A** as shown in Fig. (35). Let the surface encloses a volume V. Now the whole volume

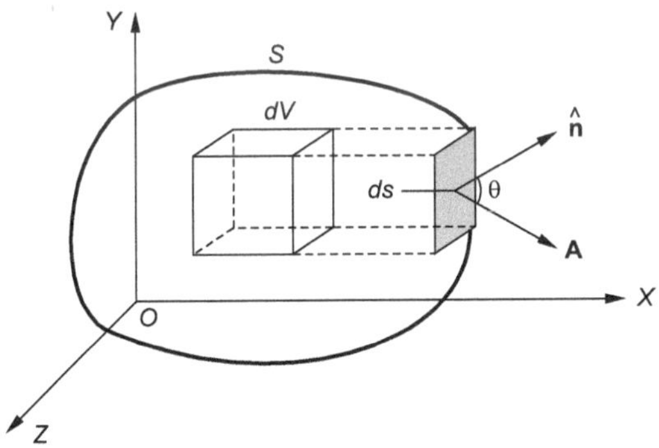

Fig. (35)

may be assumed to be divided into a very large number of cubical volume elements adjoining to each other. Consider a small cubical volume element as shown in Fig. (35). We know that div **A** represents the amount of flux diverging per unit volume and hence the flux diverging from the element of volume dV will be div **A** dV. So the total flux coming out from entire volume is given by

$$\iiint_V \text{div } \mathbf{A}\, dV \qquad ...(1)$$

Now we consider a small element of area dS on the surface S as shown in Fig. (35). Let $\hat{\mathbf{n}}$ represents the unit vector drawn normal to area dS. It should be remembered that outward drawn normal on a surface is taken as positive. If the field vector $\mathbf{A}$ and outward normal $\hat{\mathbf{n}}$ are at an angle θ, then the component of $\mathbf{A}$ along $\hat{\mathbf{n}}$ is

$$A \cos \theta = \mathbf{A} \cdot \hat{\mathbf{n}}$$

The flux of $\mathbf{A}$ through the surface element dS is given by

$$(\mathbf{A} \cdot \hat{\mathbf{n}})\, dS = \mathbf{A} \cdot d\mathbf{S}$$

($\because$ Flux is defined as the product of normal component of vector and surface area)

So the total flux through the entire surface S is given by

$$\iint_S \mathbf{A} \cdot d\mathbf{S} \qquad ...(2)$$

This must be equal to the total flux diverging from the whole volume V enclosed by the surface S. Hence from eqs. (1) and (2) we get

$$\iint_S \mathbf{A} \cdot d\mathbf{S} = \iiint_V \text{div}\, \mathbf{A}\, dV \qquad ...(3)$$

This Gauss theorem of divergence.

This Gauss theorem may also be written as

$$\boxed{\iint_S (\mathbf{A} \cdot \hat{\mathbf{n}})\, dS = \iiint_V (\nabla \cdot \mathbf{A})\, dV} \qquad ...(4)$$

1.24 GREEN'S THEOREM

If ϕ and ψ are two scalar point functions such that these functions and their first derivatives are continuously differentiable, in a region bounded by a closed surface S, then we have

$$\iiint_V (\phi\, \nabla^2\psi + \nabla\phi \cdot \nabla\psi)\, dV = \iint_S (\phi\nabla\psi) \cdot d\mathbf{S} \qquad ...(1)$$

and

$$\iiint_V (\phi\, \nabla^2\psi - \psi\nabla^2\phi)\, dV = \iint_S (\phi\nabla\psi - \psi\nabla\phi) \cdot d\mathbf{S} \qquad ...(2)$$

These equations are known as first and second form of Green's theorem respectively.

Proof of Green's theorem: Let us take the following mathematical form of Gauss's divergence theorem

$$\iiint_V \text{div}\, \mathbf{A}\, dV = \iint_S \mathbf{A} \cdot d\mathbf{S} \qquad ...(3)$$

Because $\nabla\psi$ is a vector quantity, hence $\phi\nabla\psi$ will also be a vector quantity. Let this is represented by vector $\mathbf{A}$. Thus

$$\mathbf{A} = \phi\nabla\psi$$

or

$$\hat{\mathbf{a}}_x A_x + \hat{\mathbf{a}}_y A_y + \hat{\mathbf{a}}_z A_z = \phi\left(\hat{\mathbf{a}}_x \frac{\partial\psi}{\partial x} + \hat{\mathbf{a}}_y + \frac{\partial\psi}{\partial y}\, \hat{\mathbf{a}}_z \frac{\partial\psi}{\partial z}\right)$$

From this equation we can see that

$$A_x = \phi\frac{\partial\psi}{\partial x}, \quad A_y = \phi\frac{\partial\psi}{\partial y} \quad \text{and} \quad a_z = \phi\frac{\partial\psi}{\partial z}$$

Now $$\text{div } \mathbf{A} = \frac{\partial A_x}{\partial x} + \frac{\partial A_y}{\partial y} + \frac{\partial A_z}{\partial z}$$

Substituting the values of A_x, A_y and A_z, we get

$$\text{div } \mathbf{A} = \frac{\partial}{\partial x}\left(\phi \frac{\partial \psi}{\partial x}\right) + \frac{\partial}{\partial y}\left(\phi \frac{\partial \psi}{\partial y}\right) + \frac{\partial}{\partial z}\left(\phi \frac{\partial \psi}{\partial z}\right)$$

$$= \left(\phi \frac{\partial^2 \psi}{\partial x^2} + \frac{\partial \phi}{\partial x} \cdot \frac{\partial \psi}{\partial x}\right) + \left(\phi \frac{\partial^2 \psi}{\partial y^2} + \frac{\partial \phi}{\partial y} \cdot \frac{\partial \psi}{\partial y}\right) + \left(\phi \frac{\partial^2 \psi}{\partial z^2} + \frac{\partial \phi}{\partial z} \cdot \frac{\partial \psi}{\partial z}\right)$$

$$= \phi\left(\frac{\partial^2 \psi}{\partial x^2} + \frac{\partial^2 \psi}{\partial y^2} + \frac{\partial^2 \psi}{\partial z^2}\right) + \frac{\partial \phi}{\partial x} \cdot \frac{\partial \psi}{\partial x} + \frac{\partial \phi}{\partial y} \cdot \frac{\partial \psi}{\partial y} + \frac{\partial \phi}{\partial z} \cdot \frac{\partial \psi}{\partial z}$$

$$= \phi \nabla^2 \psi + \nabla \phi \cdot \nabla \psi \qquad ..(4)$$

Substituting the value of div **A** from eq. (4) in eq. (3), we get

$$\boxed{\iiint_V (\phi \nabla^2 \psi + \nabla \phi \cdot \nabla \psi)\, dV = \iint_S (\phi \nabla \psi) \cdot \nabla \mathbf{S}} \qquad ...(a)$$

This is known as Green's first theorem.

Interchanging ϕ and ψ in eq. (a), we have

$$\iiint_V (\psi \nabla^2 \phi + \nabla \psi \cdot \nabla \phi)\, dV = \iint_S (\psi \nabla \phi) \cdot \nabla \mathbf{S} \qquad ...(5)$$

Subtracting eq. (5) from eq. (a), we get

$$\boxed{\iiint_V (\phi \nabla^2 \psi - \psi \nabla^2 \phi)\, dV = \iint_S (\phi \nabla \psi - \psi \nabla \phi) \cdot d\mathbf{S}} \qquad ...(b)$$

This is known as Green's second theorem.

1.25 HELMHOLTZ'S THEOREM

This theorem states that a vector field can be completely specified by its divergence and curl. If div **F** and curl **F** are known every where within the region and the field **F** is known on the boundary of the region, then the field can be splited up into two parts $\mathbf{F}_1$ and $\mathbf{F}_2$.

The first part $\mathbf{F}_1$ is a ***solenoidal field***. This result from div. $\mathbf{F}_1 = 0$ and curl $\mathbf{F}_1$ = curl **F**. This part can be derived from a vector potential **A** such that $\mathbf{F}_1$ = curl **A**.

The second part $\mathbf{F}_2$ is an ***irrotational field***. This part result from the source of **F**, *i.e.*, curl $\mathbf{F}_2 = 0$ and div. $\mathbf{F}_2$ = div. **F**. This part can be derived from scalar potential V such that $\mathbf{F}_2$ = grad V.

Few *Worked* EXAMPLES

EXAMPLE 1. *Using Stokes theorem, prove that*

$$\oint_C \mathbf{r} \cdot d\mathbf{l} = 0, \text{ where } \mathbf{r} \text{ is position vector.}$$

Solution. By Stokes theorem, we have

$$\oint_C \mathbf{A} \cdot d\mathbf{l} = \iint_S \text{curl } \mathbf{r} \cdot d\mathbf{S}$$

Replacing the vector **A** by position vector **r**, we get

$$\oint_C \mathbf{r}\cdot d\mathbf{l} = \iint_S \text{curl}\,\mathbf{r}\cdot d\mathbf{S} = \iint_S 0\cdot d\mathbf{S} \qquad (\because \text{curl}\,\mathbf{r} = 0)$$

Hence,
$$\oint_C \mathbf{r}\cdot d\mathbf{l} = 0$$

EXAMPLE 2. *Given* $\mathbf{A} = 2\rho\cos\phi\,\hat{\mathbf{a}}_r + \rho\hat{\mathbf{a}}_\phi$ *in cylindrical coordinate for the contour shown in Fig. (36). Verify the Stokes theorem.*

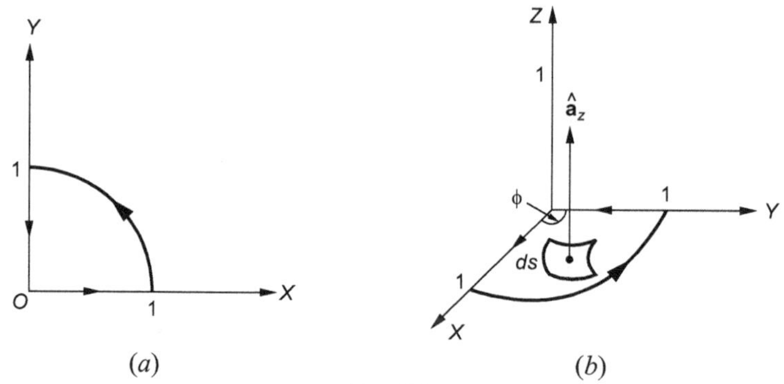

Fig. (36)

Solution. Let us consider the right hand side of Stoke's theorem, *i.e.*,

$$\iint_S (\vec{\nabla}\times\mathbf{A})\cdot d\mathbf{S}$$

The value $(\vec{\nabla}\times\mathbf{A})$ in cylindrical coordinate system is given by

$$\vec{\nabla}\times\mathbf{A} = \frac{1}{\rho}\begin{vmatrix}\hat{\mathbf{a}}_\rho & \rho\hat{\mathbf{a}}_\phi & \hat{\mathbf{a}}_z \\ \dfrac{\partial}{\partial\rho} & \dfrac{\partial}{\partial\phi} & \dfrac{\partial}{\partial z} \\ A_\rho & \rho A_\phi & A_z\end{vmatrix}$$

Given $\mathbf{A} = 2\rho\cos\phi\,\hat{\mathbf{a}}_r + \rho\,\hat{\mathbf{a}}_\phi$, *i.e.*, $A_\rho = 2\rho\cos\phi$, $A_\phi = \rho$ and $A_z = 0$
Substituting these values, we get

$$\vec{\nabla}\times\mathbf{A} = \frac{1}{\rho}\begin{vmatrix}\hat{\mathbf{a}}_\rho & \rho\hat{\mathbf{a}}_\phi & \hat{\mathbf{a}}_z \\ \dfrac{\partial}{\partial\rho} & \dfrac{\partial}{\partial\phi} & \dfrac{\partial}{\partial z} \\ 2\rho\cos\phi & \rho^2 & 0\end{vmatrix}$$

$$= \frac{1}{\rho}\left[\hat{\mathbf{a}}_\rho\left\{0 - \frac{\partial}{\partial\rho}(\rho^2)\right\} + \rho\,\hat{\mathbf{a}}_\phi\left\{\frac{\partial}{\partial z}(2\rho\cos\phi) - \frac{\partial}{\partial\rho}(0)\right\}\right.$$
$$\left. + \hat{\mathbf{a}}_z\left\{\frac{\partial}{\partial\rho}(\rho^2) - \frac{\partial}{\partial\phi}(2\rho\cos\phi)\right\}\right]$$

$$= \frac{1}{\rho}[\hat{\mathbf{a}}_z\,\{2\rho + 2\rho\sin\phi\}] = 2[\hat{\mathbf{a}}_z\,(1+\sin\phi)]$$

For the area under consideration, the differential surface

$$d\mathbf{S} = \rho\, d\rho\, d\phi\, \hat{\mathbf{a}}_z$$

$$\therefore \quad \iint_S (\vec{\nabla} \times \mathbf{A}) \cdot d\mathbf{S} = \int_0^1 \int_{\phi=0}^{\pi/2} 2(1 + \sin\phi)\, \hat{\mathbf{a}}_z \cdot (\rho\, d\rho\, d\phi\, \hat{\mathbf{a}}_z) = \left(1 + \frac{\pi}{2}\right)$$

Now, consider the surface integral

$$\oint_C \mathbf{A} \cdot d\mathbf{l} = \int_{\rho=0}^{1} \mathbf{A} \cdot d\mathbf{l} + \int_{\phi=0}^{\pi/2} \mathbf{A} \cdot d\mathbf{l} + \int_{\rho=1}^{0} \mathbf{A} \cdot d\mathbf{l}$$

$$= \int_{\rho=0}^{1} 2\rho\cos\phi\, d\rho + \int_{\phi=0}^{\pi/2} \rho^2\, d\phi + \int_{\rho=1}^{0} 2\rho\cos\phi\, d\rho$$

at $\phi = 0$ at $\rho = 1$ at $\phi = \pi/2$

$$= 1 + \frac{\pi}{2} + 0 = 1 + \frac{\pi}{2}$$

Therefore L.H.S. = R.H.S. This verifies te Stoke's theorem.

EXAMPLE 3. *Given* $\mathbf{A} = (10\,\rho^3)/4\ \hat{\mathbf{a}}_\rho$ *in cylindrical coordinate system. Verify Gauss theorem of divergence for the volume enclosed by* $\rho = 2$, $z = 0$ *to* 10.

Solution. According to Gauss divergence theorem

$$\iint_S \mathbf{A} \cdot d\mathbf{S} = \iiint_V (\vec{\nabla} \cdot \mathbf{A})\, dV$$

Consider the left hand side

$$\iint_S \mathbf{A} \cdot d\mathbf{S} = \int_0^{10} \int_0^{2\pi} \frac{10\rho^3}{4} \rho\, d\phi\, dz \Bigg|_{\rho=2}$$

$$= \frac{10\rho^4}{4}\Bigg|_{\rho=2} \int_0^{10} dz \int_0^{2\pi} d\phi = \frac{10(2)^4}{4} \times 10 \times 2\pi = 800\,\pi$$

Consider the R.H.S.

$$\iiint_V (\vec{\nabla} \cdot \mathbf{A})\, dV = \iiint_V (\vec{\nabla} \cdot \mathbf{A}) \rho\, d\rho\, d\phi\, dz$$

Now, $$(\vec{\nabla} \cdot \mathbf{A}) = \frac{1}{\rho}\frac{\partial}{\partial\rho}(\rho A_r) = \frac{1}{\rho}\frac{\partial}{\partial\rho}\left(\frac{10\rho^4}{4}\right) = \frac{1}{\rho} \times \frac{10}{4} \times 4\rho^3 = 10\rho^2$$

$$\therefore \quad \iiint_V (\vec{\nabla} \cdot \mathbf{A})\, dV = \int_0^2 \int_0^{10} \int_0^{2\pi} 10\rho^2 \times \rho\, d\rho\, d\phi\, dz = \int_0^2 10\rho^3\, d\rho \int_0^{2\pi} d\phi \int_0^{10} dz$$

$$= \left[\frac{10\rho^2}{4}\right]_0^2 2\pi \times 10 = 10 \times 4 \times 2\pi \times 10 = 800\,\pi$$

Therefore, L.H.S = R.H.S.

EXAMPLE 4. *Given* $\mathbf{A} = \left(\frac{5r^2}{4}\right)\hat{\mathbf{a}}_r$ *in spherical coordinate. Verify divergence theorem for the volume enclosed between* $r = 1$ *and* $r = 2$

Solution. According to Gauss divergence theorem

$$\iint_S \mathbf{A} \cdot d\mathbf{S} = \iiint_V (\vec{\nabla} \cdot \mathbf{A}) dV$$

Given that $\mathbf{A} = \left(\frac{5r^2}{4}\right)\hat{\mathbf{a}}_r$ *i.e.*, $A_r = \frac{5r^2}{4}$, $A_\theta = 0$ and $A_\phi = 0$

$$\therefore \qquad (\vec{\nabla} \cdot \mathbf{A}) = \frac{1}{r^2}\frac{\partial}{\partial r}\left(r^2 \times \frac{5r^2}{4}\right) = \frac{1}{r^2}\frac{\partial}{\partial r}\left(\frac{5}{4}r^4\right) = \frac{1}{r^2} \times \frac{5}{4} \times 4r^3 = 5r$$

Now, $$\iiint_V (\vec{\nabla} \cdot \mathbf{A}) dV = \iiint_V 5r \cdot r^2\, dr\, d\theta \sin\theta\, d\phi = 5\int_1^2 \int_0^\pi \int_0^{2\pi} 5r^3\, dr \sin\theta\, d\theta\, d\phi$$

$$= 5\int_1^2 r^3\, dr \int_0^\pi \sin\theta\, d\theta \int_0^{2\pi} d\phi = 5\left[\frac{r^2}{4}\right]_1^2 [-\cos\theta]_0^\pi\, [2\pi]$$

$$= 5\left[\frac{16}{4} - \frac{1}{4}\right][2][2\pi] = 5 \cdot \frac{15}{4} \cdot 4\pi = 75\pi$$

Consider the left hand side of divergence theorem. Hence $d\mathbf{S}$ is radially outward for $r = 2$ and $d\mathbf{S}$ radially inward for $r = 1$.

Therefore, $$\iint_S \mathbf{A} \cdot d\mathbf{S} = \left[\int_0^\pi \int_0^{2\pi} \frac{5r^2}{4} \cdot r^2 \sin\theta\, d\theta\, d\phi\, \hat{\mathbf{a}}_r \cdot (\hat{\mathbf{a}}_r)\Big|_{r=1} + \int_0^\pi \int_0^{2\pi} \frac{5r^2}{4} \cdot r^2 \sin\theta\, d\theta\, d\phi\, \hat{\mathbf{a}}_r \cdot \hat{\mathbf{a}}_r\Big|_{r=2}\right]$$

$$= -\frac{5}{4}(1)^4 [-\cos\theta]_0^\pi\, [2\pi] + \frac{5}{2}(2)^4 [-\cos\theta]_0^\pi\, [2\pi]$$

$$= -\frac{5}{4} \times 2 \times 2\pi \times \frac{5}{4} \times 16 \times 2 \times 2\pi = 5\pi + 80\pi = 75\pi$$

The left hand side is equal to right hand side. Hence, the theorem is proved

EXAMPLE 5. *Using Stoke's theorem, prove that curl grad $\phi = 0$*

Solution. According to Stole's theorem, we have

$$\oint_C \mathbf{A} \cdot d\mathbf{l} = \iint_S \text{curl } \mathbf{A} \cdot d\mathbf{S}$$

Let $\mathbf{A} = \text{grad } \phi$, then

$$\oint_C \text{grad } \phi \cdot d\mathbf{l} = \iint_S \text{curl grad } \phi \cdot \hat{\mathbf{n}}\, dS$$

Now, $$\text{grad } \phi \cdot d\mathbf{l} = \left(\hat{\mathbf{a}}_x \frac{\partial \phi}{\partial x} + \hat{\mathbf{a}}_y \frac{\partial \phi}{\partial y} + \hat{\mathbf{a}}_z \frac{\partial \phi}{\partial z}\right) + (\hat{\mathbf{a}}_x\, dx + \hat{\mathbf{a}}_y\, dy + \hat{\mathbf{a}}_z\, dz)$$

$$= \frac{\partial \phi}{\partial x}dx + \frac{\partial \phi}{\partial y}dy + \frac{\partial \phi}{\partial z}dz = d\phi$$

$$\therefore \qquad \oint_C d\phi = [\phi]_A^A \text{ where } \mathbf{A} \text{ is any point on } S$$

Hence, $$\iint_S \text{curl grad}\,\phi\cdot\hat{\mathbf{n}}\,dS = 0$$

This is true for all surface elements S, *i.e*,

$$\text{curl grad}\,\phi = 0$$

EXAMPLE 6. *Verify Stoke's theorem for the vector* $\mathbf{A} = x(\hat{\mathbf{a}}_x x + \hat{\mathbf{a}}_y y)$ *integrated round the square, in the plane* $z = 0$, *whose sides are along the lines* $x = 0$, $y = 0$, $x = a$, $y = a$.

Solution. The situation is shown in Fig. (37).

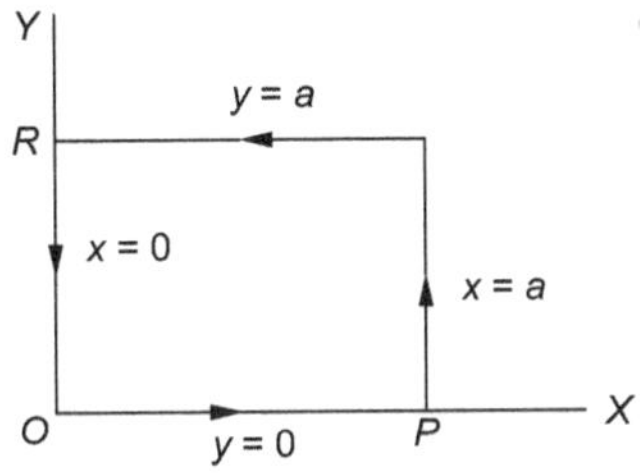

Fig. (37)

Here $$\oint_C \mathbf{A}\cdot d\mathbf{r} = \int_{OP}\mathbf{A}\cdot d\mathbf{l} + \int_{PQ}\mathbf{A}\cdot d\mathbf{l} + \int_{QR}\mathbf{A}\cdot d\mathbf{l} + \int_{RO}\mathbf{A}\cdot d\mathbf{l}$$

where $$\int_{OP}\mathbf{A}\cdot d\mathbf{l} = \int_0^a x(\hat{\mathbf{a}}_x x + \hat{\mathbf{a}}_y y)\cdot\hat{\mathbf{a}}_x\,dx = \int_0^a x^2\,dx = \frac{a^3}{3}$$

$$\int_{PQ}\mathbf{A}\cdot d\mathbf{l} = \int_0^a x(\hat{\mathbf{a}}_x x + \hat{\mathbf{a}}_y y)\cdot\hat{\mathbf{a}}_y\,dy = \int_0^a a\,y\,dy = \frac{a^3}{3}$$

$$\int_{QR}\mathbf{A}\cdot d\mathbf{l} = \int_a^0 x(\hat{\mathbf{a}}_x x + \hat{\mathbf{a}}_y y)\cdot\hat{\mathbf{a}}_x\,dx = \int_a^0 x^2\,dx = -\frac{a^3}{3} 0$$

and $$\int_{RO}\mathbf{A}\cdot d\mathbf{l} = \int_a^0 x(\hat{\mathbf{a}}_x x + \hat{\mathbf{a}}_y y)\cdot\hat{\mathbf{a}}_y\,dy = \int_a^0 0(\hat{\mathbf{a}}_x x + \hat{\mathbf{a}}_y y)\cdot\hat{\mathbf{a}}_y\,dy = 0$$

$$\therefore \quad \int_C \mathbf{A}\cdot d\mathbf{r} = \frac{a^3}{3} + \frac{a^3}{2} - \frac{a^3}{3} + 0 = \frac{a^3}{2} \qquad \text{...(1)}$$

Further, $$\iint_S \text{curl}\,\mathbf{A}\cdot d\mathbf{S} = \iint_S \text{curl}\,x(\hat{\mathbf{a}}_x x + \hat{\mathbf{a}}_y y)\,d\mathbf{S}$$

$$= \int_0^a\int_0^a \hat{\mathbf{a}}_z\,y\cdot\hat{\mathbf{a}}_z\,dx\,dy \qquad (\because \text{curl}\,x(\hat{\mathbf{a}}_x x + \hat{\mathbf{a}}_y y) = \hat{\mathbf{a}}_z\,y)$$

$$= \int_0^a\int_0^a y\,dx\,dy = \int_0^a dx\int_0^a y\,dy = a\times\frac{a^2}{2} = \frac{a^3}{2} \qquad \text{...(2)}$$

From eqs. (1) and (2) Stoke's theorem is verified.

EXAMPLE 7. *Evaluate* $\iint \mathbf{r}\cdot\hat{\mathbf{n}}\,dS$, *where S is a closed surface.*

Solution. By Gauss' theorem

$$\iint_S \mathbf{r}\cdot\hat{\mathbf{n}}\,dS = \iiint_V \text{div}\,\mathbf{r}\,dV = \iiint_V\left(\hat{\mathbf{a}}_x\frac{\partial}{\partial x} + \hat{\mathbf{a}}_y\frac{\partial}{\partial y} + \hat{\mathbf{a}}_z\frac{\partial}{\partial z}\right)(\hat{\mathbf{a}}_x x + \hat{\mathbf{a}}_y y + \hat{\mathbf{a}}_z z)\,dV$$

$$= \iiint_V \left(\frac{\partial x}{\partial x} + \frac{\partial y}{\partial y} + \frac{\partial z}{\partial z} \right) dV = 3\iiint_V dV = 3V$$

where V is the volume enclosed by surface S.

EXAMPLE 8. *Find* $\iint_S \mathbf{F} \cdot \hat{\mathbf{n}}\, d\mathbf{S}$ *where* $\mathbf{F} = \hat{\mathbf{a}}_x 4xz - \hat{\mathbf{a}}_y y^2 + \hat{\mathbf{a}}_z yz$ *and S is the surface of a cube bounded by* $x = 0,\ x = 1,\ y = 0,\ y = 1,\ z = 0,\ z = 1$.

Solution. According to Gauss' theorem

$$\iint_S \mathbf{F} \cdot \hat{\mathbf{n}}\, d\mathbf{S} = \iiint_V \text{div}\, \mathbf{F}\, dV = \iiint_V \left(\frac{\partial F_x}{\partial x} + \frac{\partial F_y}{\partial y} + \frac{\partial F_z}{\partial z} \right) dx\, dy\, dz$$

$$= \iiint_V \left[\frac{\partial}{\partial x}(4xz) + \frac{\partial}{\partial y}(-y^2) + \frac{\partial}{\partial z}(yz) \right] dx\, dy\, dz$$

$$= \iiint_V (4z - 2y + y)\, dx\, dy\, dz = \int_0^1\int_0^1\int_0^1 (4z - y)\, dx\, dy\, dz$$

$$= \int_0^1\int_0^1 [2z^2 - yz]_0^1\, dx\, dy = \int_0^1\int_0^1 (2 - y)\, dx\, dy$$

$$= \int_0^1 \left[2y - \frac{y^2}{2} \right]_0^1 dx = \int_0^1 \frac{3}{2} dx = \left[\frac{3}{2} x \right]_0^1 = \frac{3}{2}$$

EXAMPLE 9. *A region V bounded by a surface S has a continuous charge distribution of density* ρ. *If* ϕ *be the potential at any point, then prove that*

$$\iint_S \mathbf{E} \cdot d\mathbf{S} = 4\pi \iiint_V \rho\, dV$$

where $\mathbf{E} = -\nabla\phi$ and $\nabla^2 \phi = -4\pi\rho$ (*Poisson's equation*).

Solution. We know that $\iint_S \mathbf{E} \cdot d\mathbf{S} = \iiint_V \vec{\nabla} \cdot \mathbf{E}\, dV$

$$\therefore \quad \iint_S \mathbf{E} \cdot d\mathbf{S} = \iiint_V \vec{\nabla} \cdot (-\vec{\nabla}\phi)\, dV \qquad (\because\ \mathbf{E} = -\vec{\nabla}\phi)$$

$$= -\iiint_V \nabla^2 \phi\, dV = -\iiint_V (-4\pi\rho)\, dV = 4\pi \iiint_V \rho\, dV$$

EXAMPLE 10. *Use the divergence theorem to show that*

$$\iint_S \mathbf{F} \cdot \hat{\mathbf{n}}\, dS = \frac{12}{5}\pi r^5$$

where S is the surface of sphere of radius r and

$$\mathbf{F} = \hat{\mathbf{a}}_x x^3 + \hat{\mathbf{a}}_y y^3 + \hat{\mathbf{a}}_z z^3$$

Solution. According to Gauss divergence theorem

$$\iint_S \mathbf{F}\cdot\hat{\mathbf{n}}\,dS = \iiint_V (\nabla\cdot\mathbf{F})\,dV = \iiint_V [\nabla\cdot(\hat{\mathbf{a}}_x x^3 + \hat{\mathbf{a}}_y y^3 + \hat{\mathbf{a}}_z z^3)]\,dV$$

$$= \iiint_V \left(\frac{\partial}{\partial x}x^3 + \frac{\partial}{\partial y}y^3 + \frac{\partial}{\partial z}z^3\right)dV$$

$$= \iiint_V 3\,(x^2+y^2+z^2)\,dV = \iiint_V 3r^2\,dV \qquad (\because x^2+y^2+z^2=r^2)$$

In spherical polar coordinates $dV = r^2\,dr\sin\theta\,d\theta\,d\phi$ where the limits of θ varies from 0 to π and limits of ϕ varies from 0 to 2π

$$= \int_0^r\int_0^\pi\int_0^{2\pi} 3r^2\cdot r^2\;dr\sin\theta\,d\theta\,d\phi$$

$$= \int_0^r 3r^4\cdot dr \times 4\pi = 12\pi\times\frac{r^5}{5} = \frac{12}{5}\pi r^5 \qquad \left(\because \int_0^\pi\int_0^{2\pi}\sin\theta\,d\theta\,d\phi = 4\pi\right)$$

EXAMPLE 11. *If* $\mathbf{F} = \hat{\mathbf{a}}_x ax + \hat{\mathbf{a}}_y by\,\hat{\mathbf{a}}_z cz$, *where a, b, c are constant show that*

$$\iint_S \mathbf{F}\cdot\hat{\mathbf{n}}\,dS = \frac{4}{3}\pi(a+b+c)$$

where S is the surface area of a unit sphere

Solution. According to Gauss theorem

$$\iint_S \mathbf{F}\cdot\hat{\mathbf{n}}\,dS = \iiint_V \text{div}\cdot\mathbf{F}\,dV$$

where V is the volume enclosed by S.

Now
$$\iint_S \mathbf{F}\cdot\hat{\mathbf{n}}\,dS = \iiint_V \nabla\cdot(\hat{\mathbf{a}}_x ax + \hat{\mathbf{a}}_y by + \hat{\mathbf{a}}_z cz)\,dV$$

$$= \iiint_V \left[\frac{\partial}{\partial x}(ax) + \frac{\partial}{\partial y}(by) + \frac{\partial}{\partial x}(cz)\right]dV$$

$$= \iiint_V (a+b+c)\,dV = (a+b+c)\iiint_V dV$$

$$= (a+b+c)\cdot\frac{4}{3}\pi$$

[$\because$ volume V is enclosed by sphere of radius one $\frac{4}{3}\cdot\pi\cdot(1)^3 = \frac{4}{3}\pi$]

$$= \frac{4}{3}\pi\;(a+b+c).$$

EXAMPLE 12. *Prove* $\iint_S \textit{curl } F\cdot d\mathbf{S} = 0$ *for any closed surface.*

Solution. According to Gauss theorem

$$\iint_S \text{curl}\,\mathbf{F}\cdot d\mathbf{S} = \iiint_V \vec{\nabla}\cdot(\text{curl}\,\mathbf{F})\,dV = \iiint_V div\cdot\text{curl}\,\mathbf{F}\,dV$$

We know that div. of a curl is zero, hence

$$\iint_S \text{curl}\,\mathbf{F}\cdot d\mathbf{S} = 0$$

EXAMPLE 13. *The electric field* **E** *in a certain space is given by*

$$E_x = (a\,x + b\,y + c),\ E_y = 0 \text{ and } E_z = 0$$

use Gausses' theorem to evaluate the charge enclosed in a cube with one of its sides of length L along the x-axis, the two faces of the cube perpendicular to the X-axis lying in the planes $x = l$ *and* $x = l + L$.

Solution. According to Gauss's theorem

$$\iint_S \mathbf{E}\cdot d\mathbf{S} = \iiint_V \text{div}\,\mathbf{E}\,dV = \iiint_V \left(\frac{\partial E_x}{\partial x} + \frac{\partial E_y}{\partial y} + \frac{\partial E_z}{\partial z}\right) dV$$

$$= \iiint_V \left[\frac{\partial}{\partial x}(a\,x + b\,y + c) + \frac{\partial}{\partial y}(0) + \frac{\partial}{\partial z}(0)\right] dV$$

$$= \iiint_V a\,dV = a\,dV = a\,L^3 \qquad (\because\ V = L^3) \quad \text{...(1)}$$

According to Gauss' law in electrostatic

$$\iint_S \mathbf{E}\cdot d\mathbf{S} = \frac{q}{\varepsilon_0}. \qquad \text{...(2)}$$

where q is the charge within the surface S which enclosed the volume V and ε_0 is the permittivity of medium.

Comparing eqs. (1) and (2), we get

$$\frac{q}{\varepsilon_0} = a\,L^3 \quad \text{or} \quad q = \varepsilon_0\,a\,L^3$$

Exercises & Problems

EXERCISES

1. Explain cartesian or rectangular coordinate system.
2. Explain the cylindrical coordinate system.
3. Describe the spherical coordinate system.
4. How will you express the differential length, differential area and differential volume in rectangular coordinate system?
5. How will you express the differential length, differential area and differential volume in cylindrical coordinate system?

6. How will you express the differential length, differential area and differential volume in spherical coordinate system?
7. Give the basic concept of transformation of one coordinate system to another.
8. Describe the conversion from the cylindrical coordinate of a point to its cartesian coordinate and vice-versa.
9. Describe the conversion from spherical coordinates of a point to its cartesian coordinates and vice-versa.
10. Discuss how rectangular, cylindrical and spherical coordinate system can be correlated with each other.
11. What are partial derivatives? Obtain an expression for the gradient of a scalar function.
12. Define gradient of a vector field.
13. Write short notes on (*a*) the Del operator and (*b*) Laplacian operator.
14. Explain the divergence and curl of a vector field with examples.
15. Explain curl of a vector with examples.
16. Prove that the gradient of a scalar field is vector and explain physical significance of gradient.
17. Define the convergence, divergence and curl of a vector.
18. Define curl of a vector field. Write its determinant form.
19. Explain the physical meaning of divergence of vector and define the curl of a vector.
20. Explain the terms (*i*) Div., (*ii*) curl, (*iii*) gradient with examples.
21. What do you understand by the gradient of a scalar field? Explain its physical significance.
22. Explain the physical meaning of the divergence of a vector field.
23. Explain curl of a vector field. Give its physical significance.
24. Write a short note on physical interpretation of gradient, divergence and curl.
25. Define and explain the physical significance of the following:
    (*i*) divergence, (*ii*) gradient and (*iii*) curl.
26. What do you mean by solenoidal and irrotational vector field?
27. What is the difference between an irrotational field and a solenoidal field?
28. Explain the physical interpretation of curl. Write mathematical expressions for gradient, divergence and curl in cylindrical and spherical coordinate systems.
29. Given the concept of line integral surface integral and volume integral.
30. What is line integral? What is its physical significance?
31. State and prove Stokes theorem.
32. State and prove Gauss divergence theorem
33. State Gauss and Green theorems of vectors and prove the Green theorem.
34. What is the curl of a vector? Explain physical significance. State and prove Stokes theorem.
35. Explain curl of a vector field. State and prove Stokes' theorem.
36. State Gauss divergence theorem and its mathematical form.
37. State and prove Gauss theorem. Explain gradient of a scalar field and divergence of a vector field.

PRACTICE PROBLEMS

1. If $\phi(x, y, z) = 3x^2y - y^3z^2$, find the value of grad ϕ at point (1, – 2, – 1).

[**Ans.** $-\hat{\mathbf{a}}_x 12 - \hat{\mathbf{a}}_y 9 - \hat{\mathbf{a}}_z 16$]

[**Hint.** $\text{grad}\,\phi = \nabla\phi = \left(\hat{\mathbf{a}}_x \frac{\partial}{\partial x} + \hat{\mathbf{a}}_y \frac{\partial}{\partial y} + \hat{\mathbf{a}}_z \frac{\partial}{\partial z}\right)(3x^2 y - y^3 z^2)$]

$$= \hat{\mathbf{a}}_x 6xy + \hat{\mathbf{a}}_y (3x^2 - 3y^2 z^2) - \hat{\mathbf{a}}_z 2y^3 z$$

Now, substitute $x = 1, y = -2$ and $z = -1$]

2. Show that $\nabla\phi$ is a vector perpendicular to the surface $\phi(x, y, z) = C$, where C is a constant.

[**Hint.** $\phi(x, y, z) =$ constant so that $d\phi = 0$

$$\therefore \quad \frac{\partial\phi}{\partial x}dx + \frac{\partial\phi}{\partial y}dy + \frac{\partial\phi}{\partial z}dz = 0$$

or $$\left(\frac{\partial\phi}{\partial x}\hat{\mathbf{a}}_x + \frac{\partial\phi}{\partial y}\hat{\mathbf{a}}_y + \frac{\partial\phi}{\partial z}\hat{\mathbf{a}}_z\right)\cdot(dx\,\hat{\mathbf{a}}_x + dy\,\hat{\mathbf{a}}_x + dz\,\hat{\mathbf{a}}_z) = 0$$

$$\nabla\phi \cdot d\mathbf{r} = 0$$

Thus, $\nabla\phi$ is perpendicular to $d\mathbf{r}$ and hence perpendicular to the surface also].

3. If $\mathbf{A} = \hat{\mathbf{a}}_x 3x^2 + \hat{\mathbf{a}}_y 5xy^2 + \hat{\mathbf{a}}_z xyz^2$, find div **A** at point (1, 2, 3).

[**Ans.** **38**]

[**Hint.** $\text{div}\,\mathbf{A} = \frac{\partial A_x}{\partial x} + \frac{\partial A_y}{\partial y} + \frac{\partial A_z}{\partial z} = \frac{\partial}{\partial x}(3x^2) + \frac{\partial}{\partial y}(5xy^2) + \frac{\partial}{\partial z}(xyz^2)$

$$= 6x + 10xy + 2xyz$$

Substitute $x = 1, y = 2$ and $z = 3$]

4. If $A = x^2 z\,\hat{\mathbf{a}}_x - 2y^3 z^2\,\hat{\mathbf{a}}_y + xy^2 z\,\hat{\mathbf{a}}_z$, find ∇ **A** at the point. (1, – 1, 1)

[**Ans.** 3]

[**Hint.** $\mathbf{A} = A_x\,\hat{\mathbf{a}}_x + A_y\,\hat{\mathbf{a}}_y + A_z\,\hat{\mathbf{a}}_z$

$\therefore \quad A_x = x^2 z,\ A_y = -2y^3 z^2$ and $A_z = xy^2 z$

We know that

$$\nabla\mathbf{A} = \frac{\partial}{\partial x}A_x + \frac{\partial}{\partial y}A_y + \frac{\partial}{\partial z}A_z$$

or $$\nabla\mathbf{A} = \frac{\partial}{\partial x}(x^2 z) + \frac{\partial}{\partial y}(-2y^3 z^2) + \frac{\partial}{\partial z}(xy^2 z)$$

$$= 2xz - 6y^2 z^2 + xy^2 = 2(1)(1) - 6(-1)^2(1)^2 + 1(-1)^2]$$

5. Calculate div **A** when $\mathbf{A} = \hat{\mathbf{a}}_x \frac{x}{r} + \hat{\mathbf{a}}_y \frac{y}{r} + \hat{\mathbf{a}}_z \frac{z}{r}$ where

$$x^2 + y^2 + z^2 = r^2.$$

[**Ans.** $(2/r)$]

[**Hint.** $$\mathbf{A} = \hat{\mathbf{a}}_x \frac{x}{\sqrt{[(x^2+y^2+z^2)]}} + \hat{\mathbf{a}}_y \frac{y}{\sqrt{[(x^2+y^2+z^2)]}} + \hat{\mathbf{a}}_z \frac{z}{\sqrt{[(x^2+y^2+z^2)]}}]$$

$$\text{div A} = \frac{\partial A_x}{\partial x} + \frac{\partial A_y}{\partial y} + \frac{\partial A_z}{\partial z}$$

$$= \frac{\partial}{\partial x}\left[\frac{x}{\sqrt{[(x^2+y^2+z^2)]}}\right] + \frac{\partial}{\partial y}\left[\frac{y}{\sqrt{[(x^2+y^2+z^2)]}}\right] + \frac{\partial}{\partial z}\left[\frac{z}{\sqrt{[(x^2+y^2+z^2)]}}\right]$$

$$= \frac{y^2+z^2}{(x^2+y^2+z^2)^{3/2}} + \frac{x^2+z^2}{(x^2+y^2+z^2)^{3/2}} + \frac{x^2+y^2}{(x^2+y^2+z^2)^{3/2}}\Bigg]$$

6. Find curl **A**, where $\mathbf{A} = 2x\,\hat{\mathbf{a}}_x + 2y\,\hat{\mathbf{a}}_y$. Is the motion irrotational?
[**Ans.** zero, yes]

7. Transform the given vector into cylindrical corodinates

$$\mathbf{A} = z\,\hat{\mathbf{a}}_x + (1-x)\hat{\mathbf{a}}_y + \frac{y}{x}\hat{\mathbf{a}}_z$$

[**Ans.** $\mathbf{A} = [z\cos\phi + (1-\rho\cos\phi)\sin\phi]\,\hat{\mathbf{a}}_\rho + [-z\sin\phi + (1-\rho\cos\phi)\,\hat{\mathbf{a}}_\phi] + \tan\phi\,\hat{\mathbf{a}}_z$]

8. Given a vector **A** as $\mathbf{A} = y\hat{\mathbf{a}}_x \pm (x+z)\,\hat{\mathbf{a}}_y$ in cartesian coordinate system at point $P(-2, 6, 3)$. Convert the vector A in cylindrical coordinate system.

[**Ans.** $P'(6.32, 108.43°, 3)$ and

$$\mathbf{A} = (2\rho\sin\phi\cos\phi + z\sin\phi)\,\hat{\mathbf{a}}_\rho + [\rho(\cos^2\phi - \sin^2\phi) + z\cos\phi]\,\hat{\mathbf{a}}_\phi]$$

9. Two points are given as $P(2, -1, -3)$ and $Q(1, 3, 4)$. Give the vector that extends from P to Q in (*a*) cartesian coordinates, (*b*) cylindrical coordinates and (*c*) spherical coordinates.

[**Ans.** (*a*) $-\hat{\mathbf{a}}_x + 4\hat{\mathbf{a}}_y + 7\hat{\mathbf{a}}_z$, (*b*) $-2.69\,\hat{\mathbf{a}}_\rho + 3.13\,\hat{\mathbf{a}}_\phi + 7\hat{\mathbf{a}}_z$ and

(*c*) $-7.22\,\hat{\mathbf{a}}_r - 2.03\,\hat{\mathbf{a}}_\theta + 3.13\,\hat{\mathbf{a}}_\phi$]

10. Express the vector $\mathbf{A} = 2xyz\,\hat{\mathbf{a}}_x - 5(x+y+z)\,\hat{\mathbf{a}}_x$ in cylindrical coordinates and calculate $|\mathbf{A}|$ at point $P(\rho = 2, \phi = 60°$ and $z = 3)$.

[**Ans.** $\mathbf{A} = [\rho^2 z\sin 2\phi\,(\cos\phi\,\hat{\mathbf{a}}_\rho - \sin\phi\,\hat{\mathbf{a}}_\phi) - 5(\rho\cos\phi + \rho\sin\phi$
$- 5\,(\rho\cos\phi + \rho\sin\phi + z)\,\hat{\mathbf{a}}_z)]$

$|\mathbf{A}| = 30.5$]

11. (*a*) Given the cartesian coordinates of point $C(\rho = 4.4, \phi = -115°, z = 2)$
(*b*) Given the cylindrical coordinate of point $D(x = -3.1, y = 2.6, z = -3)$.
(*c*) Calculate the distance between C and D.

[**Ans.** (*a*) (– 1.859, – 3.9877, 2), (*b*) (4.05, 140°, – 3), (*c*) 8.364]

12. Transform the vector $\mathbf{A} = 10\hat{\mathbf{a}}_x - 8\hat{\mathbf{a}}_y + 6\hat{\mathbf{a}}_z$ at $P(10, -8, 6)$ into cylindrical coordinates.

[**Ans.** $\mathbf{A} = 12.80\,\hat{\mathbf{a}}_\rho + 6\hat{\mathbf{a}}_z$]

13. Transform the vector $\mathbf{A} = (2x + y)\,\hat{\mathbf{a}}_x - (y - 4x)\,\hat{\mathbf{a}}_y$ at point $Q(\rho, \phi, z)$ in cylindrical coordinates.

[**Ans.** $2\rho(1 + \sin^2\phi)\hat{\mathbf{a}}_\rho + (4\rho\cos^2\phi - \rho\sin^2\phi - 3\rho\cos\phi\sin\phi\,\hat{\mathbf{a}}_\phi)$]

14. Transform the following vector to spherical coordinates at the given points:

(*a*) $10\,\hat{\mathbf{a}}_x$ at $P(-3, 2, 4)$

(*b*) $10\,\hat{\mathbf{a}}_y$ at $Q(\rho = 5, \phi = 30°, z = 4)$

(*c*) $10\,\hat{\mathbf{a}}_z$ at $M(r = 4, \theta = 110°, \phi = 120°)$.

[**Ans.** (*a*) $10\,\hat{\mathbf{a}}_x = -5.57\,\hat{\mathbf{a}}_r - 6.18\,\hat{\mathbf{a}}_\theta + 5.55\,\hat{\mathbf{a}}_\phi$]

(*b*) $10\,\hat{\mathbf{a}}_y = 3.91\,\hat{\mathbf{a}}_r + 3.12\,\hat{\mathbf{a}}_\theta + 8.66\,\hat{\mathbf{a}}_\phi$

(*c*) $10\,\hat{\mathbf{a}}_z = -3.42\,\hat{\mathbf{a}}_r - 9.39\,\hat{\mathbf{a}}_\theta$]

15. Two given points are: $M(r = 5, \theta = 20°$ and $\phi = 120°)$ and $N(r = 2, \theta = 80°$ and $\phi = 30°)$.

(*a*) Find the distance from *M* to *N*.

(*b*) Give a unit vector in coordinate at *M* that is directed towards *N*.

(*c*) Give unit vector in spherical coordinates at *M* that is directed towards *N*.

[**Ans.** (*a*) 5.073 unit (*b*) $0.505\,\hat{\mathbf{a}}_x - 0.098\,\hat{\mathbf{a}}_y - 0.858\,\hat{\mathbf{a}}_z$

(*c*) $-0.921\,\hat{\mathbf{a}}_r - 0.0023\,\hat{\mathbf{a}}_\theta - 0.388\,\hat{\mathbf{a}}_\phi$]

16. Express $\mathbf{A} = 2y\,\hat{\mathbf{a}}_x - 2\,\hat{\mathbf{a}}_y + x\,\hat{\mathbf{a}}_z$ in spherical coordinates

[**Ans.**
$$\mathbf{A} = [(2\,r\sin^2\theta\sin\phi\cos\phi - 2\sin\theta\sin\phi + r\sin\theta\cos\theta\cos\phi)\,\hat{\mathbf{a}}_r\,]$$
$$+ \left(\frac{r}{2}\sin 2\theta\sin 2\phi - 2\cos\theta\sin\phi - r\sin^2\theta\cos\theta\right)\hat{\mathbf{a}}_\theta$$
$$- 2\,r\sin\theta\sin^2\phi + 2\cos)\,\hat{\mathbf{a}}_\phi\,]$$

17. Express the vector $\mathbf{A} = 5\,\hat{\mathbf{a}}_r + 2\sin\phi\,\hat{\mathbf{a}}_\theta + 2\cos\theta\,\hat{\mathbf{a}}_\phi$ into cartesian coordinates.

[**Ans.** $\mathbf{A} = (2\cos\phi - x\tan\phi)\,\hat{\mathbf{a}}_x + (x + 2\sin\phi)\,\hat{\mathbf{a}}_y$]

18. If *S* is any closed surface enclosing a volume *V* and $\mathbf{F} = x\,\hat{\mathbf{a}}_x + 3\hat{\mathbf{a}}_y + 2z\,\hat{\mathbf{a}}_z$, show that
$$\iint_S \mathbf{F}\cdot d\mathbf{S} = 6V$$

[**Hint.**
$$\nabla\cdot\mathbf{F} = \left(\hat{\mathbf{a}}_x\frac{\partial}{\partial x} + \hat{\mathbf{a}}_y\frac{\partial}{\partial y} + \hat{\mathbf{a}}_z\frac{\partial}{\partial z}\right)\cdot(x\hat{\mathbf{a}}_x + 3y\hat{\mathbf{a}}_y + 2z\hat{\mathbf{a}}_z)$$
$$= \frac{\partial}{\partial x}(x) + \frac{\partial}{\partial y}(3y) + \frac{\partial}{\partial z}(2z) = 1 + 3 + 2 = 6$$

According to Gauss' theorem

$$\iint_S \mathbf{F}\cdot d\mathbf{S} = \iiint_V (\nabla\cdot\mathbf{F})dV = \iiint_V 6\,dV = 6V\Big]$$

19. If $\hat{\mathbf{n}}$ is the unit outward drawn normal to any closed surface of area S, show that

$$\iiint_V div\,\hat{\mathbf{n}}\,dV = \mathbf{S}$$

[**Hint.**

$$\iiint_V div\,\hat{\mathbf{n}}\,dV = \iint_S \hat{\mathbf{n}}\cdot d\mathbf{S} \text{ by divergence theorem}$$

$$= \iint_S \hat{\mathbf{n}}\cdot\hat{\mathbf{n}}\,d\mathbf{S} = \iint_S dS = S]$$

20. Evaluate by Stoke's theorem $\oint_C (yz\,dx + zx\,dy + xy\,dz)$ where C is the curve $x + y^2 = 1$ and $z = y^2$.

[**Ans.** zero]

[**Hint.**

$$\oint_C (yz\,dx + zx\,dy + xy\,dz)$$

$$\oint_C (yz\,\hat{\mathbf{a}}_x + zx\,\hat{\mathbf{a}}_y + xy\,\hat{\mathbf{a}}_z)\cdot(\hat{\mathbf{a}}_x\,dx + \hat{\mathbf{a}}_y\,dy + \hat{\mathbf{a}}_z\,dz)$$

$$\oint_C (\mathbf{A}\cdot d\mathbf{r}) \text{ where } \mathrm{A} = yz\,\hat{\mathbf{a}}_x + zx\,\hat{\mathbf{a}}_y + xy\,\hat{\mathbf{a}}_z$$

Now

$$\nabla\times\mathbf{A} = \begin{vmatrix} \hat{\mathbf{a}}_x & \hat{\mathbf{a}}_y & \hat{\mathbf{a}}_z \\ \partial/\partial x & \partial/\partial y & \partial/\partial z \\ yz & zx & xy \end{vmatrix}$$

According to Stoke's theorem

$$\oint_C \mathbf{A}\cdot d\mathbf{r} = \iint_S \text{curl}\,\mathbf{A}\cdot d\mathbf{S} = \iint_S (\nabla\times\mathbf{A})\cdot d\mathbf{S}$$

$\therefore$

$$\oint_C \mathbf{A}\cdot d\mathbf{r} = \iint_S 0\cdot d\mathbf{S} = 0]$$

21. Prove $\iint_S \hat{\mathbf{n}}\,d\mathbf{S} = 0$, for any closed surface S.

[**Hint.** Let **A** be any arbitrary constant vector. Now we have

$$\mathbf{A}\iint_S \hat{\mathbf{n}}\,d\mathbf{S} = \iint_S \mathbf{A}\cdot\hat{\mathbf{n}}\,d\mathbf{S} = \iiint_V \text{div}\,\mathbf{A}\,dV \quad \text{(Gauss's theorem)}$$

$$= \iiint_V \left(\frac{\partial A_x}{\partial x} + \frac{\partial A_y}{\partial y} + \frac{\partial A_z}{\partial z}\right)dV = 0,$$

because the scalar components of a constant vector **A** are zero.

As $A \neq 0$, therefore $\iint_S \hat{\mathbf{n}}\,dS = 0$]

22. If S is any closed surface enclosing a volume V, prove that

$$\iint_S (3x\,\hat{\mathbf{a}}_x + 5y\,\hat{\mathbf{a}}_y + 7z\,\hat{\mathbf{a}}_z)\cdot d\mathbf{S} = 15\,V$$

[**Hint.** $\iint_S (3x\,\hat{\mathbf{a}}_x + 5y\,\hat{\mathbf{a}}_y + 7z\,\hat{\mathbf{a}}_z)\cdot d\mathbf{S}$

$$= \iiint_V \text{div}(3x\,\hat{\mathbf{a}}_x + 5y\,\hat{\mathbf{a}}_y + 7z\,\hat{\mathbf{a}}_z)dV$$

$$= \iiint_V \left\{\frac{\partial}{\partial x}(3x) + \frac{\partial}{\partial y}(5y) + \frac{\partial}{\partial z}(7z)\right\}dV$$

$$= \iiint_V (3+5+7)dV = 15 \iiint_V V = 15\,V]$$

23. A vector field $\mathbf{A}(\rho, \phi, z) = 30e^{-\rho}\,\hat{\mathbf{a}}_\rho - 2z\,\hat{\mathbf{a}}_z$. Verify the divergence theorem for the volume enclosed by $\rho = 2, z = 0, z = 5$.

[**Ans.** 129.44]

24. Given $\mathbf{A} = 30e^{-\rho}\,\hat{\mathbf{a}}_\rho - 2\,z\,\hat{\mathbf{a}}_z$ in cylindrical coordinate. Evaluate both sides of div. theorem for the volume enclosed by $\rho = 2, z = 0$ and $z' = 5$ and hence, prove the divergence theorem.

[**Ans.** $\iint_S \mathbf{A}\cdot d\mathbf{S} = 129.4$ and $\iiint_V \vec{\nabla}\cdot\mathbf{A})dV = 129.4$]

As the values are same and hence, proves the div. theorem]

25. Give $\mathbf{A} - \dfrac{10\rho^3}{4}\hat{\mathbf{a}}_\rho$ in cylindrical coordinates. Evaluate both sides of div. theorem for the volume enclosed by $\rho = 1$ and $\rho = 2$ m, $z = 0$ and $z = 10$ m.

[**Ans.** L.H.S = R.H.S. = 750π]

❑❑❑

2

Electrostatic Field and Energy

INTRODUCTION

In this chapter, electric field and energy associated with static charges *i.e.*, stationary charges are discussed. Therefore, term electrostatic field defined.

In order to evaluate electrostatic fields, Coulomb's and Gauss's Laws are used. Coulomb's law is applicable only for point charges may be defined as charges whose dimension does not changes or impact the value of electronic fields. Gauss's law is applicable in analysing electrostatic fields due to charges other than point charges. Here, we shall study electric field intensity, electric flux density($\vec{D}$), energy and potential ($\vec{E}$).

2.1 COULOMB'S LAW

Like charges repel each other while unlike charges attract. This shows that a force acts between two charges. This force is known as '*electric force*'. The electric force is *attractive* in case of unlike charges while *repulsive* in case of like charges. In 1785, Coulomb gave two laws for the force of attraction or repulsion between two electrically charged bodies separated from each other by a definite distance. The laws are stated as follows :

(*i*) The force between two electric charges is directly proportional to the product of two charges.

(*ii*) The force between two electric charges is inversely proportional to the square of the distance between them. This is known as '*inverse square law*'.

The force between the two charges always acts along a straight line joining the two point charges.

Thus, Coulomb's law states that the force between two stationary point charges is directly proportional to the product of the two charges and inversely proportional to the square of the distance between them. The force acts along the straight line joining the charges.

If q_1 and q_2 be the two point charges separated from each other by a distance r (see Fig. 1) (in vacuum or air), then the force F acting between them is given by

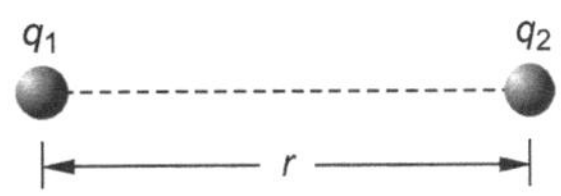

Fig. (1)

$$F \propto q_1 q_2 \text{ and } F \propto \frac{1}{r^2}$$

or
$$F \propto \frac{q_1 q_2}{r^2} = \frac{1}{4\pi\varepsilon_0} \cdot \frac{q_1 q_2}{r^2} \qquad ...(1)$$

where $\left(\frac{1}{4\pi\varepsilon_0}\right)$ is proportionality constant. Its value depends upon the medium between charges, units for charge, distance and force. If q_1 and q_2 are in coulomb, r in meter and force in newton, then

$$\frac{1}{4\pi\varepsilon_0} = 9.0 \times 10^9 \text{ newton-metre}^2/\text{coulomb}^2$$

Hence, the force between two point charges placed in vacuum (or air) is

$$\boxed{F = (9.0) \times 10^9 \frac{q_1 q_2}{r^2} \text{ newton}} \qquad ...(2)$$

The constant ε_0 is called the permittivity of free space. Its values is 8.85×10^{-12} coulomb^2/newton-meter ^{2}.

Since force is a vector quantity (having magnitude and direction), eq. (2) in vector form can be expressed as [see Fig. (2)]

$$\boxed{\mathbf{F} = \frac{1}{4\pi\varepsilon_0} \cdot \frac{q_1 q_2}{r_{12}^2} \hat{\mathbf{a}}_r} \qquad ...(3)$$

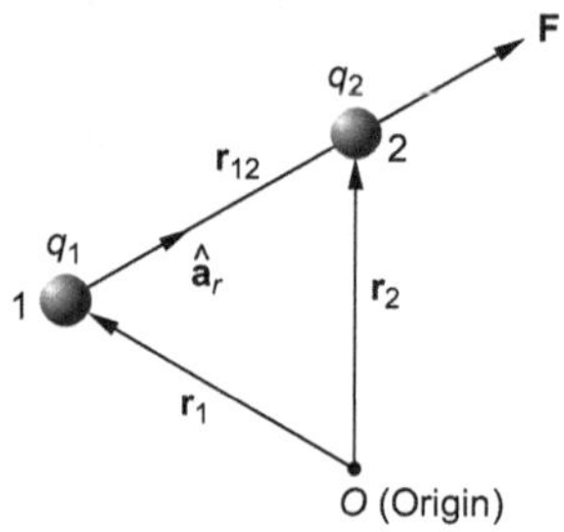

Fig. (2)

where $\hat{\mathbf{a}}_r$ is a unit vector from q_1 to q_2.

$$\hat{\mathbf{a}}_r = \frac{\mathbf{r}_{12}}{|\mathbf{r}_{12}|} = \frac{\mathbf{r}_2 - \mathbf{r}_1}{|\mathbf{r}_2 - \mathbf{r}_1|}$$

Eq. (3) can also be expressed as

$$\boxed{\mathbf{F} = \frac{1}{4\pi\varepsilon_0} \cdot \frac{q_1 q_2}{|\mathbf{r}_2 - \mathbf{r}_1|^3} (\mathbf{r}_2 - \mathbf{r}_1)} \qquad ...(4)$$

Note:

1. The unit vector $\hat{\mathbf{a}}_r$ is in the direction of line joining the two charges. In case of like charges, the unit vector $\hat{\mathbf{a}}_r$ and force $\mathbf{F}$ are in same direction [Fig. (3*a*)] while in case of unlike charges, they are in opposite direction [Fig. (3*b*)].

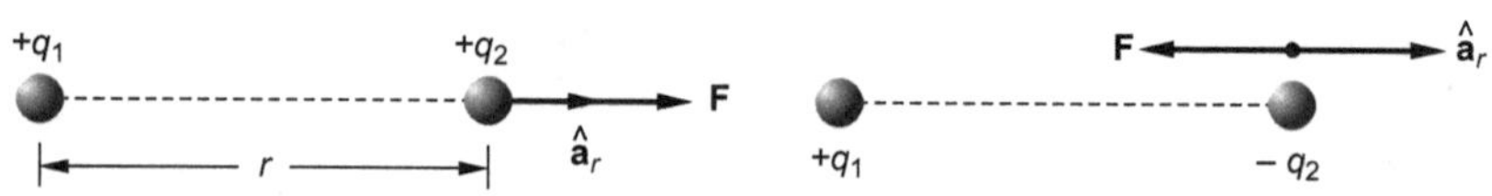

(*a*) (*b*)

Fig. (3)

2. If there are a number of charges acting on a test charge q, then the net force will be the vector sum of individual forces, $\mathbf{F}_1$, $\mathbf{F}_2$, $\mathbf{F}_3$, ... etc.
3. If instead of vacuum (or air) some insulating material (as glass, wax, paper, oil, etc.) is placed between the charge, then force F acting between them is given by

$$F = \frac{1}{4\pi\varepsilon_0 k} \cdot \frac{q_1 q_2}{r^2} \qquad \text{...(3)}$$

where k = dielectric constant of the medium (a dimensionless quantity). The insulating material is called as dielectric.

In the above formula $\varepsilon_0 k$ can be replaced by $\varepsilon(= \varepsilon_0 k)$ where ε is called as the permittivity of the dielectric. Hence,

$$\boxed{F = \frac{1}{4\pi\varepsilon} \cdot \frac{q_1 q_2}{r^2}} \qquad \text{...(4)}$$

The different forms of Coulomb's law given below in table.

Table 2.1 Different forms of Coulomb's law

General from of Coulomb's law	$F = \dfrac{q_1 q_2}{4\pi\varepsilon_0 r^2} = K\dfrac{q_1 q_2}{r^2}$
Vector from of Coulomb's law	$\mathbf{F} = \dfrac{q_1 q_2}{4\pi\varepsilon_0 r^2}\,\hat{\mathbf{a}}_r$
Coulomb's law in a medium	$\mathbf{F} = \dfrac{q_1 q_2}{4\pi\varepsilon r^2}\,\hat{\mathbf{a}}_r$

2.2 ELECTRIC FIELD INTENSITY

The region or surrounding, in which a charge experiences force is called as electric field. *The intensity of electric field at a point in the field is defined as the force experienced by a unit positive charge placed at that point* [see Fig. (3)]

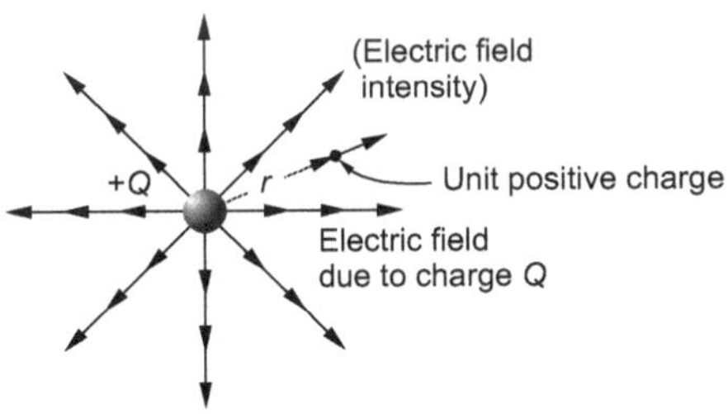

Fig. (3) Showing electric field due to charge Q and electric field intensity

Let **F** be the force experienced by a test charge q_0 placed at a point in the electric field, then the intensity of electric field **E** at that point is given by

$$\boxed{\mathbf{E} = \frac{\mathbf{F}}{q_0} \frac{\text{newton}}{\text{coulomb}}} \qquad \text{...(1)}$$

It is obvious that intensity of electric field is a *vector quantity*. Its direction will be the same as the direction of force, *i.e.*, the direction in which the positive charge placed in the field tends to move. The unit of electric field is *newton per coulomb*.

Let us consider that a unit positive charge is placed at a distance r from a point charge Q in air. From Coulomb's law, the force on unit positive charge is given by

$$F = \frac{1}{4\pi\varepsilon_0} \cdot \frac{Q \times 1}{r^2}$$

$$\therefore \quad \boxed{E = \frac{F}{1} = \frac{1}{4\pi\varepsilon_0} \frac{Q}{r^2}} \quad \text{...(2)}$$

Vector form

Consider a Cartesian coordinate system in which a charge Q is located at a distance $\mathbf{r}_1$ from the origin as shown in Fig. (4). Let us find the electric field intensity at point P, which is at a distance $\mathbf{r}$ from the origin.

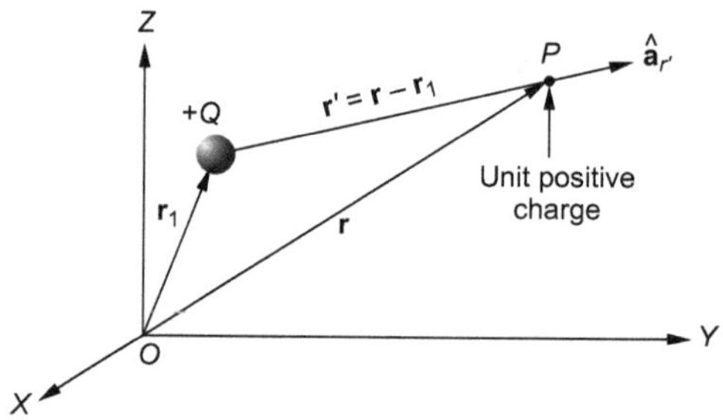

Fig. (4) Field at point P due to charge +Q

The electric field intensity at P is given by

$$\mathbf{E} = \frac{Q}{4\pi\varepsilon_0 |\mathbf{r} - \mathbf{r}_1|^2} \hat{\mathbf{a}}_{r'}$$

where $\hat{\mathbf{a}}_{r'}$ is the unit vector in the direction of $(\mathbf{r} - \mathbf{r}_1)$.

$$\hat{\mathbf{a}}_{r'} = \frac{\mathbf{r} - \mathbf{r}_1}{|\mathbf{r} - \mathbf{r}_1|}$$

$$\therefore \quad \boxed{\mathbf{E} = \frac{Q(\mathbf{r} - \mathbf{r}_1)}{4\pi\varepsilon_0 |\mathbf{r} - \mathbf{r}_1|^3}} \quad \text{...(3)}$$

For a system of charges, we have

$$\mathbf{E} = \frac{Q_1(\mathbf{r} - \mathbf{r}_1)}{4\pi\varepsilon_0 |\mathbf{r} - \mathbf{r}_1|^3} + \frac{Q_2(\mathbf{r} - \mathbf{r}_2)}{4\pi\varepsilon_0 |\mathbf{r} - \mathbf{r}_2|^3} + \ldots + \frac{Q_n(\mathbf{r} - \mathbf{r}_n)}{4\pi\varepsilon_0 |\mathbf{r} - \mathbf{r}_n|^3}$$

$$\therefore \quad \boxed{E = \sum_{i=1}^{n} \frac{Q_i(\mathbf{r} - \mathbf{r}_i)}{4\pi\varepsilon_0 |\mathbf{r} - \mathbf{r}_i|^3}}$$

2.3 ELECTRIC FIELD INTENSITY DUE TO A SYSTEM OF CHARGES

Before considering the electric field intensity due to a system of charges, let us consider line charge density (ρ_L), surface charge density (ρ_S) and volume charge density (ρ_V).

1. **Line charge density (ρ_L):** Consider that a charge Q is uniformly distributed on a line conductor. The charge on the conductor per unit length is called as line charge density. This is denoted by ρ_L. The total charge on a conductor can be be obtained by integral. If L is the length of the conductor, then

$$Q = \int_0^L \rho_L \, dL \quad \text{...(1)}$$

2. **Surface charge density (ρ_S):** When the charge is distributed on a surface (2-dimensional body) of a conductor, then the charge per unit surface area is called surface charge density. This is denoted by ρ_S. The net charge enclosed by the surface is given by

$$Q = \iint_S \rho_S \, dS \quad \text{...(2)}$$

3. **Volume charge density (ρ_V):** When the charge is distributed around the volume of a conductor, then charge contained per unit volume of the conductor is called as volume charge density This is denoted by ρ_V. Consider any three dimensional body having volume charge density ρ_V as shown in fig (5). Consider a small volume dV of the body. The charge contained by small volume dV is given by

$$dQ = \rho_V \, dV$$

Therefore, the total charge enclosed by the body is given by

$$Q = \iiint_V \rho_V \, dV \quad \text{...(3)}$$

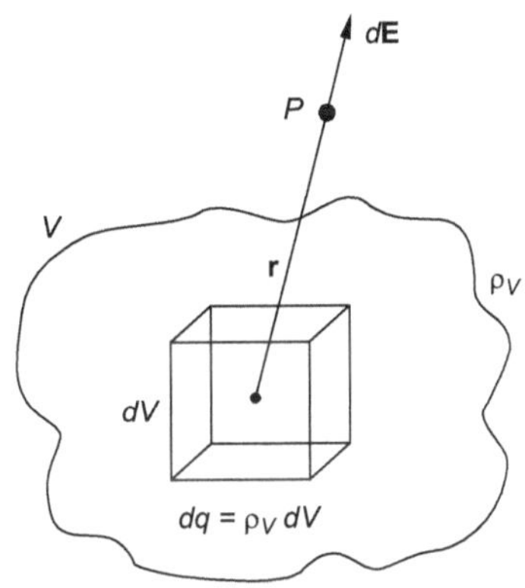

Fig. (5)

Expressions for Electric Field Intensity

1. Electric field intensity due to charge distribution is given by

$$\mathbf{E} = \int_L \frac{\rho_L \, dL}{4\pi \varepsilon_0 |\mathbf{r}|^2} \hat{\mathbf{a}}_r \quad \text{...(4)}$$

2. Electric field intensity due to surface charge distribution is given by

$$\mathbf{E} = \iint_S \frac{\rho_S \, dS}{4\pi \varepsilon_0 |\mathbf{r}|^2} \hat{\mathbf{a}}_r \quad \text{...(5)}$$

3. Electric field intensity due to volume charge distribution is given by

$$\mathbf{E} = \iiint_V \frac{\rho_V \, dV}{4\pi \varepsilon_0 |\mathbf{r}|^2} \hat{\mathbf{a}}_r \quad \text{...(6)}$$

2.4 ELECTRIC FIELD INTENSITY DUE TO INFINITELY LONG CHARGED WIRE

(1) Cartesian coordinate system

Consider the case of an infinitely long charged wire of negligible thickness as shown in Fig. (6). Let ρ_L be the density of charge per unit length. Our aim is to calculate the electric field at a point P,

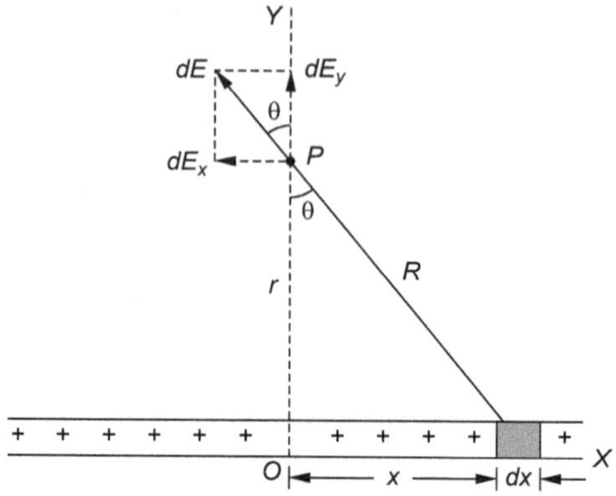

Fig. (6)

a distance r from the wire. Since, we know Coulomb's law which is applicable to point charge only. Therefore, let us divider line charge into infinitesmall point charges. Now, consider one small element of length dx at a distance x from O. The charge on this element is $dq = \rho_L\, dx$. The field $d\mathbf{E}$ at a point P due to this charge dq

$$dE = \frac{1}{4\,\pi\varepsilon_0} \times \frac{dq}{R^2} = \frac{1}{4\,\pi\varepsilon_0} \times \frac{\rho_L\, dx}{R^2} \qquad \text{(where } AP = R\text{)...(1)}$$

The x and y components of dE are

$$dE_x = -\,dE \sin\theta \quad \text{and} \quad dE_y = dE\cos\theta$$

Further consider an element of length dx at a distance x from O towards left. In this case, the components of dE along X and Y axis will be $dE \sin\theta$ and $dE\cos\theta$ respectively. Due to symmetry, the horizontal components of the field intensity will cancel each other and only vertical components remain. Hence, the total intensity of electric field at P will be $2\, dE\cos\theta$ in Y-direction.

$$\therefore \qquad E = \int_0^{+\infty} 2\, dE\cos\theta \qquad (\therefore \text{ now the wire extends from 0 to } +\infty)$$

or

$$E = \int_0^{\infty} \frac{1}{4\pi\varepsilon_0} \times \frac{\rho_L\, dx}{R^2} \times 2\cos\theta = \frac{\sigma}{2\pi\varepsilon_0}\int_0^{\infty} \frac{dx\cos\theta}{R^2} \qquad ...(2)$$

From fig $\quad \dfrac{x}{r} = \tan\theta$

$$\therefore \qquad \frac{dx}{r} = \sec^2\theta\, d\theta$$

or $\quad dx = r\sec^2\theta\, d\theta \quad$ and $\quad R^2 = r^2 + x^2 = r^2 + r^2 + \tan^2\theta$

when $x = 0$, $\theta = 0$ and when $x = \infty$, $\theta = \dfrac{\pi}{2}$

Substituting these values, we get

$$E = \frac{\rho_L}{2\pi\varepsilon_0}\int_0^{\pi/2} \frac{r\sec^2\theta\, d\theta\cdot\cos\theta}{(r^2\tan^2\theta + r^2)} = \frac{\rho_L}{2\pi\varepsilon_0}\int_0^{\pi/2} \frac{r\sec^2\theta\cos\theta\, d\theta}{r^2(1+\tan^2\theta)}$$

$$= \frac{\rho_L}{2\pi\varepsilon_0}\int_0^{\pi/2} \frac{\sec^2\theta\cos\theta\, d\theta}{r\sec^2\theta} \qquad (\because\ 1 + \tan^2\theta = \sec^2\theta)$$

$$= \frac{\rho_L}{2\pi\varepsilon_0 r}\int_0^{\pi/2}\cos\theta\, d\theta = \frac{\rho_L}{2\pi\varepsilon_0 r}[\sin\theta]_0^{\pi/2}$$

$$= \frac{\rho_L}{2\pi\varepsilon_0 r}[1-0] = \frac{\rho_L}{2\pi\varepsilon_0 r}$$

$$\therefore \qquad \boxed{E = \frac{\rho_L}{2\pi\varepsilon_0 r}} \qquad ...(3)$$

The direction of electric field is outward if the wire carries a positive charge and inward if the wire carries a negative charge.

In vector form

$$\boxed{\mathbf{E} = \frac{\rho_L}{2\pi\varepsilon_0 r}\hat{\mathbf{a}}_r} \qquad ...(4)$$

(2) Cylindrical Coordinate System

Consider an infinitely long line charge having charge density ρ_L placed along Z-axis as shown in Fig. (7).

In order to find out the electric field intensity at P, we consider an infinitesimal element dz at a distance $\mathbf{r}'$ from origin along Z-axis. The electric field intensity $d\mathbf{E}$ at P due to dz is given by

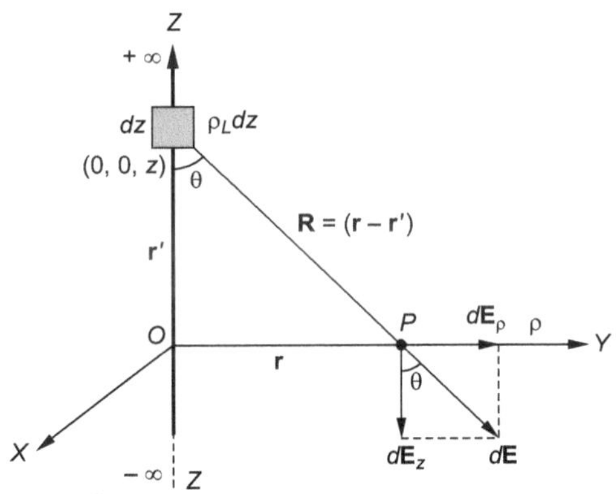

Fig. (7) Electric field intensity due to infinite line charge

$$d\mathbf{E} = \frac{\rho_L\, dz}{4\pi\varepsilon_0}\frac{(\mathbf{r}-\mathbf{r}')}{|\mathbf{r}-\mathbf{r}'|^3}$$

Here, $\mathbf{r} = y\hat{\mathbf{a}}_y = \rho\hat{\mathbf{a}}_\rho$ and $\mathbf{r}' = z\hat{\mathbf{a}}_z$

$$\therefore \qquad \mathbf{r}-\mathbf{r}' = \rho\hat{\mathbf{a}}_\rho - z\hat{\mathbf{a}}_z$$

Therefore, $$d\mathbf{E} = \frac{\rho_L\, dz\,(\rho\hat{\mathbf{a}}_\rho - z\hat{\mathbf{a}}_z)}{4\pi\varepsilon_0(\rho^2+z^2)^{3/2}} \qquad [\because |\mathbf{r}-\mathbf{r}'| = \sqrt{(\rho^2+z^2)}]$$

No element of the charge produces ϕ component of electric field intensity, *i.e.*, $E_\phi = 0$. However, each element produces E_ρ and E_z components. The contribution to E_z by elements which are above and below the z-axis at point P will cancel. Therefore, only E_ρ component contributes to electric field intensity. Now, we have

$$dE_\rho = \frac{\rho_L \, \rho \, dz}{4\pi\varepsilon_0 (\rho^2 + z^2)^{3/2}}$$

Therefore, total electric field intensity at P is given by

$$E_\rho = \int_{-\infty}^{+\infty} \frac{\rho_L \, \rho \, dz}{4\pi\varepsilon_0 (\rho^2 + z^2)^{3/2}}$$

$$= \frac{\rho_L \, \rho}{4\pi\varepsilon_0} \left[\frac{1}{\rho^2} \cdot \frac{z}{\sqrt{\rho^2 + z^2}} \right]_{-\infty}^{+\infty} = \frac{\rho_L \rho}{4\pi\varepsilon_0} \left[\frac{2}{\rho^2} \right]$$

$$\therefore \quad \boxed{E_\rho = \frac{\rho_L}{2\pi\varepsilon_0 \cdot \rho}}$$

2.5 ELECTRIC FIELD DUE TO A FINITE LINE OF CHARGE

(1) Cartesian Coordinate System

Fig. (8) shows a uniformly charged wire of length L and carries a total charge q spread uniformly about it. Let $\rho_L = (\rho/L)$ be the density of charge per unit length of the wire. Here, we shall calculate the electric field at point P on a perpendicular bisector of wire. Let the distance of P from the centre O be r. Consider two symmetrical elements dx of the wire at a distance x from the centre. One such element is shown in the Fig. (8). Due to symmetry, the horizontal components of electric field intensity will be cancelled. Their remains only vertical components. So, the resultants electric field due to two elements dx will be along Y-axis. Now,

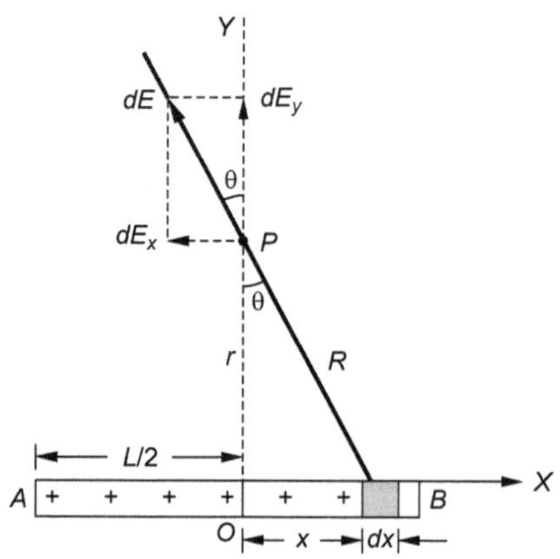

Fig. (8)

$$dE = \frac{1}{4\pi\varepsilon_0} \times \frac{\rho_L \, dx}{R^2}$$

or

$$dE = \frac{1}{4\pi\varepsilon_0} \cdot \frac{\rho_L \, dx}{(r^2 + x^2)} \qquad [\because R = \sqrt{(r^2 + x^2)}\,]$$

$$\therefore \qquad dE_y = \frac{1}{4\pi\varepsilon_0} \times \frac{\rho_L\, dx}{(r^2+x^2)} \cos\theta \qquad (\because dE_y = dE\cos\theta) \quad ...(1)$$

The total electric field due to wire of lengths L can be obtained by integrating above expression (1) within the limits $x = -\frac{L}{2}$ to $x = +\frac{L}{2}$. Thus,

$$E = \int_{-L/2}^{L/2} dE_y = \frac{\rho_L}{4 r\varepsilon_0} \int_{-L/2}^{L/2} \frac{dx}{(r^2+x^2)} \cos\theta$$

From figure $\qquad \cos\theta = \frac{r}{(r^2+x^2)^{1/2}} \qquad [\because R = (r^2+x^2)^{1/2}]$

Substituting the value of $\cos\theta$ in above expression, we get

$$E = \frac{2\rho_L r}{4\pi\varepsilon_0} \int_0^{L/2} \frac{dx}{(r^2+x^2)^{3/2}} \qquad ...(2)$$

$$= \frac{\rho_L r}{2\pi\varepsilon_0} \left[\frac{1}{r^2} \cdot \frac{x}{(r^2+x^2)^{1/2}} \right]_0^{L/2} = \frac{\rho_L r}{2\pi\varepsilon_0} \left[\frac{1}{r^2} \cdot \frac{\frac{L}{2}}{\left\{ r^2 + \left(\frac{L}{2}\right)^2 \right\}^{1/2}} \right]$$

or $$\boxed{E = \frac{\rho_L}{2\pi\varepsilon_0 r} \times \frac{L}{(L^2+4r^2)^{1/2}}} \qquad ...(3)$$

(2) Cylindrical Coordinate System

Let us consider a line charge distribution along the line between $(0, 0, -a)$ and $(0, 0, a)$ in cylindrical coordinates. Here the cylindrical coordinates and denoted as (r, ϕ, z). The charge density is ρ_L coulomb/m. The aim of this article is to obtain an expression for electric field intensity at a point $P(r, \phi, 0)$ as shown in Fig. (9).

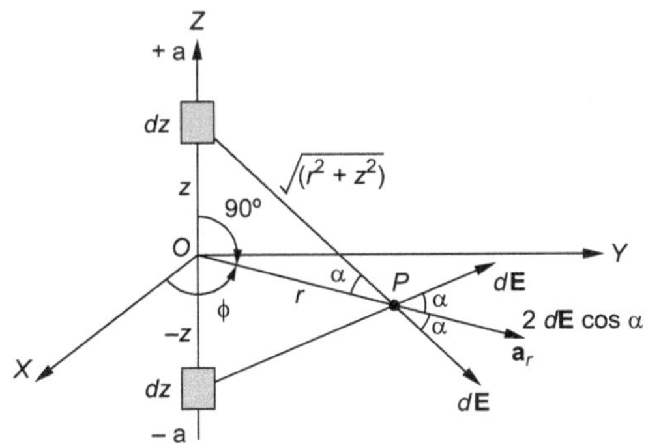

Fig. (9)

As the situation has a symmetry and hence, the electric field at P has only r component. Consider a small element dz on the charge line at a distance z from the origin. The electric field intensity dE at P due to element dz is given by

$$dE = \frac{\rho_L\, dz}{4\pi\varepsilon_0 (r^2+z^2)}$$

Similarly, we can consider the electric field intensity at point P due to element dz at a distance $-z$ from the origin.

The resultant intensity $d\mathbf{E}$ in the direction of r is given by

$$d\mathbf{E} = 2 \cdot \frac{\rho_L \, dz \cos \alpha}{4 \pi \varepsilon_0 (r^2 + z^2)} \hat{\mathbf{a}}_r$$

where $\hat{\mathbf{a}}_r$ is unit vector along the direction of r.

The total electric intensity at P due to finite line charge is given by

$$\mathbf{E}(r, \phi, 0) = 2 \int_{z=0}^{a} \frac{\rho_L \, dz \cos \alpha}{4 \pi \varepsilon_0 (r^2 + z^2)} \hat{\mathbf{a}}_r$$

$$\therefore \qquad \mathbf{E}(r, \phi, 0) = 2 \int_{z=0}^{a} \frac{\rho_L \, dz \, r}{4 \pi \varepsilon_0 (r^2 + z^2)^{3/2}} \hat{\mathbf{a}}_r \qquad \left[\because \cos \alpha = \frac{r}{\sqrt{r^2 + z^2}} \right]$$

$$= 2 \cdot \frac{\rho_L \, r}{4 \pi \varepsilon_0} \left[\frac{z}{r^2 (r^2 + z^2)^{1/2}} \right]_{z=0}^{a} \hat{\mathbf{a}}_r$$

or

$$\boxed{\mathbf{E}(r, \phi, 0) = \frac{\rho_L \, a}{2 \pi \varepsilon_0 \, r (r^2 + a^2)^{1/2}}}$$

This is the desired expression.

2.6 ELECTRIC FIELD INTENSITY DUE TO INFINITE SHEET OF CHARGE

Consider an infinite sheet of charge of density ρ_S coulomb per metre^2 placed in Y-Z plane as shown as shown in Fig. (10). The aim is to calculate the electric field intensity at a point $P(x, 0, 0)$.

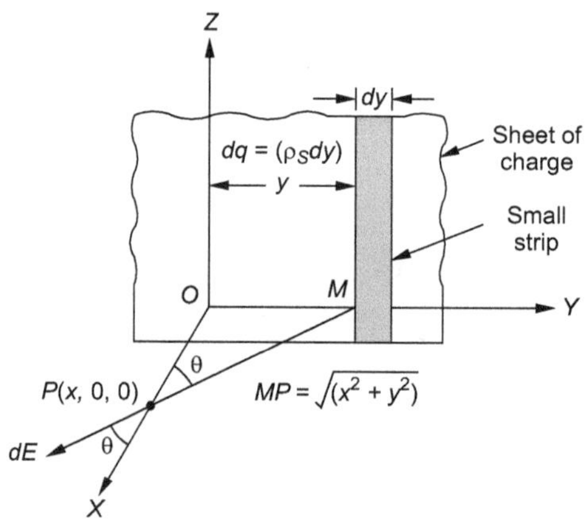

Fig. (10) Showing a sheet of charge

In order to consider the electric field intensity at a point P due to this sheet of charge, we divide this sheet into infinite number of strips. Consider one such strip of thickness dy at a distance y.

The contribution to E_x at P due to this strip is given by

$$dE_x = \frac{\rho_S \, dy}{2 \pi \varepsilon_0} \cdot \frac{1}{\sqrt{x^2 + y^2}} \cdot \cos \theta \qquad \text{...(1)}$$

From figure, $\cos\theta = \dfrac{x}{\sqrt{(x^2+y^2)}}$

$$\therefore \quad dE_x = \frac{\rho_S\, dy}{2\pi\varepsilon_0}\cdot\frac{1}{\sqrt{(x^2+y^2)}}\cdot\frac{x}{\sqrt{(x^2+y^2)}} = \frac{\rho_S\, x\, dy}{2\pi\varepsilon_0\,(x^2+y^2)} \qquad ...(2)$$

The total electric field strength can be obtained by integrating the above expression within the limits $-\infty$ to $+\infty$. Therefore,

$$E_x = \frac{\rho_S\, x}{2\pi\varepsilon_0}\int_{-\infty}^{+\infty}\frac{dy}{(x^2+y^2)} \qquad ...(3)$$

In order to solve this integral, put $y = x\tan\theta$, *i.e.*, $dy = x\sec^2\theta\, d\theta$

$$\therefore \quad E_x = \frac{\rho_S\, x}{2\pi\varepsilon_0}\int_{-\pi/2}^{+\pi/2}\frac{x\sec^2\theta\, d\theta}{(x^2+x^2\tan^2\theta)} = \frac{\rho_S\, x^2}{2\pi\varepsilon_0}\int_{-\pi/2}^{+\pi/2}\frac{\sec^2\theta\, d\theta}{x^2(1+\tan^2\theta)}$$

$$= \frac{\rho_S}{2\pi\varepsilon_0}\int_{-\pi/2}^{+\pi/2}\frac{\sec^2\theta\, d\theta}{\sec^2\theta)} = \frac{\rho_S}{2\pi\varepsilon_0}\cdot\int_{-\pi/2}^{+\pi/2} d\theta = \frac{\rho_S}{2\pi\varepsilon_0}[\theta]_{-\pi/2}^{\pi/2}$$

$$= \frac{\rho_S}{2\pi\varepsilon_0}\left[\frac{\pi}{2}+\frac{\pi}{2}\right] = \frac{\rho_S\,\pi}{2\pi\varepsilon_0} = \frac{\rho_S}{2\pi\varepsilon_0}$$

Thus, $$\boxed{E_x = \frac{\rho_S}{2\varepsilon_0}} \quad \text{for} \quad x > 0 \qquad ...(4)$$

If the point is taken on the negative X-axis, then

$$\boxed{E_x = -\frac{\rho_S}{2\varepsilon_0}} \quad \text{for} \quad x < 0 \qquad ...(5)$$

In general $$\boxed{\mathbf{E} = \frac{\rho_S}{2\varepsilon_0}\hat{\mathbf{a}}_n}$$

where $\hat{\mathbf{a}}_n$ is unit vector outward normal to the sheet.

Eq. (6) shows that the electric field strength is constant in magnitude as well as direction.

Solved EXAMPLES

EXAMPLE 1. *Find* **E** *at* (0, 3, 4) *due to a point charge* $q = 0.5\ \mu C$ *placed at origin.*

Solution. The situation is shown in Fig. (11).

The electric field intensity at P is given by

$$\mathbf{E} = \frac{q}{4\pi\varepsilon_0 r^2}\hat{\mathbf{a}}_r \qquad ...(1)$$

where $\hat{\mathbf{a}}_r$ is unit vector along the direction of $\mathbf{r}$.

$$\hat{\mathbf{a}}_r = \frac{\mathbf{r}}{|\mathbf{r}|}$$

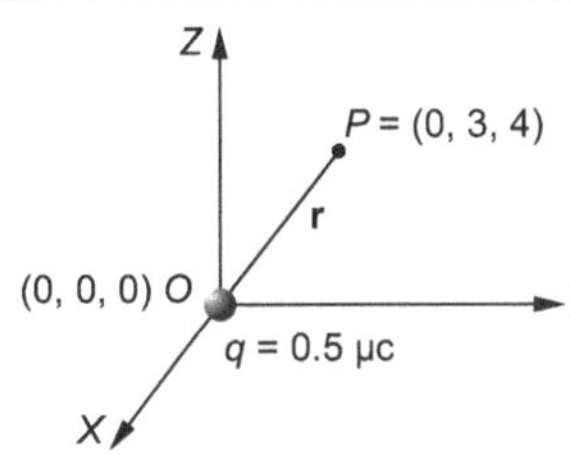

Fig. (11)

Here $\mathbf{r} = (3-0)\hat{\mathbf{a}}_y + (4-0)\hat{\mathbf{a}}_z = 3\hat{\mathbf{a}}_y + 4\hat{\mathbf{a}}_z$

and $$|\mathbf{r}| = \sqrt{[(3)^2 + (4)^2]} = \sqrt{(9+16)} = 5$$

For free space, $\varepsilon_0 = 8.85 \times 10^{-12}$ F/m

Substituting these values in eq. (1), we get

$$\mathbf{E} = \frac{0.5 \times 10^{-6}}{4\pi \times (8.85 \times 10^{-12})(5)^2} \cdot \frac{3\hat{\mathbf{a}}_y + 4\hat{\mathbf{a}}_z}{5}$$

$$= 35.97(3\hat{\mathbf{a}}_y + 4\hat{\mathbf{a}}_z) = 107.9\hat{\mathbf{a}}_y + 143.87\hat{\mathbf{a}}_z$$

EXAMPLE 2. *Fine electric field intensity* **E** *at a point* $P(1, 1, 1)$ *due to four* identical 3×10^{-9} C *charges located* at points $P_1(1, 1, 0)$, $P_2(-1, 1, 0)$, $P_3(-1, -1, 0)$ *and* $P_4(1, -1, 0)$.

Solution. The situation is shown in Fig. (13).

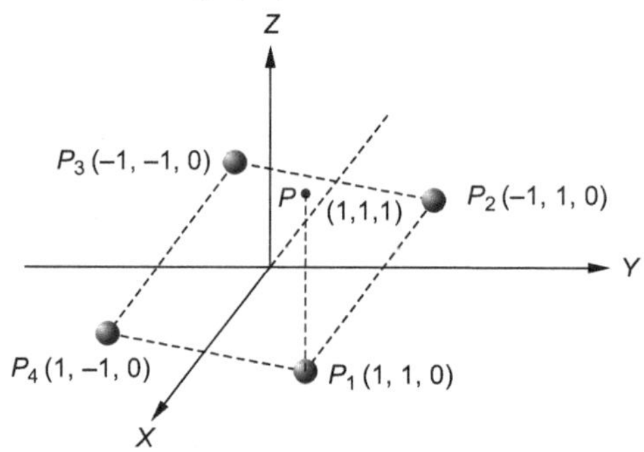

Fig. (13)

Electric field at point $P(1, 1, 1)$ will be the sum of electric fields due to charges at P_1, P_2, P_3, and P_4 as shown in Fig. (13). Therefore,

$$\mathbf{E}_{P(1,1,1)} = \mathbf{E}_{P_1} + \mathbf{E}_{P_2} + \mathbf{E}_{P_3} + \mathbf{E}_{P_4}$$

$$= \frac{1}{4\pi\varepsilon_0} \cdot \frac{3 \times 10^{-9}\ \text{C}}{\left[\sqrt{(1-1)^2 + (1-1)^2 + (1-0)^2}\right]^3} \cdot [(1-1)\hat{\mathbf{a}}_x + (1-1)\hat{\mathbf{a}}_y + (1-0)\hat{\mathbf{a}}_z]$$

$$+ \frac{1}{4\pi\varepsilon_0} \cdot \frac{3 \times 10^{-9}}{\left[\sqrt{(1+1)^2 + (1-1)^2 + (1-0)^2}\right]^3} \cdot \left[(1+1)\hat{\mathbf{a}}_x + (1-1)\hat{\mathbf{a}}_y + (1-0)\hat{\mathbf{a}}_z\right]$$

$$+ \frac{1}{4\pi\varepsilon_0} \cdot \frac{3 \times 10^{-9}}{\left[\sqrt{(1+1)^2 + (1+1)^2 + (1-0)^2}\right]^3} \cdot \left[(1+1)\hat{\mathbf{a}}_x + (1+1)\hat{\mathbf{a}}_y + (1-0)\hat{\mathbf{a}}_z\right]$$

$$+ \frac{1}{4\pi\varepsilon_0} \cdot \frac{3 \times 10^{-9}}{\left[\sqrt{(1-1)^2 + (1+1)^2 + (1-0)^2}\right]^3} \cdot [(1-1)\hat{\mathbf{a}}_x + (1+1)\hat{\mathbf{a}}_y + (1-0)\hat{\mathbf{a}}_z]$$

$$= \frac{3 \times 10^{-9}}{4\pi\varepsilon_0}\left[\hat{\mathbf{a}}_z + \left(\frac{2\hat{\mathbf{a}}_x + \hat{\mathbf{a}}_z}{(5^{3/2})}\right) + \left(\frac{2\hat{\mathbf{a}}_x + 2\hat{\mathbf{a}}_y + \hat{\mathbf{a}}_z}{(3)^3}\right) + \left(\frac{2\hat{\mathbf{a}}_y + \hat{\mathbf{a}}_z}{5^{3/2}}\right)\right] \frac{\text{Volt}}{\text{metre}}$$

$$= (6.82\,\hat{\mathbf{a}}_x + 6.82\,\hat{\mathbf{a}}_y + 32.8\,\hat{\mathbf{a}}_z)\text{V/m}$$

EXAMPLE 3. *Fine the total electric field intensity at point* (0, 6, 5)m *due to a charge* 20 μC *located at* (2, 0, 6)m, *a charge of* 60 μC *located at* (0, – 1, 2)m *and a charge* 100 μC *located at* (2, 3, 4)m.

Solution. The situation is shown in Fig. (12).

The electric field intensity at point P_0 is given by

$$\mathbf{E} = \mathbf{E}_1 + \mathbf{E}_2 + \mathbf{E}_3$$

$$= \frac{1}{4\pi\varepsilon_0}\left[\frac{q_1\,\mathbf{a}_{r_1}}{r_1^2} + \frac{q_2\,\mathbf{a}_{r_2}}{r_2^2} + \frac{q_3\,\mathbf{a}_{r_3}}{r_3^2}\right]$$

$$= \frac{1}{4\pi\varepsilon_0}\left[\frac{q_1\,\mathbf{r}_1}{r_1^3} + \frac{q_1\,\mathbf{r}_2}{r_2^3} + \frac{q_3\,\mathbf{r}_3}{r_3^3}\right]$$

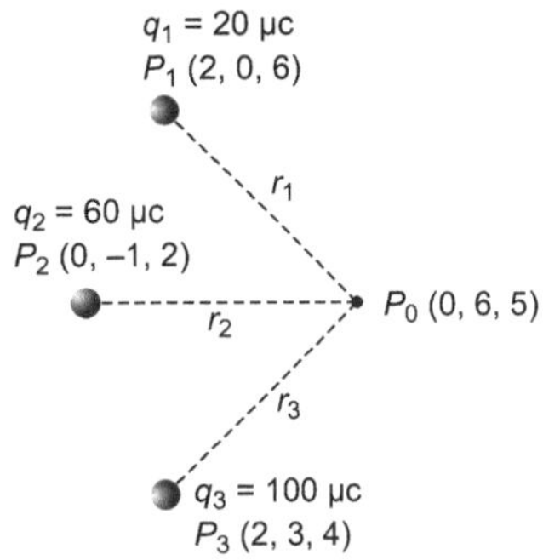

Fig. (12)

where $\mathbf{r}_1 = -2\,\hat{\mathbf{a}}_x + 6\,\hat{\mathbf{a}}_y - \hat{\mathbf{a}}_z$

$\mathbf{r}_2 = 7\,\hat{\mathbf{a}}_y + 3\,\hat{\mathbf{a}}_z$

and $\mathbf{r}_3 = -2\,\hat{\mathbf{a}}_x + 3\,\hat{\mathbf{a}}_y + \hat{\mathbf{a}}_z$

From these vectors, we have

$$\mathbf{r}_1 = |\mathbf{r}_1| = \sqrt{(4+36+1)} = 6.40 \text{ m}$$

$$\mathbf{r}_2 = |\mathbf{r}_2| = \sqrt{(49+9)} = 7.615 \text{ m}$$

$$\mathbf{r}_3 = |\mathbf{r}_3| = \sqrt{(4+9+1)} = 3.74 \text{ m}$$

$$\therefore \quad \mathbf{E} = (9\times 10^9)\left[\frac{20\times10^{-6}}{(6.40)^3}(-2\,\hat{\mathbf{a}}_x + 6\,\hat{\mathbf{a}}_y - \hat{\mathbf{a}}_z) + \frac{60\times10^{-6}}{(7.615)^3}(7\,\hat{\mathbf{a}}_y + 3\,\hat{\mathbf{a}}_z) + \frac{100\times10^{-6}}{(3.74)^2}(-2\,\hat{\mathbf{a}}_x + 3\,\hat{\mathbf{a}}_y + \hat{\mathbf{a}}_z)\right]$$

or

$$\mathbf{E} = [-35781.15\,\hat{\mathbf{a}}_x + 64291.82\,\hat{\mathbf{a}}_y + 20185.92\,\hat{\mathbf{a}}_z] \text{ V/m}$$

EXAMPLE 4. *A charge of* – 0.3 μC *is located at* A(25, – 30, 15) cm *and a second charge of* 0.5 μC *is at* B(– 10, 8, 12) cm. *Find* **E** *at* (*a*) *origin* (0, 0, 0) *and* (*b*) P(15, 20, 50) cm.

Solution. (*a*) The electric field at origin due to both charges is given by

$$\mathbf{E} = \mathbf{E}_1 + \mathbf{E}_2$$

$$= \frac{1}{4\pi\varepsilon_0}\cdot\frac{-0.3\times10^{-6}\left[(0-0.25)\,\hat{\mathbf{a}}_x + (0+0.30)\,\hat{\mathbf{a}}_y + (0-0.15)\hat{\mathbf{a}}_z\right]}{\left[\sqrt{(0.25-0)^2 + (-0.30-0)^2 + (0.15-0)^2}\right]^3}$$

$$+ \frac{1}{4\pi\varepsilon_0}\cdot\frac{0.5\times10^{-6}[(0+0.1)\hat{\mathbf{a}}_x + (0-0.08)\hat{\mathbf{a}}_y + (0-0.12)\hat{\mathbf{a}}_z]}{\left[\sqrt{(0+0.1)^2 + (0-0.08)^2 + (0-0.12)^2}\right]^3}$$

$$= \frac{2700}{0.073}(0.25\,\hat{\mathbf{a}}_x - 0.3\,\hat{\mathbf{a}}_y + 0.15\,\hat{\mathbf{a}}_z) + \frac{4500}{0.0054}(0.1\,\hat{\mathbf{a}}_x - 0.08\,\hat{\mathbf{a}}_y - 0.12\,\hat{\mathbf{a}}_z)$$

$$= (9246.57\,\hat{\mathbf{a}}_x - 11095.89\,\hat{\mathbf{a}}_y + 5547.94\,\hat{\mathbf{a}}_z)$$

$$+ (83333.33\,\hat{\mathbf{a}}_x - 66666.66\,\hat{\mathbf{a}}_y - 10^5\,\hat{\mathbf{a}}_z)\frac{\text{V}}{\text{m}}$$

$$\therefore \quad \mathbf{E} = (92.3\,\hat{\mathbf{a}}_x - 77.6\,\hat{\mathbf{a}}_y - 105.3\,\hat{\mathbf{a}}_z)\frac{\text{kV}}{\text{m}}$$

(*b*) The electric field at $P(15, 20, 50)$ due to both charges is given by

$$\mathbf{E} = \frac{-0.3\times10^{-6}}{4\,\pi\varepsilon_0}\cdot\frac{\left[(0.15-0.25)\,\hat{\mathbf{a}}_x + (0.2+0.3)\,\hat{\mathbf{a}}_y + (0.5-0.15)\hat{\mathbf{a}}_z\right]}{\left[\sqrt{(0.15-0.25)^2 + (0.2+0.3)^2 + (0.5-0.15)^2}\right]^3}$$

$$+\frac{0.5\times10^{-6}}{4\,\pi\varepsilon_0}\cdot\frac{\left[(0.15+0.10)\,\hat{\mathbf{a}}_x + (0.2+0.08)\,\hat{\mathbf{a}}_y + (0.5-0.12)\,\hat{\mathbf{a}}_z\right]}{\left[\sqrt{(0.15+0.10)^2 + (0.2-0.08)^2 + (0.5-0.12)^2}\right]^3}$$

Solving we get

$$\mathbf{E} = (32.9\,\hat{\mathbf{a}}_x + 5.94\,\hat{\mathbf{a}}_y + 19.69\,\hat{\mathbf{a}}_z)\frac{\text{kV}}{\text{m}}.$$

EXAMPLE 5. *A charge is distributed along Z-axis between* ± 6 m *with uniform charge density* 25 nC/m. *Calculate electric field intensity at point* (2, 0, 0)m *in free space.*

Solution. The situation in Fig. (14).

We know that electric field intensity due to a finite line charge over a straight line is given by

$$\mathbf{E} = \frac{\rho_L}{2\,\pi\varepsilon_0\,r}\cdot\frac{a}{(r^2+a^2)^{1/2}}$$

where r is the distance of field point from line charge.

Substituting the values, we get

$$|\,\mathbf{E}\,| = \frac{25\times10^{-9}}{2\pi\times8.854\times10^{-12}\times2}\times\frac{6}{\sqrt{[(2)^2+(6)^2]}}$$

$$= 213.16 \text{ V/m}$$

The direction of **E** is along X-axis. Therefore,

$$\mathbf{E} = 213.16\,\hat{\mathbf{a}}_x$$

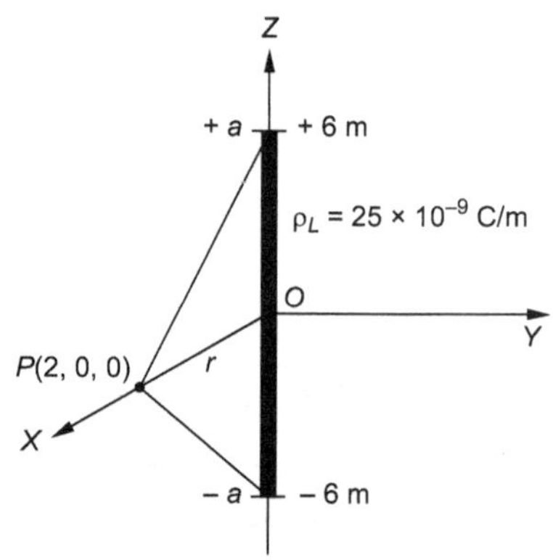

Fig. (14)

EXAMPLE 6. *An infinitely long uniform line charge is located at* $y = 3$, $z = 5$. *If* $\rho_L = 30\times10^{-9}$ C/m, *find electric field intensity at* (*i*) $P_1(0, 0, 0)$ *and* (*ii*) $P_2(0, 6, 1)$.

Solution. The situation is shown in fig (15).

The electric field intensity due to line charge is given by

$$\mathbf{E} = \frac{\rho_L}{2\,\pi\varepsilon_0 r}\,\hat{\mathbf{a}}_r$$

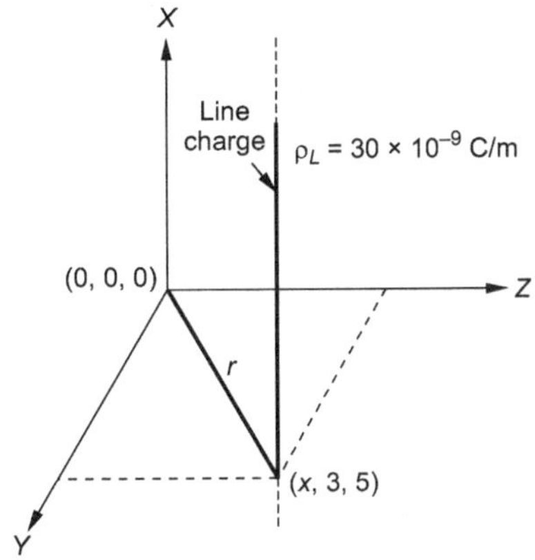

Fig. (15)

(*i*) At point $P_1(0, 0, 0)$, we have

$$\mathbf{r} = (y-3)\,\hat{\mathbf{a}}_y + (z-5)\hat{\mathbf{a}}_z$$

$$\therefore \qquad \hat{\mathbf{a}}_r = \frac{(0-3)\,\hat{\mathbf{a}}_y + (0-5)\,\hat{\mathbf{a}}_z}{\sqrt{[(3)^2+(5)^2]}} = \frac{-3\,\hat{\mathbf{a}}_y - 5\,\hat{\mathbf{a}}_z}{\sqrt{34}}$$

Therefore,
$$\mathbf{E} = \frac{30\times10^{-9}}{2\pi\,(8.854\times10^{-12})\times\sqrt{34}} \times \frac{(-3\,\hat{\mathbf{a}}_y - 5\,\hat{\mathbf{a}}_z)}{\sqrt{34}}$$

$$= 15.86\,(-3\,\hat{\mathbf{a}}_y - 5\,\hat{\mathbf{a}}_z) = -47.58\,\hat{\mathbf{a}}_y - 79.3\,\hat{\mathbf{a}}_z$$

(*ii*) At point P_2 (0,6,1). Here

$$\mathbf{r} = (6-3)\,\hat{\mathbf{a}}_y + (1-5)\,\hat{\mathbf{a}}_z = 3\,\hat{\mathbf{a}}_y - 4\,\hat{\mathbf{a}}_z$$

$$\therefore \qquad \hat{\mathbf{a}}_r = \frac{3\hat{\mathbf{a}}_y - 4\hat{\mathbf{a}}_z}{\sqrt{[(3)^2+(4)^2]}} = \frac{3\,\hat{\mathbf{a}}_y - 4\,\hat{\mathbf{a}}_z}{5}$$

Now,
$$\mathbf{E} = \frac{30\times10^{-9}}{2\pi\,(8.854\times10^{-12})\times5} \times \frac{3\hat{\mathbf{a}}_y - 4\hat{\mathbf{a}}_z}{5}$$

$$= 21.57\,(3\,\hat{\mathbf{a}}_y - 4\,\hat{\mathbf{a}}_z) = 64.71\,\hat{\mathbf{a}}_y - 86.28\,\hat{\mathbf{a}}_z$$

EXAMPLE 7. *A sheet of charge* $\rho_S = 2$ nC/m^2 *is present at the plane* $x = 3$ *in free space and line charge* $\rho_L = 20$ nC/m *is located at* $P(1, 0, 4)$. *Find the magnitude of electric field intensity* **E** *at the origin.*

Solution. The electric field $\mathbf{E}_1$ at origin due to charge sheet is given by

$$\mathbf{E}_1 = \frac{\rho_S}{2\varepsilon_0}(-\hat{\mathbf{a}}_x) = \frac{2\times10^{-9}}{2\times(8.85\times10^{-12})}(-\hat{\mathbf{a}}_x) = -113\,\hat{\mathbf{a}}_x$$

The electric field $\mathbf{E}_2$ at origin due to line charge is given by

$$\mathbf{E}_2 = \frac{\rho_L}{2\pi\varepsilon_0\cdot r}\,\hat{\mathbf{a}}_r$$

Here $\hat{\mathbf{a}}_r = \dfrac{-\hat{\mathbf{a}}_x - 4\hat{\mathbf{a}}_z}{\sqrt{(1)^2 + (4)^2}} = \dfrac{-\hat{\mathbf{a}}_x - 4\hat{\mathbf{a}}_z}{\sqrt{17}}$

$$\therefore \quad \mathbf{E}_2 = (2 \times 9 \times 10^9) \times \frac{20 \times 10^{-9}}{\sqrt{17}} \times \frac{(-\hat{\mathbf{a}}_x - 4\hat{\mathbf{a}}_z)}{\sqrt{17}}$$

$$\left(\because \frac{1}{4\pi\varepsilon_0} = 9 \times 10^9 \text{ or } \frac{1}{2\pi\varepsilon_0} = 2 \times 9 \times 10^9\right)$$

or $\mathbf{E}_2 = -21.17\,\hat{\mathbf{a}}_x - 84.70\,\hat{\mathbf{a}}_z$

The resultant intensity **E** is given by

$$\mathbf{E} = \mathbf{E}_1 + \mathbf{E}_2 = -113\,\hat{\mathbf{a}}_x + [-21.17\,\hat{\mathbf{a}}_x - 84.70\,\hat{\mathbf{a}}_z]$$
$$= -134.17\,\hat{\mathbf{a}}_x - 84.70\,\hat{\mathbf{a}}_z$$

The magnitude of $|\mathbf{E}|$ is give by

$$|\mathbf{E}| = \sqrt{[(134.17)^2 + (84.70)^2]} = \mathbf{158.52\ V/m}$$

EXAMPLE 8. *Three infinite uniform sheet charges are located in free space as shown in Fig. (16). The charges are:*

3 nC/m^2 *at* $z = -4$, 6 nC/m^2 *at* $z = 1$ *and* -8 nC/m^2 *at* $z = 4$.

Find **E** *at points (i)* $P_A(2, 5, -5)$, *(ii)* $P_B(4, 2, -3)$, *(iii)* $P_C(-1, -5, 2)$ *and (iv)* $P_D(-2, 4, 5)$.

Solution. The situation is shown in Fig. (16).

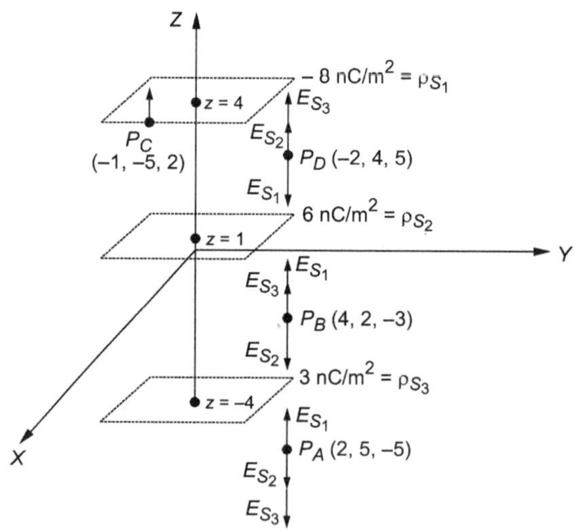

Fig. (16)

(*a*) Net electric field at $P_A(2, 5, -5)$

$$\mathbf{E} = \mathbf{E}_{S_1} + \mathbf{E}_{S_2} + \mathbf{E}_{S_3} = \frac{\rho_{S_1}}{2\varepsilon_0}\hat{\mathbf{a}}_z + \frac{\rho_{S_2}}{2\varepsilon_0}(-\hat{\mathbf{a}}_z) + \frac{\rho_{S_3}}{2\varepsilon_0}(-\hat{\mathbf{a}}_z)$$

$$= \frac{8\times10^{-9}}{2\times8.854\times10^{-12}}\hat{\mathbf{a}}_z - \frac{6\times10^{-9}}{2\times8.854\times10^{-12}}\hat{\mathbf{a}}_z - \frac{3\times10^{-9}}{2\times8.854\times10^{-12}}\hat{\mathbf{a}}_z$$

$$= \frac{10^{-9}}{2\times8.854\times10^{-12}}(8-6-3)\hat{\mathbf{a}}_z = -56.47\,\hat{\mathbf{a}}_z \text{ V/m}$$

(*b*) Net electric field at $P_B(4, 2, -3)$

$$\mathbf{E} = \mathbf{E}_{S_1} + \mathbf{E}_{S_2} + \mathbf{E}_{S_3}$$

$$= \frac{8\times10^{-9}}{2\times8.854\times10^{-12}}\hat{\mathbf{a}}_z + \frac{6\times10^{-9}}{2\times8.854\times10^{-12}}(-\hat{\mathbf{a}}_z) + \frac{3\times10^{-9}}{2\times8.854\times10^{-12}}(\hat{\mathbf{a}}_z)$$

$$= \frac{10^{-9}}{2\times8.854\times10^{-12}}[8\hat{\mathbf{a}}_z - 6\hat{\mathbf{a}}_z + 3\hat{\mathbf{a}}_z]$$

$$= 56.47\,(5\,\hat{\mathbf{a}}_z) = 282.36\,\hat{\mathbf{a}}_z \text{ V/m}$$

(*c*) Net electric field at $P_C(-1, -5, 2)$

$$\mathbf{E} = \mathbf{E}_{S_1} + \mathbf{E}_{S_2} + \mathbf{E}_{S_3} = \frac{\rho_{S_1}}{2\varepsilon_0}(\hat{\mathbf{a}}_z) + \frac{\rho_{S_2}}{2\varepsilon_0}(\hat{\mathbf{a}}_z) + \frac{\rho_{S_3}}{2\varepsilon_0}(\hat{\mathbf{a}}_z)$$

$$= \frac{10^{-9}}{2\times8.854\times10^{-12}}(8+6+3)\hat{\mathbf{a}}_z = 960\,\hat{\mathbf{a}}_z \text{ V/m}$$

(*d*) Net electric field at $P_D(-2, 4, 5)$

$$\mathbf{E} = \mathbf{E}_{S_1} + \mathbf{E}_{S_2} + \mathbf{E}_{S_3}$$

$$= \frac{10^{-9}}{2\times8.854\times10^{-12}}(-8\,\hat{\mathbf{a}}_z + 6\,\hat{\mathbf{a}}_z + 3\,\hat{\mathbf{a}}_z) = 56.47\,\hat{\mathbf{a}}_z \text{ V/m}$$

EXAMPLE 9. *A ring of radius a metre is located on X-Y plane with its origin at centre. The line charge density is* ρ_L *which is uniformly distributed over the ring. Calculate electric field strength at P(0, 0, d).*

Solution. The situation is show in Fig. (17).

Consider a small segment dL of the ring. The small charge dq on this element is $\rho_L\, dL$. The electric field intensity at a point P distant $\mathbf{r}$ is given by

$$d\mathbf{E} = \frac{\rho_L dL}{4\pi\varepsilon_0 r^2}\hat{\mathbf{a}}_r$$

where $\hat{\mathbf{a}}_r$ is the unit vector in the direction of $\mathbf{r}$.

Here $r = \sqrt{(a^2 + d^2)}$

The magnitude of $d\mathbf{E}$, *i.e.*, $|d\mathbf{E}|$ is given by

$$|d\mathbf{E}| = \frac{\rho_L dL}{4\pi\varepsilon_0(a^2+d^2)}\hat{\mathbf{a}}_r \cdot \hat{\mathbf{a}}_z = \frac{\rho_L dL\cos\theta}{4\pi\varepsilon_0(a^2+d^2)}$$

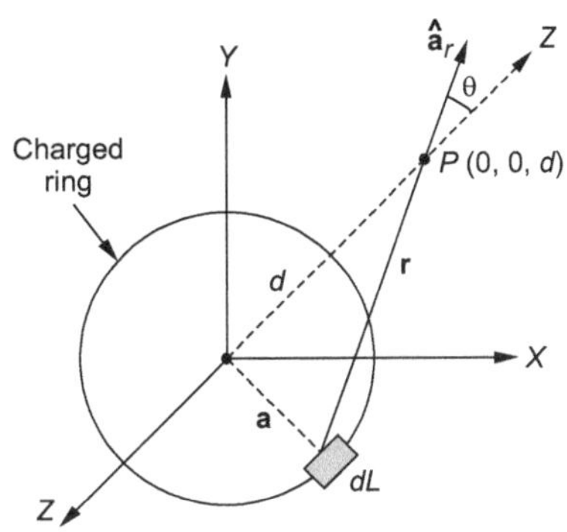

Fig. (17)

The total electric field strength can be obtained by integrating the above expression between the limits 0 to 2 πa because the length of the ring is 2 πa.

$$\therefore \quad |\mathbf{E}| = \int_0^{2\pi a} \frac{\rho_L\, dL\, d}{4\pi\varepsilon_0 (a^2+d^2)^{3/2}} = \frac{d\,\rho_L}{4\pi\varepsilon_0(a^2+d^2)^{3/2}} \cdot \int_0^{2\pi a} dL$$

$$= \frac{d\,\rho_L}{4\pi\varepsilon_0(a^2+d^2)^{3/2}} \times 2\,\pi a = \frac{d\,\rho_L\, a}{2\varepsilon_0(a^2+d^2)^{3/2}}$$

Therefore, the electric field intensity at point P in vector form is given as

$$\mathbf{E} = \frac{d\,\rho_L\, a}{2\varepsilon_0(a^2+d^2)^{3/2}}\,\hat{\mathbf{a}}_z$$

EXAMPLE 10. *A circular disc of radius a is situated in X-Y plane at z = 0 with its centre at the origin. The charge density on the disc is ρ_S C/m^2. Calculate the electric field intensity at a point P(0, 0, h) in cylindrical coordinate system.*

Solution. The situation is shown in Fig. (18).

Consider a small elemental area dS of the disc. Small charge dq on this element will be $\rho_s\, \rho\, d\rho\, d\phi$. The electric field intensity at a point P due to this small element will be

$$d\mathbf{E} = \frac{\rho_S\, dS}{4\,\pi\varepsilon_0(\rho^2+h^2)} \cdot \frac{(-\rho\,\hat{\mathbf{a}}_\rho + h\,\hat{\mathbf{a}}_z)}{\sqrt{(\rho^2+h^2)}}$$

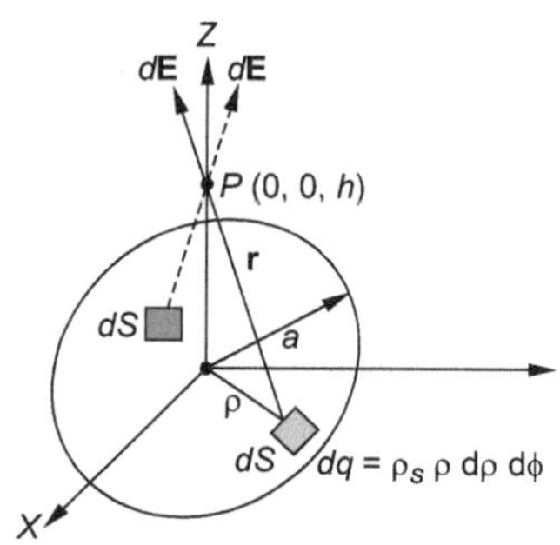

Fig. (18)

Similarly, if we consider an elemental area as (shown dotted) on another side, the component of $d\mathbf{E}$ will be in opposite direction. The radial components of electric field intensity will be cancelled. Only, the z components will be added. Therefore, the total electric field strength will be in Z-direction. So, the value of $\mathbf{E}$ at P is given by

$$\mathbf{E} = \iint_S d\mathbf{E} = \frac{\rho s\, dS}{4\,\pi\varepsilon_0(\rho^2+h^2)} \cdot \frac{h\hat{\mathbf{a}}_z}{\sqrt{(\rho^2+h^2)}} = \iint_S \frac{\rho_S\,\rho\, d\rho\, d\phi \cdot h}{4\,\pi\varepsilon_0(\rho^2+h^2)^{3/2}}\,\hat{\mathbf{a}}_z$$

$$= \frac{\rho_S\, h}{4\pi\varepsilon_0} \cdot \int_{\rho=0}^{a} \frac{\rho\, d\rho}{(\rho^2+h^2)^{3/2}}\,\hat{\mathbf{a}}_z \int_0^{2\pi} d\phi = \frac{\rho_S\, h}{2\varepsilon_0} \int_{\rho=0}^{a} \frac{\rho\, d\rho}{(\rho^2+h^2)^{3/2}}\,\hat{\mathbf{a}}_z$$

$$\therefore \quad \mathbf{E} = \frac{\rho_S\, h}{2\varepsilon_0}\left[\frac{-1}{\sqrt{a^2+h^2}} + \frac{1}{h}\right]\hat{\mathbf{a}}_z$$

Note: For an infinite uniform plane sheet $a \to \infty$ .In this case

$$\mathbf{E} = \frac{\rho_S h}{2\varepsilon_0} \times \frac{1}{h}\,\hat{\mathbf{a}}_z = \frac{\rho_S}{2\varepsilon_0}\,\hat{\mathbf{a}}_z$$

2.7 ELECTRIC FLUX DENSITY

To understand the electric flux density, consider a sphere with charge Q placed inside a large hollow sphere without touching it. The outer sphere is momentarily earthed and isolated from the inner sphere. It is observed that an equal and opposite charge gets develop on the outer sphere. It is, therefore, concluded that there is a displacement of electric charge from inner sphere to outer sphere through the medium between them. The displacement is regarded as displacement flux or simply as *electric flux*. This is denoted by ψ and is equal in magnitude to the charge that produces it, *i.e.*,

$$\psi = Q \text{ coulomb} \quad \text{...(1)}$$

Let us now consider *electric displacement density or electric flux density* **D**. The electric field intensity at a distance r from a point charge Q in free space is given by

$$\mathbf{E} = \frac{Q}{4\pi\varepsilon_0 r^2}\hat{\mathbf{a}}_r \quad \text{or} \quad |\mathbf{E}| = \frac{Q}{4\pi\varepsilon_0 r^2} \quad \text{...(2)}$$

From eq. (2)

$$\varepsilon_0 E = \frac{Q}{4\pi r^2} = \frac{\psi}{4\pi r^2} = \frac{\text{Charge}}{\text{Area}} \quad \text{...(3)}$$

It is clear from eq. (3) thar $\varepsilon_0\, E$ has the dimensions of charge per unit area, *i.e.*, charge density. If electric field intensity is taken in vector from (*i.e.*, **E**) in free space, then the product $\varepsilon_0\, \mathbf{E}$ is represented by symbol **D** and is called as *electric flux density.*

Therefore,

$$\boxed{\mathbf{D} = \varepsilon_0 \mathbf{E}} \text{ in free space} \quad \text{...(4)}$$

$$\boxed{\mathbf{D} = \varepsilon \mathbf{E} = \varepsilon_0 \varepsilon_r \mathbf{E}} \text{ in dielectric medium} \quad \text{...(5)}$$

Electric flux density (**D**) is a vector quantity. The direction is always taken normal to the surface.

Electric Flux Density in Terms of an Isolated Point Charge

Fig. (19) shows an electric field of an isolated charge Q.

Consider a small surface area $d\mathbf{S}$ at a distance **r**. If **D** be the electric flux density then small charge $d\psi$ or dQ can be represented as

$$d\psi = \mathbf{D} \cdot d\mathbf{S}$$

The vector **D** is taken as normal to the surface.

The total flux over the complete surface is given by

$$\boxed{\psi = \oint\!\!\oint_S \mathbf{D} \cdot d\mathbf{S}} \quad \text{...(6)}$$

where the integral is taken over the complete surface.

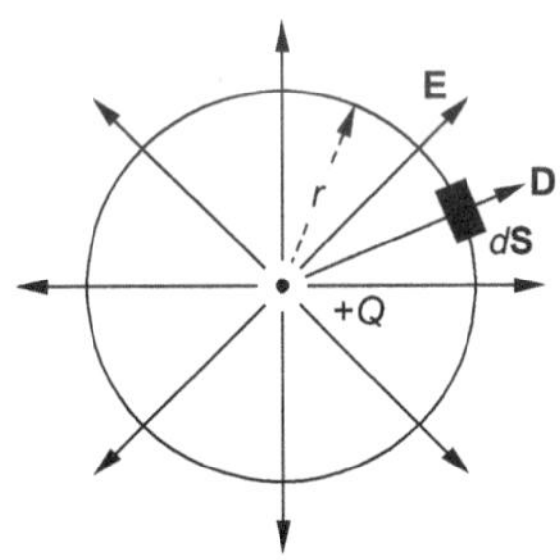

Fig. (19) Showing electric flus density in terms of isolated charge

D for Line Charge Distribution.

In case of line charge distribution

$$\mathbf{E} = \int_L \frac{\rho_L\, dL}{4\pi\varepsilon_0 r^2}\hat{\mathbf{a}}_r \quad \text{or} \quad \varepsilon_0 \mathbf{E} = \int_L \frac{\rho_L\, dL}{4\pi r^2}\hat{\mathbf{a}}_r$$

$$\therefore \quad \boxed{\mathbf{D} = \int_L \frac{\rho_L\, dL}{4\pi r^2}\hat{\mathbf{a}}_r} \quad \text{...(7)}$$

D for Surface Charge Distribution

In case of surface charge distribution

$$\mathbf{E} = \iint_S \frac{\rho_S \, dS}{4\pi\varepsilon_0 r^2} \hat{\mathbf{a}}_r \quad \text{or} \quad \varepsilon_0 \mathbf{E} = \iint_S \frac{\rho_S \, dS}{4\pi r^2} \hat{\mathbf{a}}_r$$

$$\therefore \qquad \boxed{\mathbf{D} = \iint_S \frac{\rho_S \, dS}{4\pi r^2} \hat{\mathbf{a}}_r}$$

D for Volume Charge Distribution

In case of volume charge distribution

$$\mathbf{E} = \iiint_V \frac{\rho_V \, dV}{4\pi\varepsilon_0 r^2} \hat{\mathbf{a}}_r \quad \text{or} \quad \varepsilon_0 \mathbf{E} = \iiint_V \frac{\rho_V \, dV}{4\pi r^2} \hat{\mathbf{a}}_r$$

$$\therefore \qquad \boxed{\mathbf{D} = \iiint_V \frac{\rho_V \, dV}{4\pi r^2} \hat{\mathbf{a}}_r}$$

EXAMPLE 1. *Find the flux density at a point* $P(6, 4, -5)$ *caused by a point charge of* 20 mC *at the origin.*

Solution. The electric field **E** at a point due to point charge is given by

$$\mathbf{E} = \frac{Q}{4\pi\varepsilon_0 r^2} \hat{\mathbf{a}}_r$$

$\therefore$ Electric flux density **D** is given by

$$\mathbf{D} = \varepsilon_0 \mathbf{E} = \frac{Q}{4\pi r^2} \hat{\mathbf{a}}_r$$

$$= \frac{20 \times 10^{-3}}{4\pi \times 77} \cdot \frac{6\hat{\mathbf{a}}_x + 4\hat{\mathbf{a}}_y - 5\hat{\mathbf{a}}_z}{\sqrt{77}}$$

$$= (14.13\,\hat{\mathbf{a}}_x + 9.42\,\hat{\mathbf{a}}_y - 11.78\,\hat{\mathbf{a}}_z)\ \mu\text{C/m}^2.$$

EXAMPLE 2. *Find the electric flux density at a point* $P(6, 4, -5)$ *caused by a uniform line charge* $\rho_L = 20\ \mu\text{C/m}$ *on Z-axis.*

Solution. The electric field due to a line charge is given by

$$\mathbf{E} = \frac{\rho_L}{2\pi\varepsilon_0 r} \hat{\mathbf{a}}_r$$

$$\mathbf{D} = \frac{\rho_L}{2\pi r} \hat{\mathbf{a}}_r$$

Here $\qquad \mathbf{r} = 6\hat{\mathbf{a}}_x + 4\hat{\mathbf{a}}_y$ and $r = \sqrt{(6)^2 + (4)^2} = \sqrt{52}$

Therefore, $\qquad \mathbf{D} = \dfrac{20 \times 10^{-6}}{2\pi\sqrt{52}} \times \dfrac{6\hat{\mathbf{a}}_x + 4\hat{\mathbf{a}}_y}{\sqrt{52}} = (0.37\hat{\mathbf{a}}_x + 0.25\hat{\mathbf{a}}_y)\dfrac{\mu\text{C}}{\text{m}^2}$

EXAMPLE 3. *A 60 μC charge is located at origin. Find the total electric flux passing through (a) portion of the sphere r* = 26 cm *bounded by* $0 < \theta < \pi/2$, $0 < \phi < \pi/2$ *and (b) the closed surface defined by* $\rho = 26$ cm *and* $z = \pm 26$ cm.

Solution. (*a*) We know that

$$\mathbf{D} = \frac{q}{4\pi r^2}\hat{\mathbf{a}}_r = \frac{60\times10^{-6}}{4\pi r^2}\hat{\mathbf{a}}_r$$

Now,
$$\psi = \iint_S \mathbf{D}\cdot d\mathbf{S} = \int_0^{\pi/2}\int_0^{\pi/2}\left(\frac{60\times10^{-6}}{4\pi r^2}\hat{\mathbf{a}}_r\right)\cdot(r^2\sin\theta\, d\theta\, d\phi\,\hat{\mathbf{a}}_r)$$

$$= \frac{60\times10^{-6}}{4\pi}\int_0^{\pi/2}\sin\theta\, d\theta\int_0^{\pi/2}d\phi = \frac{60\times10^{-6}}{4\pi}[-\cos\theta]_0^{\pi/2}[\phi]_0^{\pi/2}$$

$$= \frac{60\times10^{-6}}{4\pi}\times1\times\frac{\pi}{2} = 7.5\times10^{-6}\text{ C} = \mathbf{7.5\ \mu C}$$

(*b*) The situation is shown in Fig. (20). Here

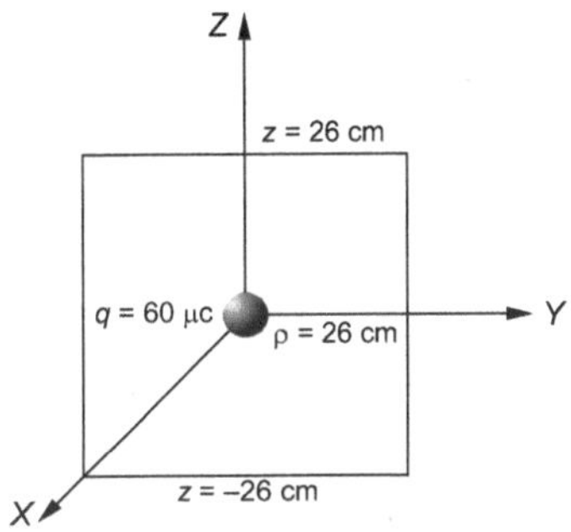

Fig. (20)

$$\mathbf{D} = \frac{60}{4\pi\rho^2}\hat{\mathbf{a}}_\rho\ \frac{\mu\text{C}}{\text{m}^2}$$

$$\therefore\quad \psi = \iint_S \mathbf{D}\cdot d\mathbf{S}$$

$$= \iint_S\left(\frac{60}{4\pi\rho^2}\hat{\mathbf{a}}_\rho\right)\cdot(\rho\, d\phi\, dz\,\hat{\mathbf{a}}_\rho)$$

$$= \int_{-26}^{26}\int_0^{2\pi}\frac{60}{4\pi\rho}d\phi\, dz\Big|_{\rho=26}$$

$$= \frac{60}{4\pi\times26}\int_0^{2\pi}d\phi\int_{-26}^{26}dz = \frac{60}{4\pi\times26}2\pi\times(26+26)$$

$$= \frac{60\times52}{2\times26} = \mathbf{60\ \mu C}$$

2.8 GAUSS'S LAW

The Gauss's law states that "***the surface integral of the normal component of electric flux density* D *over any closed surface surrounding charges is equal to the amount of charge enclosed*.

Let ψ be the electric flux through a closed surface having volume *V* and charge *Q*, then the Gauss's law can be expressed as

$$\boxed{\psi = \oint_S \mathbf{D}\cdot d\mathbf{S} = Q(\text{charge enclosed})} \quad ...(1)$$

If the charge is distributed throughout the volume with a charge density ρ (C/m^3), the Gauss's law can be expressed as

$$\boxed{\oint_S \mathbf{D}\cdot d\mathbf{S} = \int_V \rho\, dV} \qquad (\because\ \psi = \int_V \rho\, dV) \quad ...(2)$$

Let us consider the surface of a sphere of radius r. The sphere encloses a charge Q at its origin. The electric field intensity **E** and electric flus density **D** are shown in Fig. (21).

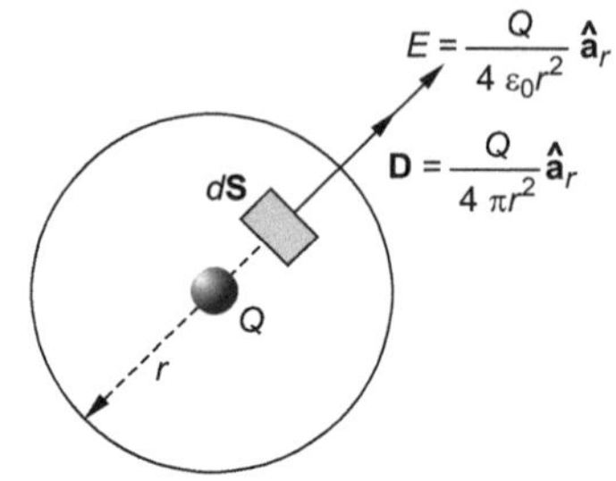

Fig. (21)

Here $$d\mathbf{S} = r^2 \sin\theta \, d\theta \, d\phi \, \hat{\mathbf{a}}_r$$

Now, $$\mathbf{D}.\, d\mathbf{S} = \left(\frac{Q}{4\pi r^2}\hat{\mathbf{a}}_r\right)\cdot(r^2 \sin\theta d\theta\, d\phi)\,\hat{\mathbf{a}}_r = \frac{Q}{4\pi}\cdot\sin\theta\, d\theta\cdot d\phi$$

$\therefore$ $$\oint_S \mathbf{D}\cdot d\mathbf{S} = \frac{Q}{4\pi}\int_0^{\pi}\sin d\theta \int_0^{2\pi} d\phi = \frac{Q}{4\pi}(2)(2\pi) = Q$$

$\therefore$ $$\boxed{\oint \mathbf{D}\cdot d\mathbf{S} = Q}$$

2.9 APPLICATIONS OF GAUSS'S LAW

We shall consider the application of Gauss's law in the following cases:

1. **Infinite Line Charge**

   Consider a uniform line charge distribution ρ_L lying along the Z-axis and extending from $-\infty$ to $+\infty$ as shown in Fig. (22). The aim is to calculate electric field intensity **E**

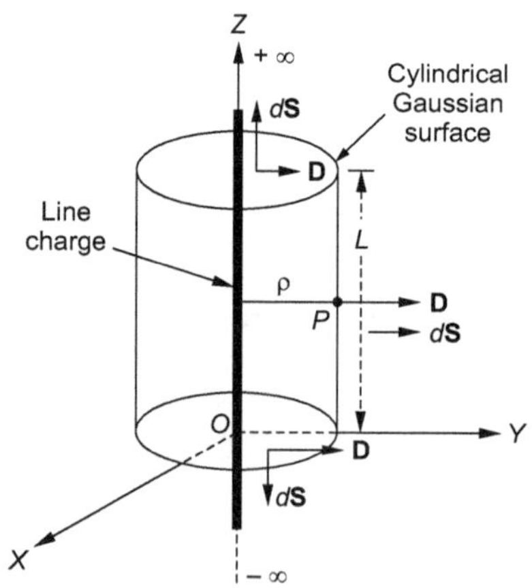

Fig. (22) Line charge with Gaussian surface

   and electric flux density **D** at a distance ρ. For this purpose, we consider a cylindrical Gaussian surface of radius ρ and length L. The cylindrical surface is selected because D_ρ is everywhere normal to the surface.

The total charge enclosed in the Gaussian surface is given by

$$Q = \int \rho_L \, dL = \rho_L L \qquad ...(1)$$

According to Gauss' s law

$$Q = \oint_{\text{Cyl.}} \mathbf{D} \cdot d\mathbf{S}$$

The Gaussian surface (cylindrical surface) has three surfaces, *i.e.*, upper surface, side curved surface and bottom surface.

Since electric flux flows radially outward for infinitely line charge. Therefore,

$$\therefore \quad \oint_{\text{cyl.}} \mathbf{D} \cdot d\mathbf{S} = \iint_{\text{top}} D \, dS \cos 90° + \iint_{\text{sides}} D \, dS \cos 0 + \iint_{\text{bottom}} D \, dS \cos 90°$$

$$= 0 + \int_{z=0}^{L} \int_{\phi=0}^{2\pi} D \cdot (\rho \, d\phi \, dz) + 0 = D\rho \int_0^L dz \int_0^{2\pi} d\phi = D\rho L 2\pi \qquad ...(2)$$

From eqs. (1) and (2), we get

$$\rho_L L = D\rho L \, 2\pi$$

or

$$\boxed{D = \frac{\rho_L}{2\pi\rho}} \qquad ...(3)$$

and

$$\boxed{E = \frac{\rho_L}{2\pi\varepsilon_0 \rho}} \qquad ...(4)$$

In vector form,

$$\mathbf{D} = \frac{\rho_L}{2\pi\rho} \hat{\mathbf{a}}_\rho \quad \text{and} \quad \mathbf{E} = \frac{\rho_L}{2\pi\varepsilon_0 \rho} \hat{\mathbf{a}}_\rho \qquad ...(5)$$

2. Coaxial Cable

A coaxial cable is formed by two cylindrical conductors as shown in Fig. (23). The inner radius of the coaxial cable is a and outer radius is b. Let L is the length of the cable.

The surface charge distribution of inner conductor is P_S. The aim is to calculator D at a point distant ρ where $a < \rho < b$. For this purpose we consider a Gaussian surface as right circular cylinder of length L and radius ρ where $a < \rho < b$.

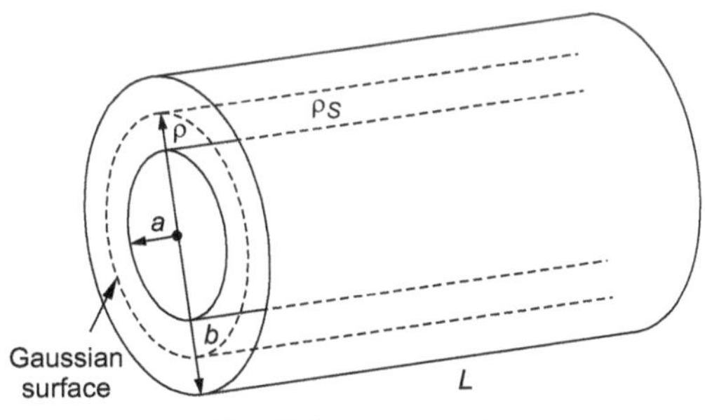

Fig. (23) Coaxial cable

The total charge on inner conductor is

$$Q = \int_{z=0}^{L} \int_0^{2\pi} \rho_S \cdot (a \, d\phi \, dz) = \rho_S \, a \int_{z=0}^{L} dz \int_0^{2\pi} d\phi = \rho_S \, a \, L \, 2\pi \qquad ...(1)$$

$$\text{Now} \quad \oint_S \mathbf{D} \cdot d\mathbf{S} = \oint_S D\rho \, dS \cos 0° = D\rho \int_0^L \int_0^{2\pi} \rho \, d\phi \, dz = D_\rho \, \rho \int_0^L dz \int_0^{2\pi} d\phi$$

$$= D_\rho \, \rho \cdot L \cdot 2\pi \qquad ...(2)$$

According to Gauss's law

$$\oint_S \mathbf{D}\cdot d\mathbf{S} = Q \qquad ...(3)$$

Substituting the values form eqs. (1) and (2) in eq. (3), we get

$$D_\rho\, \rho\, L 2\pi = \rho_S\, a\, L\, 2\,\pi$$

$$\therefore \qquad D_\rho = \frac{\rho_S\, a}{\rho}$$

In vector form $\qquad \mathbf{D}_\rho = \dfrac{\rho_S\, a}{\rho}\hat{\mathbf{a}}_\rho$

In general $\qquad \boxed{\mathbf{D} = \dfrac{\rho_S\, a}{\rho}\hat{\mathbf{a}}_\rho} \qquad ...(4)$

We can also express this result in terms of charge per unit length because the inner conductor has $2\,\pi\, a\, \rho_S$ coulomb charge per unit length, *i.e.*, $\rho_L = 2\,\pi\, a\, \rho_S$ or $\rho_S = \rho_L/2\pi a$

$$\therefore \qquad \boxed{\mathbf{D} = \frac{\rho_L}{2\pi\rho} = \hat{\mathbf{a}}_\rho} \qquad ...(5)$$

For $\rho > b$ or $\rho < a$, $\mathbf{D} = 0$, because charge enclosed $Q = 0$

For $\rho < a$, $(\vec{D}) = 0$

Since, there is no enclosed charge. For $\rho < a$, $(\vec{D}) = 0$ because, net enclosed charge $(+\theta, -\theta)$ will be zero.

2.10 ELECTRIC POTENTIAL

Now we proceed to define electric potential.

Consider the electric field due to a positive charge $+q$ as shown in Fig. (24). Let there be a positive test charge $+q_0$ in this field. The test charge q_0 experiences an electric repulsive force **F** due to charge $+q$ as shown in the figure. Now consider that the test charge is moved from point B to another point A by an external agent. Some work has to be done against the force of repulsion. The ratio of workdone in taking a test charge from one point to another point in an electric field to the magnitude of test charge is defined as electric potential difference between these points. Therefore,

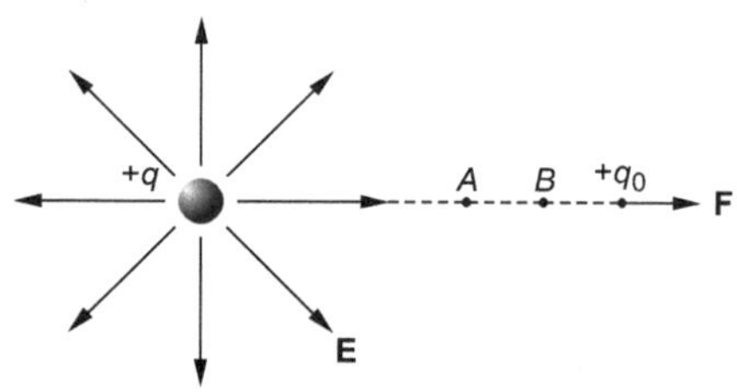

Fig. (24)

$$V_A - V_B = \frac{\text{Work done}}{\text{Magnitude of test charge}} = \frac{W}{q_0}$$

If point B is chosen at infinity, then electric potential V_B is arbitrarily taken as zero.

Then, $\qquad V_A = \dfrac{W}{q_0}$

where, W is workdone by external agent to carry the test charge q_0 from infinity to point A.

Thus, electric potential at a point in electric field is defined as the work done by an external agent in carrying a unit positive test charge from infinity to that point against the electric force of the field.

2.11 RELATION BETWEEN ELECTRIC POTENTIAL DIFFERENCE AND ELECTRIC FIELD

Let a uniform electric field **E** is set up by certain stationary charge (not shown) as shown in Fig. (25). Suppose a positive test charge q_0 is moved by an external agent without acceleration between two points A and B in the electric field. The electric force on the charge q_0 is $q_0\,\mathbf{E}$. The direction of this force is in the direction of electric field strength **E** which points downwards. Now, to move the charge in the upward direction by external agent, an equal and opposite force $-\,q_0\,\mathbf{E}$ must be applied. Let the charge q_0 is moved through a small distance dl by the external agent. The workdone dW by the external agent is given by

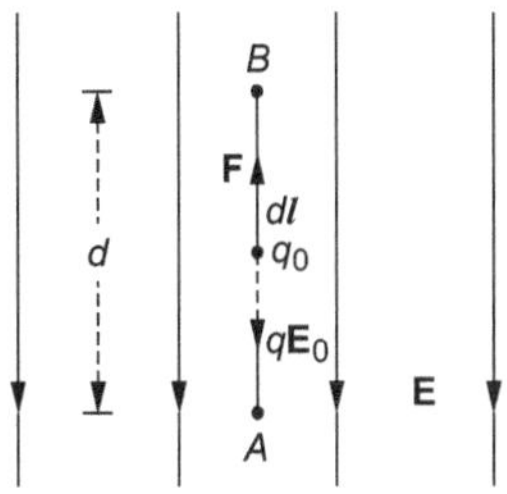

Fig. (25)

$$dW = \mathbf{F} \cdot dl = -\,q_0\,\mathbf{E} \cdot dl \qquad ...(1)$$

Therefore, the total workdone W_{AB} in carrying q_0 from A to B will be

$$W_{AB} = \int_A^B -\,q_0\,\mathbf{E} \cdot dl$$

$$= -\,q_0 \int_A^B \mathbf{E} \cdot dl \qquad ...(2)$$

Further, $$V_B - V_A = \frac{W_{AB}}{q_0} \qquad ...(3)$$

∴ Potential difference between two points A and B and will be

$$V_B - V_A = -\int_A^B \mathbf{E} \cdot dl \qquad ...(4)$$

If the reference point A is taken at infinity, then $V_A = 0$.

Hence, $$V_B - V_A = -\int_\infty^B \mathbf{E} \cdot dl \qquad ...(5)$$

The integral in eq. (5) is called as line integral. ***So, the electric potential at a point in the electric field can be expressed as a line integral of the electric field.***

2.12 FIELD AS THE GRADIENT OF POTENTIAL

Here, we shall consider how to calculate **E** when V is known. Consider two surfaces of potentials V and $V + dV$ in an electric field as shown in Fig. (26). Let a positive test charge q_0 at point A on the surface V is moved to the point B on the surface $V + dV$ along the vector displacement dl by any external agent. The force experienced by the test charge q_0 at A due to electric field will be $q_0\,\mathbf{E}$. This force is in the direction of **E** which is at right angle to the surface V. In order

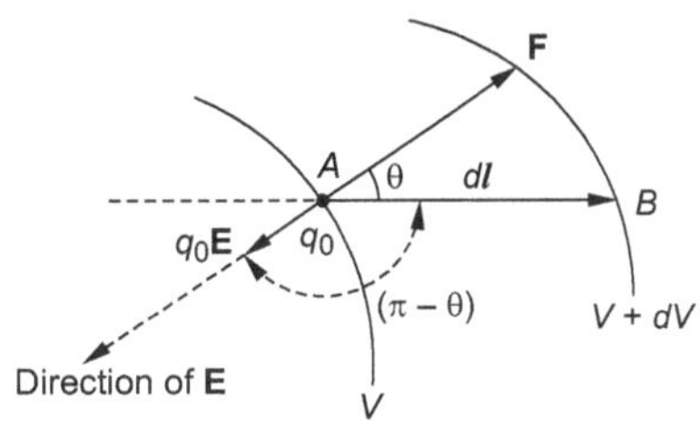

Fig. (26)

to move the test charge q_0 without acceleration, the external agent must apply an equal and opposite force **F** as shown in figure. Thus,

$$\mathbf{F} = q_0\, \mathbf{E} \qquad \text{...(1)}$$

Further, the workdone by the external agent to move the test charge from A to B along dl is given by

$$dW = \mathbf{F} \cdot dl \quad \text{or} \quad dW = -q_0\, \mathbf{E} \cdot dl$$

$$\therefore \qquad \frac{dW}{q_0} = \mathbf{E} \cdot dl \qquad \text{...(2)}$$

We know that $\left(\frac{dW}{q_0}\right)$ is the potential difference dV between points A and B. Thus,

$$dV = -\mathbf{E} \cdot dl \qquad \text{...(3)}$$

Let the coordinates of A and B be (x, y, z) and $(x + dx, y + dy, z + dz)$ respectively. The potential difference dV between A and B can be expressed as

$$dV = \frac{\partial V}{\partial x} dx + \frac{\partial V}{\partial y} dy + \frac{\partial V}{\partial z} dz$$

$$= \left(\frac{\partial V}{\partial x} \hat{\mathbf{a}}_x + \frac{\partial V}{\partial y} \hat{\mathbf{a}}_y + \frac{\partial V}{\partial z} \hat{\mathbf{a}}_z\right) \cdot (dx\, \hat{\mathbf{a}}_x + dy\, \hat{\mathbf{a}}_y + dz\, \hat{\mathbf{a}}_z)$$

$$= (\text{grad } V) \cdot \mathbf{dl} = \vec{\nabla} V \cdot \mathbf{dl} \qquad \text{...(4)}$$

$\therefore \left(\frac{\partial V}{\partial x} \hat{\mathbf{a}}_x + \frac{\partial V}{\partial y} \hat{\mathbf{a}}_y + \frac{\partial V}{\partial z} \hat{\mathbf{a}}_z\right.$ = grad V and $dx\, \hat{\mathbf{a}}_x + dy\, \hat{\mathbf{a}}_y + dz\, \hat{\mathbf{a}}_z$ is the displacement vector **dl** between A and B).

Comparing eqs. (3) and (4), we get

$$\mathbf{E} = -\text{grad } V = -\vec{\nabla} V \qquad \text{...(5)}$$

If E_x, E_y and E_z be the components of **E**, then eq. (5) can be written as

$$E_x\, \hat{\mathbf{a}}_x + E_y\, \hat{\mathbf{a}}_y + E_z\, \hat{\mathbf{a}}_z = -\left(\frac{\partial V}{\partial x} \hat{\mathbf{a}}_x + \frac{\partial V}{\partial y} \hat{\mathbf{a}}_y + \frac{\partial V}{\partial z} \hat{\mathbf{a}}_z\right)$$

Thus, we have $\boxed{E_x = -\frac{\partial V}{\partial x},\ E_y = -\frac{\partial V}{\partial y} \text{ and } E_z = -\frac{\partial V}{\partial z}}$ ...(6)

Thus, the electric intensity at a point in the electric field is equal to the negative potential gradient in that direction.

In case of polar coordinates, the components are radial along radius vector and azimuthal perpendicular to the radius vector. These can be calculated as

$$E_r = -\left(\frac{\partial V}{\partial r}\right)_\theta \quad \text{and} \quad E_\theta = -\frac{1}{r}\left(\frac{\partial V}{\partial \theta}\right)_r \qquad \text{...(7)}$$

2.13 EQUIPOTENTIAL SURFACES

Equipotential surface in an electric field is a surface on which the potential is same at every point. In other words, ***the locus of all points which have the same electric potential is called***

equipotential surface. As the potential difference between any two points on the equipotential is zero and hence, no work is done in taking a charge from one point to another. This is only possible when the charge is taken perpendicular to the field. In this way, the equipotental surface at every point is perpendicular to the field. We know that is an electric field, the direction of field at different points is represented by lines of force. ***Hence, the lines of force at every point of the equipotential surface are perpendicular to the surface.***

In case of uniform field, where the lines of forces are straight and parallel, the equipotenital surfaces are planes perpendicular to the lines of force as shown in Fig. (27 *a*). The equipotenital

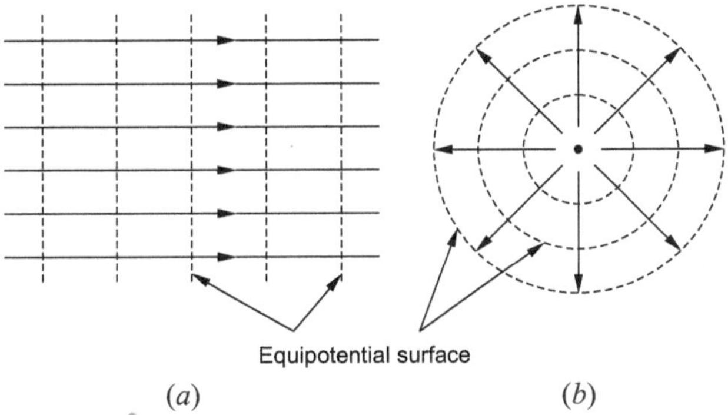

Fig. (27)

surfaces are a family of concentric spheres for a sphere of charge or a point charge as shown in Fig. (27 *b*).

Equipotenital surfaces in electrostatics are similar to wavefronts in optics. The wavefronts in optics are the locus of all points which are in the same phase. These are the planes perpendicular to the direction of rays. On the other hand, the equipotential surfaces are perpendicular to the lines of force.

Thus, we have the following points in case of equipotenital surfaces:

(*i*) The electric field along the equipotenital surface is zero.

(*ii*) The electric field is always normal to the surface.

(*iii*) The workdone in moving a charge on the equipotenital surface is zero.

$\therefore$ Workdone = potential difference × charge = 0 × charge = zero

(*iv*) Plane at infinite will be equipotential surface.

(*v*) The equipotenital surfaces act as wave-fronts in optics.

2.14 POTENTIAL DUE TO A POINT CHARGE

Consider a point charge q at the origin of spherical coordinate system as shown in Fig. (28). The aim of this article is to calculate the potential difference between two points A and B. The distances of these two points from q are r_A and r_B respectively.

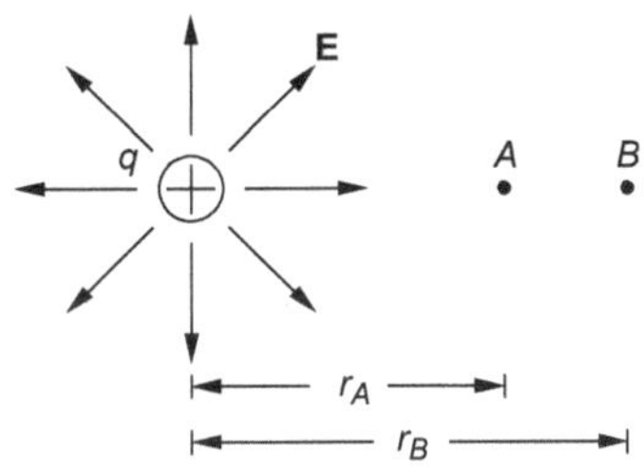

Fig. (28)

The potential difference between A and B is given by

$$V_{AB} = -\int_B^A \mathbf{E} \cdot dl = -\int_{r_B}^{r_A} \mathbf{E} \cdot dl \qquad ...(1)$$

where dl is the small element of length between points A and B.

We know that electric field due to a point charge

$$\mathbf{E} = \frac{q}{4\pi\varepsilon_0 r^2}\hat{\mathbf{a}}_r \qquad ...(2)$$

where $\hat{\mathbf{a}}_r$ is the unit vector along the direction of $\mathbf{r}$.

In eq. (1), $\quad dl = dr\,\hat{\mathbf{a}}_r \qquad ...(3)$

Substitution the values of **E** and dl from eqs. (2) and (3) in eq. (1), we get

$$V_{AB} = -\int_{r_B}^{r_A} \frac{q}{4\pi\varepsilon_0 r^2}\hat{\mathbf{a}}_r \cdot dr\,\hat{\mathbf{a}}_r = -\int_{r_B}^{r_A} \frac{q}{4\pi\varepsilon_0 r^2} dr \qquad (\because \hat{\mathbf{a}}_r \cdot \hat{\mathbf{a}}_r = 1)$$

$$= -\frac{q}{4\pi\varepsilon_0}\int_{r_B}^{r_A}\frac{dr}{r^2} = \frac{q}{4\pi\varepsilon_0}\left[\frac{1}{r}\right]_{r_B}^{r_A} = \frac{q}{4\pi\varepsilon_0}\left[\frac{1}{r_A} - \frac{1}{r_B}\right]$$

If the reference point B is taken at infinity, then the potential at point A is given by

$$V_A = \frac{q}{4\pi\varepsilon_0}\left[\frac{1}{r_A} - \frac{1}{\infty}\right] = \frac{q}{4\pi\varepsilon_0 r_A}$$

2.15 POTENTIAL DUE TO UNIFORMLY CHARGED LINE

Consider an infinitely long of charge with density ρ_L, as shown in Fig. (29).

The aim of this article is to consider the potential difference between the two points A and B.

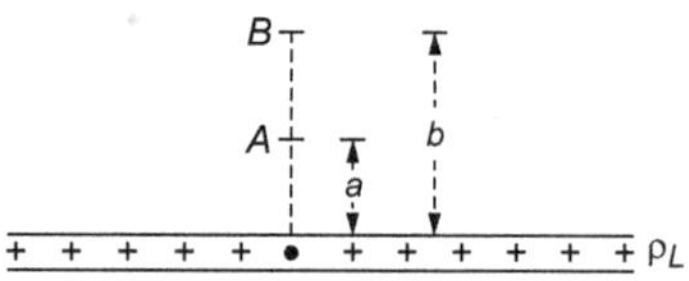

Fig. (29) Uniformly long line of charge

In case of infinitely long line of charge, the electric field strength **E** is given by

$$\mathbf{E} = \frac{\rho_L}{2\pi\varepsilon_0 r}\hat{\mathbf{a}}_r \qquad ...(1)$$

The potential difference between points A and B is given by

$$V_{AB} = -\int \mathbf{E}\cdot dl = -\int \mathbf{E}\cdot dr\,\hat{\mathbf{a}}_r$$

$$= -\int_b^a \frac{\rho_L}{2\pi\varepsilon_0 r}\hat{\mathbf{a}}_r \cdot dr\,\hat{\mathbf{a}}_r = -\int_b^a \frac{\rho_L}{2\pi\varepsilon_0 r}dr$$

$$= -\frac{\rho_L}{2\pi\varepsilon_0}\int_b^a \frac{dr}{r} = -\frac{\rho_L}{2\pi\varepsilon_0}[\log_e r]_b^a$$

$$= -\frac{\rho_L}{2\pi\varepsilon_0}[\log_e a - \log_e b]$$

$$= \frac{\rho_L}{2\pi\varepsilon_0}[\log_e b - \log_e a] = \frac{\rho_L}{2\pi\varepsilon_0}\log_e\left(\frac{b}{a}\right)$$

$$\therefore \quad \boxed{V_{AB} = \frac{\rho_L}{2\pi\varepsilon_0}\log_e\left(\frac{b}{a}\right)}$$

2.16 POTENTIAL AT A POINT BETWEEN TWO COAXIAL CYLINDERS

Consider two coaxial cylinders of radii a of inner cylinder and b of outer cylinder respectively as shown in Fig. (30). Let the inner cylinder is at higher potential and the outer cylinder is at lower potential. There is a point P at a distance r from the axis of inner cylinder. The aim of this article is to calculate the potential at point P.

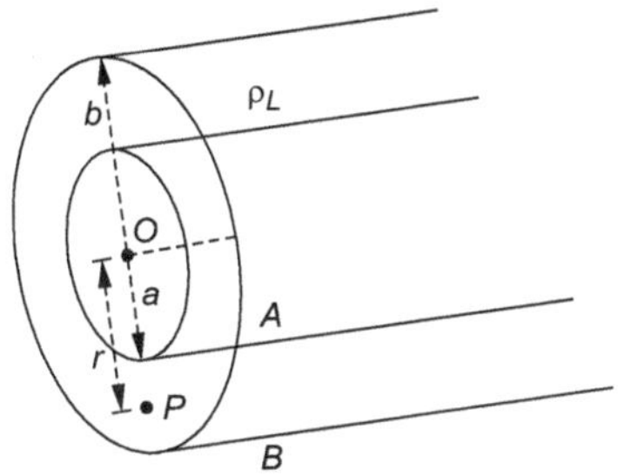

Fig. (30)

The potential difference between the two cylinders is given by

$$V_{AB} = -\int_b^a \mathbf{E}\cdot dl = -\int_b^a \frac{\rho_L\, dr}{2\pi\varepsilon_0 r} = \frac{\rho_L}{2\pi\varepsilon_0}[\log_e r]_a^b$$

$$\therefore\ V_{ab} = \frac{\rho_L}{2\pi\varepsilon_0}\log_e\left(\frac{b}{a}\right) \qquad ...(1)$$

Similarly, the potential difference V_r between two points specified by r and b is given by

$$V_r = \frac{\rho_L}{2\pi\varepsilon_0}\log_e\left(\frac{b}{r}\right) \qquad ...(2)$$

From eq. (1) $$\frac{\rho_L}{2\pi\varepsilon_0} = \frac{V_{ab}}{\log_e(b/a)} \qquad ...(3)$$

From eq. (2) $$\frac{\rho_L}{2\pi\varepsilon_0} = \frac{V_r}{\log_e(b/r)} \qquad ...(4)$$

Comparing eqs. (3) and (4), we get

$$\frac{V_r}{\log_e(b/r)} = \frac{V_{ab}}{\log_e(b/a)}$$

$$\therefore\ \boxed{V_r = \frac{V_{ab}\log_e(b/r)}{\log_e(b/a)}} \qquad ...(5)$$

Eq. (5) gives the potential difference between any point ($a < r < b$) in terms of potential difference between two cylinders.

2.17 POTENTIAL DUE TO CHARGED CONDUCTING SPHERE

(*i*) When point *P* lies outside the sphere

Consider a conducting charged with centre O and radius R as shown in Fig. (31). Let σ be the surface charge density and the total charge be q. Here it should be remembered that when a conducting sphere is give charge, the whole charge is distributed uniformly on the surface of the sphere and there will be no charge inside the

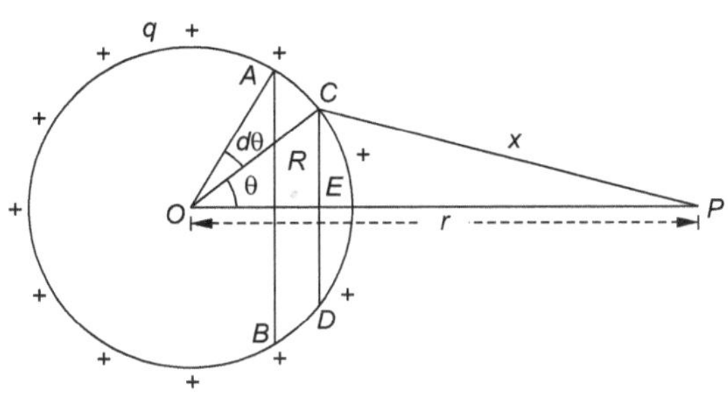

Fig. (31)

sphere. Now the problem is to find out the potential at an external point P distant r from the centre of the sphere. For this purpose we divide the sphere into a number of rings with centres on OP. Further consider one such ring $ABCD$ between two planes AB and CD. Let $CP = x$, $\angle COP = \theta$ and $\angle AOC = d\theta$.

From right angled triangle OEC, $CE = OC \sin\theta = R \sin\theta$

From sector AOC, $AC = R\, d\theta$.

The circumference of the ring $= 2\pi \times (R \sin\theta)$

$$\therefore \quad \text{Area of the ring} = 2\pi R \sin\theta \times R\, d\theta = 2\pi R^2 \sin\theta\, d\theta$$

$$\text{Charge on the ring} = \text{area of the ring} \times \text{surface density}$$

$$= 2\pi R^2 \sin\theta\, d\theta \times \sigma$$

where $\sigma = \dfrac{\text{Total charge on sphere}}{\text{Total surface area}} = \dfrac{q}{4\pi R^2}$

$$\therefore \quad \text{charge on } dq = 2\pi R^2 \sin\theta\, d\theta \times \left(\frac{q}{4\pi R^2}\right) = \frac{q \sin\theta\, d\theta}{2} \quad \text{...(1)}$$

So the potential at P due to the charge on the ring

$$dV = \frac{1}{4\pi\varepsilon_0} \cdot \frac{dq}{x} \quad \text{...(2)}$$

(Even point of the narrow ring $ABCD$ is the same distance x from the point P).

$$dV = \frac{q \sin\theta\, d\theta}{8\pi\varepsilon_0\, x} \quad \text{...(3)}$$

From figure, $x = R^2 + r^2 - 2Rr\cos\theta$

Differentiating this equation, we get

$$2x\, dx = 2Rr \sin\theta\, d\theta \quad \text{...(4)}$$

or $\sin\theta\, d\theta = \dfrac{x\, dx}{Rr}$

Substituting the value of $\sin\theta\, d\theta$ from eq. (4) in eq. (3) we have

$$dV = \frac{q\, x\, dx}{8\pi\varepsilon_0 Rrx} = \frac{q\, dx}{8\pi\varepsilon_0 Rr} \quad \text{...(5)}$$

In order to obtain the potential due to whole sphere we integrate the above equation within the limits $x = r - R$ and $x = r + R$. Hence

$$V = \int_{r-R}^{r+R} dV = \int_{r-R}^{r+R} \frac{q\, dx}{8\pi\varepsilon_0 Rr} = \frac{q}{8\pi\varepsilon_0 Rr} [x]_{x-R}^{r+R}$$

$$= \frac{q}{8\pi\varepsilon_0 Rr} [r + R - r + R] = \frac{q}{8\pi\varepsilon_0 Rr} \times 2R$$

$$\therefore \quad \boxed{V = \frac{1}{4\pi\varepsilon_0} \cdot \frac{q}{r}} \quad \text{...(6)}$$

This expression shows that the potential at an external point is the same as if the total charge were concentrated at the centre. Further, the potential decreases with distance in inverse proportion.

(*ii*) **When *P* lies on the surface**

In this case $r = R$

$$\therefore \text{ Potential at the surface} = \frac{1}{4\pi\varepsilon_0} \cdot \frac{q}{R} \quad ...(7)$$

2.18 POTENTIAL DUE TO DIPOLE

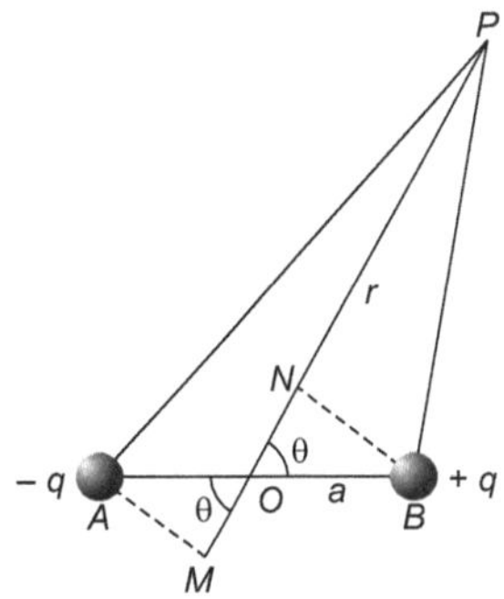

Fig. (32)

When two charges equal in magnitude but opposite in sign are separated by a small distance the system is called an electric dipole. The moment p of this dipole is $2\ a$, where $2a$ is the distance between two charge of magnitude q. The moment p is a vector quantity pointing from negative charge to positive charge. Fig. (32) shows the dipole of charge $+q$ and $-q$ of length $2\ a$. Here the aim is to determine the potential at an external point P at a distance r from the centre O of the dipole along the line OP making an angle θ with AB. Let BN and AM be the perpendicular to OP. Now

Potential at P due to charge $+q$ at B

$$V_1 = \frac{1}{4\pi\varepsilon_0}\left(\frac{q}{BP}\right) = \frac{1}{4\pi\varepsilon_0}\left(\frac{q}{NP}\right)$$

Potential at P due to charge $-q$ at A

$$V_2 = \frac{1}{4\pi\varepsilon_0}\left(\frac{-q}{AP}\right) = \frac{1}{4\pi\varepsilon_0}\left(\frac{-q}{MP}\right)$$

$\therefore$ Resultant potential at P is given by

$$V = \frac{1}{4\pi\varepsilon_0}\left(\frac{q}{NP} - \frac{q}{MP}\right) \quad ...(1)$$

From right angled triangle ONB, $ON = OB \cos\theta = a\cos\theta$

$$\therefore \quad NP = OP - ON = (r - a\cos\theta)$$

Similarly, from right angled triangle AMO,

$$OM = AO\cos\theta = a\cos\theta$$

$$\therefore \quad MP = MO + OP = (a\cos\theta + r) = (r + a\cos\theta)$$

Substituting these vlaues in eq. (1), we get

$$V = \frac{1}{4\pi\varepsilon_0}\left[\frac{q}{(r - a\cos\theta)} - \frac{q}{(r + a\cos\theta)}\right]$$

$$= \frac{q}{4\pi\varepsilon_0}\left[\frac{(r + a\cos\theta) - (r - a\cos\theta)}{(r - a\cos\theta)(r + a\cos\theta)}\right]$$

$$= \frac{q}{4\pi\varepsilon_0}\left[\frac{2a\cos\theta}{(r^2 - a^2\cos^2\theta)}\right] = \frac{p\cos\theta}{4\pi\varepsilon_0(r^2 - a^2\cos^2\theta)} \quad ...(2)$$

($\because\ 2aq = p$ = dipole moment)

As $r >> a$, $a^2\cos^2\theta$ term may be neglected in comparison with r^2. Hence,

$$\boxed{V = \frac{p\cos\theta}{4\pi\varepsilon_0 r^2}} \quad ...(3)$$

If $\theta = 90°$ (perpendicular bisector of the line joining the charges *i.e., equatorial plane,* $V = 0$. *Thus the potential vanishes everywhere in the equatorial plane.* This means that no work is done by an external agent in carrying a test charge from infinity along the perpendicular bisector of the dipole. We also note that

(*i*) equatorial plane is equipotential line

(*ii*) the potential of a dipole depends upon the dipole moment **p** and not on q or $2\,a$ separately.

Electric Field at any Point due to a Dipole (in Polar Coordinates)

In spherical coordinates

$$\mathbf{E} = -\vec{\nabla} V$$

$$\therefore \qquad \mathbf{E} = -\left(\frac{\partial V}{\partial r}\hat{\mathbf{a}}_r + \frac{\partial V}{r\partial\theta}\hat{\mathbf{a}}_\theta + \frac{1}{r\sin\theta}\frac{\partial V}{\partial\phi}\hat{\mathbf{a}}_\phi\right)$$

or

$$\mathbf{E} = \left[\left(-\frac{\partial V}{\partial r}\right)\hat{\mathbf{a}}_r + \left(-\frac{\partial V}{r\partial\theta}\right)\hat{\mathbf{a}}_\theta + \frac{1}{r\sin\theta}\left(-\frac{\partial V}{\partial\phi}\right)\hat{\mathbf{a}}_\phi\right] \qquad \text{...(4)}$$

Now, we shall calculate the individual components.

The radial component of electric field E_r along OP is given by

$$E_r = \frac{\partial V}{\partial r}$$

Substituting the value of V from eq. (3), we get

$$E_r = -\frac{\partial}{\partial r}\left(\frac{p\cos\theta}{4\pi\varepsilon_0 r^2}\right) = -\frac{p\cos\theta}{4\pi\varepsilon_0}\left(\frac{-2}{r^3}\right) = \frac{2\,p\cos\theta}{4\pi\varepsilon_0 r^3}$$

The transverse component of electric field E_θ is given by

$$E_\theta = -\frac{1}{r}\left(\frac{\partial V}{\partial\theta}\right) = -\frac{1}{r}\frac{\partial}{\partial\theta}\left(\frac{p\cos\theta}{4\pi\varepsilon_0 r^2}\right) = \frac{p\sin\theta}{4\pi\varepsilon_0 r^3} \qquad \text{...(5)}$$

Further $E_\phi = 0$

Substituting these values in eq. (4), we get

$$\mathbf{E} = \left[\frac{2p\cos\theta}{4\pi\varepsilon_0 r^3}\hat{\mathbf{a}}_r + \frac{p\sin\theta}{4\pi\varepsilon_0 r^3}\hat{\mathbf{a}}_\theta + 0\right]$$

or

$$\mathbf{E} = \left[\frac{p}{4\pi\varepsilon_0 r^3}(2\cos\theta\,\hat{\mathbf{a}}_r + \sin\theta\,\hat{\mathbf{a}}_\theta)\right]$$

Solved EXAMPLES

EXAMPLE 1. *An electric field* **E** *is given as*

$$\mathbf{E} = 6x^2\hat{\mathbf{a}}_x + 6y\hat{\mathbf{a}}_y + 4\hat{\mathbf{a}}_z$$

Find (*a*) V_{AB} *if points A and B are specified by* $A(2, 6, -1)$ *and* $B(-3, -3, 2)$ (*b*) V_A *if* $V = 0$ *at* $C(4, -2, -35)$ (*c*) V_B *if* $V = 2$ *at* $D(1, 2, -4)$.

Solution. (*a*) In cartesian coordinates,

$$V_{AB} = -\int_B^A \mathbf{E}\cdot dl \qquad ...(1)$$

where $dl = dx\,\hat{\mathbf{a}}_x + dy\,\hat{\mathbf{a}}_y + dz\,\hat{\mathbf{a}}_z$

Let us calculate the value of $\mathbf{E}\cdot dl$

$$\mathbf{E}\cdot dl = (6x^2\hat{\mathbf{a}}_x + 6y\,\hat{\mathbf{a}}_y + 4\hat{\mathbf{a}}_z)\cdot(dx\hat{\mathbf{a}}_x + dy\hat{\mathbf{a}}_y + dz\hat{\mathbf{a}}_z)$$

$$= 6x^2\,dx + 6\,y\,dy + dz$$

Substituting this value in eq. (1), we get

$$V_{AB} = -\int_{(-3,-3,2)}^{(2,6,-1)} (6x^2\,dx + 6\,y\,dy + 4dz)$$

$$= -\left[\int_{-3}^{2} 6x^2\,dx + \int_{-3}^{6} 6\,y\,dy + \int_{2}^{-1} 4\,dz\right]$$

$$= -\left[\left\{\frac{6x^3}{3}\right\}_{-3}^{2} + \left\{\frac{6y^2}{2}\right\}_{-3}^{2} + \{4z\}_2^{-1}\right]$$

$$= -\left[\frac{6}{3}(8+27) + \frac{6}{2}(36-9) + 4(-1-2)\right]$$

$$= -70 - 81 + 12 = -139$$

$\therefore$ $V_{AB} = -\mathbf{139\ V}$

(*b*) $V_{AC} = V_A - 0 = \int_{(4,-2,-35)}^{(2,6,-1)} (6x^2\,dx + 6\,y\,dy + 4\,dz)$

$\therefore$ $$V_A = \left[-6\int_4^2 x^2\,dx - 6\int_{-2}^{6} y\,dy - 4\int_{-35}^{1} dz\right]$$

Solving, we get $V_A = -\mathbf{120\ V}$

(*c*) $V_{BD} = V_B - V_D = V_B - 2$

$\therefore$ $$V_B - 2 = -\int_{(1,2,-4)}^{(-3,-3,2)} (6x^2\,dx + 6y\,dy + 4\,dz)$$

Solving, we get $V_B = \mathbf{19\ V}$

EXAMPLE 2. *A field* **E** *is given as*

$$\mathbf{E} = (-6y/x^2)\,\hat{\mathbf{a}}_x + (6/x)\,\hat{\mathbf{a}}_y + 5\,\hat{\mathbf{a}}_z \text{ V/m}$$

Calculate the potential difference $V_{AB,}$ *given that* $A(-7, 2, 1)$ *and* $B(4, 1, 2)$.

Solution. We know that

$$V_{AB} = \int_B^A \mathbf{E}\cdot dl$$

$$= -\int_B^A \left(-\frac{6y}{x^2}\,\hat{\mathbf{a}}_x + \frac{6}{x}\,\hat{\mathbf{a}}_y + 5\,\hat{\mathbf{a}}_z\right)\cdot(dx\,\hat{\mathbf{a}}_x + dy\,\hat{\mathbf{a}}_y + dz\,\hat{\mathbf{a}}_z)$$

$$= -\int_B^A \left[\frac{-6y}{x^2}\,dx + \frac{6}{x}\,dy + 5\,dz\right]$$

$$= \left[\int_{x=4,\,y=1}^{x=-7}\left(\frac{6y}{x^2}\right)dx - \int_{y=1,\,x=-7}^{y=2}\left(\frac{6}{x}\right)dy - \int_{z=2}^{1} 5\,dz\right]$$

$$= \left[\int_4^{-7}\frac{6}{x^2}\,dx + \frac{6}{7}\int_1^2 dy - 5\int_2^1 dz\right]$$

$$= \left[\left\{-\frac{6}{3x^2}\right\}_4^{-7} + \frac{6}{7}[y]_1^2 - 5\{z\}_2^1\right]$$

Solving, we get V_{AB} = **8.21 V**

EXAMPLE 3. *The potential at a point P(– 4,3,6) is given as*

$$V = 2x^2\,y - 5z$$

Find the folliwng values at P.

(i) the potential, *(ii) the electric field intensity* **E**

(iii) the direction of **E** *and* *(iv) the electric flux density* **D**.

Solution. (*i*) The potential at point $P(-4, 3, 6)$ is given by

$$V_P = 2\,(-4)^2\,(3) - 5\,(6) = \mathbf{66\ V}$$

(*ii*) $\mathbf{E} = -\vec{\nabla}\,V$

$$= -\left[\frac{\partial}{\partial x}(2x^2y - 5z)\,\hat{\mathbf{a}}_x + \frac{\partial}{\partial y}(2x^2y - 5z)\,\hat{\mathbf{a}}_y + \frac{\partial}{\partial z}(2x^2y - 5z)\,\hat{\mathbf{a}}_z\right]$$

$$= -[4x\,y\,\hat{\mathbf{a}}_x + 2x^2\,\hat{\mathbf{a}}_y - 5\,\hat{\mathbf{a}}_z] = -4x\,y\,\hat{\mathbf{a}}_x - 2x^2\,\hat{\mathbf{a}}_y + 5\,\hat{\mathbf{a}}_z$$

Now, $\mathbf{E}_p = -4(-4)(3)\,\hat{\mathbf{a}}_x - 2(-4)^2\,\hat{\mathbf{a}}_y + 5\,\hat{\mathbf{a}}_z = 48\,\hat{\mathbf{a}}_x - 32\,\hat{\mathbf{a}}_y + 5\,\hat{\mathbf{a}}_z$

or $|\mathbf{E}_p| = \sqrt{[(48)^2 + (32)^2 + (5)^2]} = \mathbf{57.9\ V/m}$

(*iii*) The direction of **E** at P is given by

$$\hat{\mathbf{a}}_{Ep} = \frac{48\,\hat{\mathbf{a}}_x - 32\,\hat{\mathbf{a}}_y + 5\hat{\mathbf{a}}_z}{|E_p|} = \frac{48\,\hat{\mathbf{a}}_x - 32\,\hat{\mathbf{a}}_y + 5\hat{\mathbf{a}}_z}{(57.9)}$$

$$= 0.829\,\hat{\mathbf{a}}_x - 0.553\,\hat{\mathbf{a}}_y + 0.086\,\hat{\mathbf{a}}_z$$

(*iv*) $\mathbf{D} = \varepsilon_0\mathbf{E}$

$$= [-4x\,y\hat{\mathbf{a}}_x - 2x^2\,\hat{\mathbf{a}}_y + 5\hat{\mathbf{a}}_z]\,\varepsilon_0$$

$$= [-4x\,y\hat{\mathbf{a}}_x - 2x^2\,\hat{\mathbf{a}}_y + 5\hat{\mathbf{a}}_z]\cdot 8.854\times10^{-12}$$

$$= -35.4\,x\,y\hat{\mathbf{a}}_x - 17.17\,x^2\,\hat{\mathbf{a}}_y + 44.3\,\hat{\mathbf{a}}_z$$

EXAMPLE 4. *The potential is expressed as*

$$V = \frac{10}{r^2}\sin\theta\cos\phi$$

Find: (*i*) *electric field strength* **E** *and*

(*ii*) *electric flux density* **D** *at* (2, π/2, 0).

Solution. (*i*) Given that $V = \frac{10}{r^2} \sin\theta \cos\phi$

We know that $\mathbf{E} = -\vec{\nabla} V$

In spherical coordinates, the value of $\vec{\nabla} V$ is given by

$$\vec{\nabla} V = \frac{\partial V}{\partial r}\hat{\mathbf{a}}_r + \frac{1}{r}\frac{\partial V}{\partial \theta}\hat{\mathbf{a}}_\theta + \frac{1}{r\sin\theta}\frac{\partial V}{\partial \phi}\hat{\mathbf{a}}_\phi$$

$$= \left(-\frac{20}{r^3}\sin\theta\cos\phi\right)\hat{\mathbf{a}}_r + \frac{1}{r}\left(\frac{10\cos\theta\cos\phi}{r^2}\right)\hat{\mathbf{a}}_\theta + \frac{1}{r\sin\theta}\left(-\frac{10\sin\theta\sin\phi}{r^2}\hat{\mathbf{a}}_\phi\right)$$

$$= -\frac{10}{r^3}[2\sin\theta\cos\phi\,\hat{\mathbf{a}}_r - \cos\theta\cos\phi\,\hat{\mathbf{a}}_\theta + \sin\phi\,\hat{\mathbf{a}}_\phi]$$

$$\therefore\ -\vec{\nabla} V = \frac{10}{r^3}[2\sin\theta\cos\phi\,\hat{\mathbf{a}}_r - \cos\theta\cos\phi\,\hat{\mathbf{a}}_\theta + \sin\phi\,\hat{\mathbf{a}}_\phi]$$

or $$\mathbf{E} = -\frac{10}{r^3}[2\sin\theta\cos\phi\,\hat{\mathbf{a}}_r - \cos\theta\cos\phi\,\hat{\mathbf{a}}_\theta + \sin\phi\,\hat{\mathbf{a}}_\phi]$$

(*ii*) The electric flux density **D** is given by

$$\mathbf{D} = \varepsilon_0 \mathbf{E}$$

$$= \frac{10}{r^3}[2\sin\theta\cos\phi\,\hat{\mathbf{a}}_r - \cos\theta\cos\phi\,\hat{\mathbf{a}}_\theta + \sin\phi\,\hat{\mathbf{a}}_\phi]$$

Let us consider the value of **D** at (2, π/2,0). We have

$$\mathbf{D} = \frac{10\varepsilon_0}{8}\left[2\sin\left(\frac{\pi}{2}\right)\cos(0)\hat{\mathbf{a}}_r - \cos\left(\frac{\pi}{2}\right)\cos(0)\hat{\mathbf{a}}_\theta + \sin(0)\hat{\mathbf{a}}_\phi\right]$$

$$= \frac{10\varepsilon_0}{8}[2\times1\times1\hat{\mathbf{a}}_r - 0 + 0]$$

$$= \frac{20}{8}\varepsilon_0\,\hat{\mathbf{a}}_r = \frac{5}{2}\varepsilon_0\,\hat{\mathbf{a}}_r$$

EXAMPLE 5. *A region is specified by the potential function given by*

$$V = 2x^2 + 5y^2 - 3z^2$$

Calculate the electric field strength at point (2, 3, 4) *in this region.*

Solution. We know that $\mathbf{E} = -\vec{\nabla} V$

$$\therefore\quad \mathbf{E} = -\left(\frac{\partial}{\partial x}\hat{\mathbf{a}}_x + \frac{\partial}{\partial y}\hat{\mathbf{a}}_y + \frac{\partial}{\partial z}\hat{\mathbf{a}}_z\right).(2x^2 + 5y^2 - 3z^2)$$

$$= [4x\hat{\mathbf{a}}_x + 10y\hat{\mathbf{a}}_y - 6z\hat{\mathbf{a}}_z]$$

At (2, 3, 4), we get

$$\mathbf{E} = -[8\hat{\mathbf{a}}_x + 30\hat{\mathbf{a}}_y - 24\hat{\mathbf{a}}_z]$$

EXAMPLE 6. *A uniform line charge density ρ_L C/m is existing from – L to + L on Y-axis. Find the potential at P (a, 0, 0).*

Solution. The situation is shown in Fig. (33). Consider a very small element *dy* at a distance *y* from the origin.

The potential at point *P* due to small element *dy* is given by

$$dV = \frac{\rho_L\, dy}{4\pi\varepsilon_0 r}$$

The total potential due to line charge is given by

$$V_P = \int_{-L}^{+L} \frac{\rho_L\, dy}{4\pi\varepsilon_0 r} = \frac{\rho_L}{4\pi\varepsilon_0}\int_{-L}^{L}\frac{dy}{\sqrt{(a^2+y^2)}}$$

$$= \frac{2\rho_L}{4\pi\varepsilon_0}\int_0^L \frac{dy}{\sqrt{(a^2+y^2)}}$$

$$= \frac{2\rho_L}{4\pi\varepsilon_0}\left[\log_e\left(y+\sqrt{(a^2+y^2)}\right)\right]_0^L$$

$$= \frac{\rho_L}{2\pi\varepsilon_0}\log_e\left[\frac{L+\sqrt{(a^2+L^2)}}{\sqrt{a^2}}\right]_0^L$$

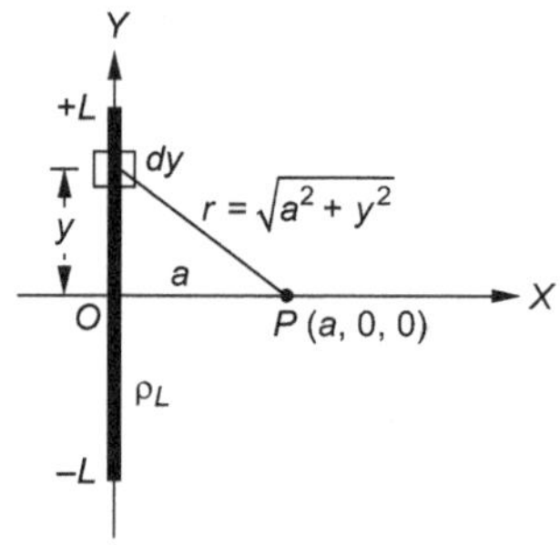

Fig. (33)

EXAMPLE 7. *Given cylindrical electric field*

$$\mathbf{E} = (5/\rho)\,\hat{\mathbf{a}}_\rho \text{ V/m} \qquad 0 \le \rho \le 2m$$

$$\mathbf{E} = 2.5\,\hat{\mathbf{a}}_\rho \text{ V/m} \qquad r \ge 2m$$

Find the potential difference V_{AB} for A (1 m, 0, 0) and *B* (4m, 0, 0).

Solution. The potential difference V_{AB} is given by

$$V_{AB} = -\int_B^A \mathbf{E}\cdot dl = -\left[\int_B^A \mathbf{E}\cdot dl + \int_2^1 \mathbf{E}\cdot dl\right] = -\int_4^2 2.5\,d\rho - \int_2^1\left(\frac{5}{\rho}\right)d\rho$$

$$= [2.5\rho]_4^2 - 5[\log_e \rho]_2^1 = 2.5\times 2 - 5\log_e(0.5)$$

or $$V_{AB} = 8.465 \text{ V}$$

EXAMPLE 8. *Calculate potential difference V_{AB} for a line charge ρ_L* = 5 nC/m *on Z-axis where A is* (2m, π/2, 0) *and B is* (4m, π, 5m).

Solution. We know that $V_{AB} = -\int_B^A \mathbf{E}\cdot dl$

For a line charge on *Z*-axis, electric field **E** is given by

$$\mathbf{E} = \frac{\rho_L}{2\pi\varepsilon_0\rho}\,\hat{\mathbf{a}}_\rho$$

In cylindrical coordinate system

$$d\mathbf{L} = d\rho\,\hat{\mathbf{a}}_\rho$$

$$\therefore \quad V_{AB} = -\int_4^2 \frac{\rho_L}{2\pi\varepsilon_0\rho}\hat{\mathbf{a}}_\rho\, d\rho\, \hat{\mathbf{a}}_\rho = -\int_4^2 \frac{\rho_L}{2\pi\varepsilon_0}\frac{d\rho}{\rho}$$

$$= \frac{\rho_L}{2\pi\varepsilon_0}[\log_e \rho]_4^2 = -\frac{5\times10^{-9}}{2\pi(8.854\times10^{-12})}\log_e\left(\frac{2}{4}\right)$$

$$= \frac{5\times10^{-9}}{2\pi(8.854\times10^{-12})}\log_e\left(\frac{4}{2}\right) = \frac{5\times10^{-9}}{2\pi(8.854\times10^{-12})}\log_e(2)$$

Solving we get V_{AB} = **62.3 V**

2.19 ENERGY STORED IN ELECTROSTATIC FIELD

Let a test charge is moved from a point of lower potential to a point of higher potential in an electric field. In this process, the external agent that moves the point charge has to do some work on it. As a result, the potential energy of the system is increased, *i.e.*, the system has a positive energy. On the other hand, if the test charge is moved from higher potential to lower potential in an electric field, then the electric field does work on the test charge. Therefore, the potential energy of the system is decreased, *i.e.*, there is a loss of potential energy. This energy is supplied to the charge.

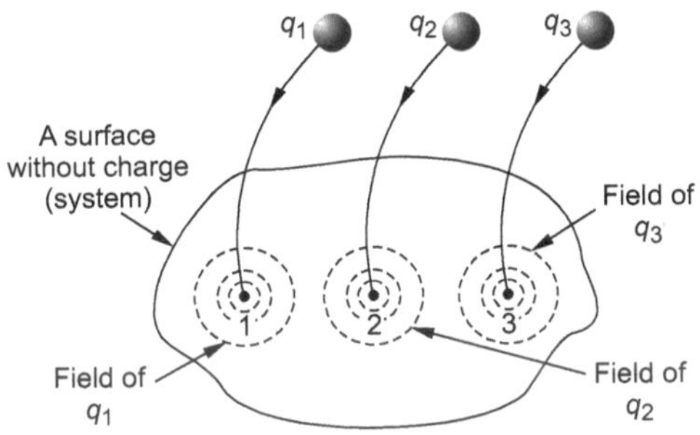

Fig. (34) Showing a set of three charges positioned on a surface without charge.

Now, we shall calculate the potential energy of a region consisting of a number (say n) of charges. First of all we shall find the potential energy of the region due to three charges, q_1, q_2 and q_3 and then generalize it for n charges. For this purpose, we have to calculate the workdone by external source in positioning these charges on a surface as shown in Fig. (34). The region is assumed to be without charge.

In bringing the charge q_1 from infinity to position 1, no work is done because the field does not exist on the system. So, $W_1 = 0$.

Now consider that charge q_2 is taken from infinity to position 2. The movement of this charge will take place in the field of charge q_1. This requires certain amount of work. The workdone is given by W_2 = charge × potential $q_2 V_{2,1}$, where $V_{2,1}$ is the potential at the location of $W_2 = q_2 \times V_{2,1}$.

Similarly, in bringing the charge q_3 from infinity to position 3, the movement will be in the field of charges q_1 and q_2. The workdone is given by

$W_3 = q_3 V_{3,1} + q_3 V_{3,2}$, where $V_{3,1}$ and $V_{3,2}$ are the potentials at position 3 due to charges q_1 and q_2 respectively.

Now, total work done = Potential energy of the field

$$W_E = 0 + (q_2 V_{2,1}) + (q_3 V_{3,1} + q_3 V_{3,2}) \qquad ...(1)$$

$$V_{2,1} = \frac{1}{4\pi\varepsilon_0} \cdot \frac{q_1}{D_{2,1}}, \quad V_{3,1} = \frac{1}{4\pi\varepsilon_0} \cdot \frac{q_1}{D_{3,1}}$$

$$q_3 V_{3,1} = q_3 \frac{1}{4\pi\varepsilon_0} \cdot \frac{q_1}{D_{3,1}} = q_1 \frac{1}{4\pi\varepsilon_0} \cdot \frac{q_1}{D_{3,1}} = q_1 V_{1,3}$$

Let us consider that the three charges are brought in the reverse order, *i.e.*, charge q_3 is taken first and then charge q_2 and finally charge q_1. The total workdone in this situation is given by

$$W_E = 0 + (q_2 V_{2,3}) + (q_1 V_{1,2} + q_1 V_{1,3}) \quad ...(2)$$

Adding eqs. (1) and (2), we get

$$2W_E = (q_1 V_{1,2} + q_1 V_{1,3}) + (q_2 V_{2,1} + q_2 V_{2,3}) + (q_3 V_{3,1} + q_3 V_{3,2})$$

or $$2W_E = q_1(V_{1,2} + V_{1,3}) + q_2(V_{2,1} + V_{2,3}) + q_3(V_{3,1} + V_{3,2})$$

or $$2W_E = q_1 V_1 + q_2 V_2 + q_3 V_3 \quad ...(3)$$

where $V_1 = V_{1,2} + V_{1,3}$ = total potential at charge q_1

$\therefore$ $$W_E = \frac{1}{2}[q_1 V_1 + q_2 V_2 + q_3 V_3] \quad ...(4)$$

If we consider a number of charges, then in general

$$\boxed{W_E = \frac{1}{2} \sum_{i=1}^{i=n} q_i V_i} \quad ...(5)$$

To get the general expression for energy stored in a region of continuous charge distribution, we replace summation by integration and substitute q by $\rho_V \, dV$

$$\boxed{W_E = \frac{1}{2} \iiint_V \rho_V V \, dV} \quad ...(6)$$

According to Gauss's law, $\vec{\nabla} \cdot \mathbf{D} = \rho_V$

So, $$W_E = \frac{1}{2} \iiint_V (\vec{\nabla} \cdot \mathbf{D}) V \, dV \quad ...(7)$$

where V is scalar and $\mathbf{D}$ is a vector field.

To obtain the value of volume integral, we use the following vector identity

$$\vec{\nabla} \cdot (V\mathbf{D}) \equiv V(\vec{\nabla} \cdot \mathbf{D}) + \mathbf{D} \cdot (\vec{\nabla} V) \quad ...(8)$$

or $$V(\vec{\nabla} \cdot \mathbf{D}) \equiv \vec{\nabla} \cdot (V\mathbf{D}) - \mathbf{D} \cdot (\vec{\nabla} V)$$

Substituting this value in eq. (7), we get

$$W_E \equiv \frac{1}{2}\left[\iiint_V (\vec{\nabla} \cdot V\mathbf{D}) dV - \iiint_V \mathbf{D} \cdot (\vec{\nabla} V) dV\right]$$

Using divergence theorem, the fist volume integral of this equation can be changed into closed surface integral. Now, we have

$$W_E = \frac{1}{2}\left[\oint_S V\mathbf{D} \, dS - \iiint_V \mathbf{D} \cdot (\vec{\nabla} V) dV\right] \quad ...(9)$$

This is known as energy density for any field **E**.

$$\therefore \qquad W_E = -\frac{1}{2}\iiint_V \mathbf{D}\cdot(\vec{\nabla}V)dV = \frac{1}{2}\iiint_V \mathbf{D}\cdot(-\vec{\nabla}V)dV$$

$$= \frac{1}{2}\iiint_V (\mathbf{D}\cdot\mathbf{E})dV \qquad (\because \mathbf{E} = -\vec{\nabla}V)$$

$$= \frac{1}{2}\iiint_V (\varepsilon_0\,\mathbf{E}\cdot E)dV \qquad (\because \mathbf{D} = \varepsilon_0\,\mathbf{E})$$

or
$$\boxed{W_E = \frac{1}{2}\iiint_V \varepsilon_0 E^2\, dV} \qquad ...(10)$$

This is known as energy density for any field **E**.

EXAMPLE 1. *A spherical conducting shell of radius a, centred at origin, has a potential field,*

$$V = \begin{cases} V_0 & r \le a \\ V_0\, a/r^2 & r > a \end{cases}$$

with zero reference at infinity. Find an expression for the stored energy that this point represents.

Solution. We know that $\quad W_E = \frac{1}{2}\iiint_V \varepsilon_0 E^2\, dV$

Here, we have $\quad W_E = \frac{\varepsilon_0}{2}\int_0^{\infty}\int_0^{\pi}\int_0^{2\pi}\left(\frac{V_0\, a}{r^2}\right)^2 r^2 \sin\theta\, d\theta\, dr\, d\phi$

$$= \frac{\varepsilon_0}{2}\left[\int_0^{a}(V_0\, a)^2 \frac{dr}{r^2}\int_0^{\pi}\sin\theta\, d\theta \int_0^{2\pi} d\phi\right] = 2\pi\varepsilon_0\, V_0^2\, a.$$

EXAMPLE 2. *Find the stored energy in a system of four identical charges* $q = 2$ nC *at the corners* 1 m *on a side.*

Solution. See Fig. (35). Here

$$W_E = \frac{1}{2}\,[q_1 V_2 + q_2 V_2 + q_3 V_3 + q_4 V_4]$$

$$= \frac{1}{2}\,[4qV] \text{ by symmetry}$$

where $\quad V = \frac{q}{4\pi\varepsilon_0}\left[\frac{1}{r_{12}} + \frac{1}{r_{13}} + \frac{1}{r_{14}}\right]$

$$= \frac{2\times 10^{-9}}{4\pi\varepsilon_0}\left[\frac{1}{1} + \frac{1}{1} + \frac{1}{\sqrt{2}}\right] = 48.66 \text{ V}$$

$$W_E = 2\times(2\times 10^{-9})\times 48.66 = \mathbf{194.64\ nJ}$$

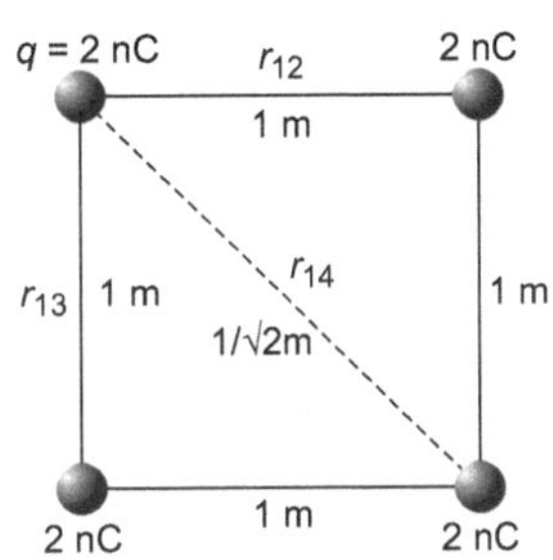

Fig. (35)

EXAMPLE 3. *Three point charges 3, 4 and 5 coulomb are situated in free space at three corners of an equilateral triangle with sides 5 cm. Find energy density due to electric field in the triangle.*

Solution. See Fig. (36). We know that

$$W_E = \frac{1}{2}\varepsilon_0 E^2$$

So, net field intensity at centre

$$\mathbf{E} = \mathbf{E}_1 + \mathbf{E}_2 + \mathbf{E}_3$$

$$\therefore \quad \mathbf{E}_1 = \frac{q_1}{4\pi\varepsilon_0 r^2}\cdot\hat{\mathbf{a}}_r$$

$$= \frac{3}{4\pi\varepsilon_0 (2.88)^2\times 10^{-4}}\;\frac{2.5\,\hat{\mathbf{a}}_x + 1.44\,\hat{\mathbf{a}}_y}{2.88}$$

$$\because \quad |\mathbf{r}_A| = |\mathbf{r}_B| = |\mathbf{r}_C| = 2.88 \text{ cm}$$

$$\mathbf{E}_2 = \frac{4}{4\pi\varepsilon_0 (2.88)^2\times 10^{-4}}\;\frac{2.5\,\hat{\mathbf{a}}_x + 1.44\,\hat{\mathbf{a}}_y}{2.88}$$

$$\mathbf{E}_3 = \frac{5}{4\pi\varepsilon_0 (2.88)^2\times 10^{-4}}(-\hat{\mathbf{a}}_y)$$

So, $$|\mathbf{E}| = |\mathbf{E}_1| + |\mathbf{E}_2| + |\mathbf{E}_3| = 1.878\times 10^{13}\text{ V/m}$$

$$\therefore \quad \text{Energy density} = \frac{1}{2}\varepsilon_0 E^2 = \mathbf{1.562\times 10^{15}\ J/m^3}$$

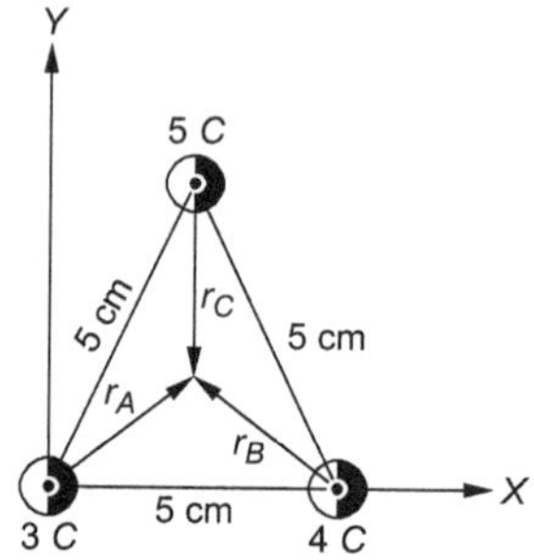

Fig. (36)

Exercises & Problems

EXERCISES

1. State and explain Coulomb's Law of electrostatics.
2. Find electric field intensity due to infinitely long line charge.
3. What is Gauss's law? Explain.
4. Write down applications of the Gauss's Law.
5. Find expression of charge enclosed in differential volume element by using Gauss's law
6. Find electric field intensity due to infinite sheet of charge.
7. Define the term potential and gradient relationship between potential and electric field intensity.
8. What is an electric dipole? Derive an expression for electric field intensity at any point due to an electric dipole.

   The different forms of Coulomb's law are given below in table.
9. What is equipotential surface? Give two examples.
10. Find electric potential at a point due to point charge and also due to group of point charge.

11. Prove the relation $\mathbf{E} = -\vec{\nabla} V$ and write formula for gradient in Cartesian, cylindrical and spherical co-ordinates.
12. Find electric potential at long distant point due to electric dipole.
13. Explain energy stored and energy density in static electric field.

PRACTICE PROBLEMS

1. Two point charge $q_1 = 50\ \mu C$ are $q_2 = 10\ \mu C$ are located at (– 1, 1, – 3) and (3, 1, 0) m respectively. Find the force on q_1 due to q_2.
2. Four point charges, each 20 μC, are on the X and Y axis at ± 4 m. Find the force on 100 μC point charge at (0, 0, 3) m.
3. Calculate electric field **E** at a point N (3, – 4, 2) in free space caused by

   (*a*) a charge $q_1 = 2\ \mu C$ at P_1 (0, 0, 0) (*b*) a charge $q_2 = 3\ \mu C$ at P_2 (– 1, 2, 3)

   (*c*) both the charges q_1 and q_2.
4. Find the value of **D** in the region about a uniform line charge of 8 nC/m lying along Z-axis in free space.
5. A circular disc of radius 4 m with a charge density $\rho_S = 12 \sin \phi\ \mu C/m^2$ is enclosed by surface S. What net flux crosses S?
6. A charge configuration in cylindrical co-ordinates is given by $\rho = 5\ re^{-2r}\ C/m^3$. Use Gauss's law to find **D**.
7. Find an approximate value for total charge enclosed in an incremental volume of $10^{-9}\ m^3$ located at origin if $\mathbf{D} = (e^{-x} \sin y\ \hat{\mathbf{a}}_x - e^{-x} \cos y\ \hat{\mathbf{a}}_y + 2z\ \hat{\mathbf{a}}_z)\ C/m^2$.
8. What is potential at centre of a square having charge of + 2, + 1, – 2 and + 3 in terms of 10^{-8} coulomb each, if side of square is 1.41 m.
9. Two identical uniform line charges lying along X and Y axis with charge densities $\rho_L = 20\ \mu C/m$. Obtain **D** at (3, 3, 3)m.
10. Given potential function $V = (2x + 4y)\ V$ in free space. Find stored energy in 1 m^3 volume centered at origin.
11. Given electric field $\mathbf{E} = -5\ e^{-r/a}\ \hat{\mathbf{a}}_r$ in cylindrical coordinate. Find energy stored in volume described by $r \le 2\ a$ and $0 \le Z \le 5\ a$.

UNIT - II

1. Conductors, Dielectrics and Capacitance
2. Poissons's and Laplace's Equations and their Applications

In this chapter, we will analyze the effects if charge in motion.

3

Conductors, Dielectrics and Capacitance

INTRODUCTION

The electric charge in motion constitute an electric current. The materials or medium carrying the current is called a conductor. In metallic conductors, the charge is carried by electrons. The aim of this chapter is first of all to define current and current density and then to consider about the conducting materials. The substances which do not contain free electrons or the number of free electrons is too low to constitute the electric current are known as dielectrics. In dielectrics, the electrons are tightly bound to the nucleus of the atom. When a slab of dielectric medium is placed between the plates of a capacitor, the capacitance increases. Therefore, after considering current and current density we shall consider about dielectrics and their uses in capacitors.

3.1 CURRENT AND CURRENT DENSITY

Current: We know that in a metallic conductor, free charges (electrons) are in random motion like gas molecules confined in a container. There is no directed motion along the conductor. Let two end of conductor be connected to a battery, *i.e.*, an electric field is maintained within the conductor. Now, the field acts on the electrons and gives them a resultant motion in the opposite direction of $-\mathbf{E}$ because a free charge in electric field experiences a force. The flow of electrons constitutes an electric current. Thus, ***when a charge flows in a conductor from one place to another place, then this flow of charge is called as electric current.*** The electric current is measured by the rate of flow of charge. Thus, if a net charge q passes through a specified surface area of the conductor in t second, then the current I is given by

$$I = \left(\frac{q}{t}\right) \qquad ...(1)$$

The current is a scalar quantity and its S.I. unit is ampere. If in the above eq. (1), $q = 1$, coulomb and $t = 1$ second, then $I = 1$ ampere. Thus,

$$1 \text{ ampere} = 1 \text{ coulomb/second}$$

When the current varies with time then at any instant t, the current is defined as

$$I = \frac{dq}{dt} \qquad ...(2)$$

It should be remembered that 1 ampere is equal to 6.15×10^{18} electrons per second. It means that in a conductor carrying a current of 1 ampere, 6.25×10^{18} electrons enters at one end and equal number of electrons leave at the other end in one second. One should not be confused that 6.25×10^{18} electrons travel per second from one end of the conductor to the other end.

It is natural to consider that the electric current flows along the conductor analogous as water flows through a pipe. Likewise, the charge flows faster where the conductor is smaller in cross-section and slower where the conductor is larger in cross-section because the charge must be conserved. It appears to define the direction of current as the direction of motion of free charges. However, the difficulty arises in case of liquids and gases where both the positively charged particles and negatively charged particles move in opposite directions. Naturally the question is that whichever should be the direction of current? As a convention, **the direction of current is** that in which **positive charges will move.** In a metallic conductor, positive charges can not move and the entire transfer of the charge is only due to flow of electrons. ***Thus, the conventional current is in opposite direction to the direction of movement of electrons.***

Current Density (J): Current I is a characteristic of a particular conductor. This is a macroscopic quantity like mass, volume, length etc. The current is a scalar quantity. A related microscopic quantity is the current density **J**. It is a vector quantity and is characteristic of a point inside the conductor rather than conductor as a whole. ***The current density is defined as the ratio of the current to the cross-sectional area.*** If a current I in a conductor be distributed uniformly over the area of cross-section A of the conductor, then current density is defined as

$$\boxed{J = \frac{I}{A}} \qquad \text{...(3)}$$

The direction of current density at a point is the direction of positive charge at that point. The unit of current density is 'ampere/metre^2'.

If the current is not uniformly distributed, then $J = \dfrac{I}{A}$ gives the average current density. However, it is of general interest to consider the current density at a point. When the current density is not uniform, we define it through point relationship as

$$J = \lim_{\Delta S \to 0} \left(\frac{\Delta I}{\Delta S} \right) \qquad \text{...(4)}$$

where ΔI is the increment of current crossing an incremental surface ΔS normal to current density.

Let us consider the relationship between current density **J** and current I. Consider a small area ΔS of the surface S through which charges are flowing as shown in Fig. (1) The amount of charge flowing across that area in unit time is given by

$$dq = \mathbf{J} \cdot \hat{\mathbf{n}}\, dS$$

where $\hat{\mathbf{n}}$ is unit vector normal to the surface. Therefore, the total charge passing per unit area through any surface is given by

$$q = \int_S \mathbf{J} \cdot d\mathbf{S}$$

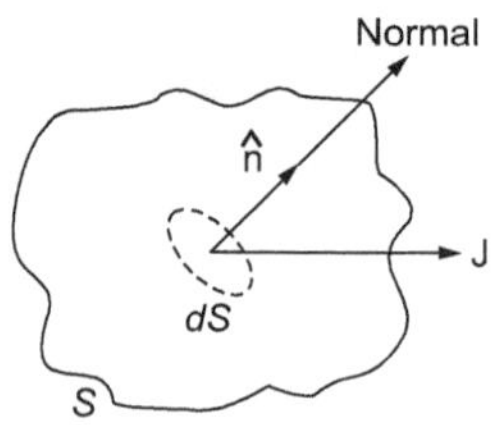

Fig. (1)

Now,

$$\boxed{I = q = \int_S \mathbf{J} \cdot d\mathbf{S}} \qquad \text{...(5)}$$

Let us now consider the relationship between current density and volume charge density. Consider an element of charge $\Delta Q = \rho_V \Delta V = \rho_V \Delta S\, \Delta L$ as shown in Fig. (2*a*).

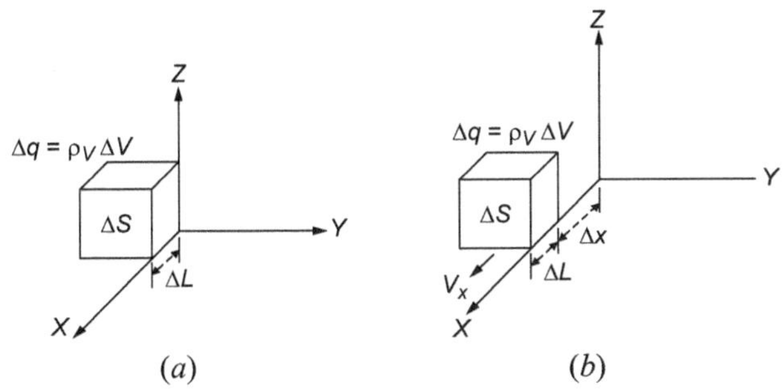

Fig. (2)

Suppose the charge is moving in *X* direction with velocity v_x. In time interval Δt, the charge has moved through a distance Δx as shown in Fig. (2*b*). Therefore, charge shifting is $\rho_V \Delta S\, \Delta x$ through a reference plane perpendicular to the direction of motion. The incremental time is Δt.

The resultant current is given by

$$\Delta I = \frac{\Delta q}{\Delta t} = \rho_V\, \Delta S \frac{\Delta x}{\Delta t} = \rho_V \Delta S\, v_x$$

where v_x represents the component of velocity along *x*-axis.

Now, $$J_x = \frac{\Delta I}{\Delta S} = \rho_V v_x$$

In general $$\boxed{\mathbf{J} = \rho_V\ \mathbf{v}} \qquad \text{...(6)}$$

This is known as convention current density.

3.2 DRIFT VELOCITY OF CHARGE CARRIERS IN A CONDUCTOR

We know that the electrons in a conductor move atrandom with sufficiently high speed in the vacant spaces between the stationary positive ions*. When an electric field is established between the two ends of the conductor, the free electrons experiences an electric force. Due to this force, the motion of electrons is accelerated. However, the speed of the electron does not increase continuously. The reason is that the electrons collide again and again with the positive ions of the conductor. As a result, they lose their energy which, of course, appears in the form of heat. Thus, the field does not give an accelerated motion to the electrons but it simply gives them a small constant velocity along the conductor which is superposed on the random motion of the electrons. This constant velocity of the electrons is called as ***drift velocity***. The drift velocity is very small in the comparison to the random motion of the electrons.

Relation Between Current Density and Drift Velocity

As shown in Fig. (3), let an electric field is maintained between the two ends of a conductor towards the left. The electrons move towards the right. Let the drift velocity of the electrons be v_d.

* When few electrons leave their atoms so that they become free, the atoms of the metal becomes positively charged and are called as positive ions.

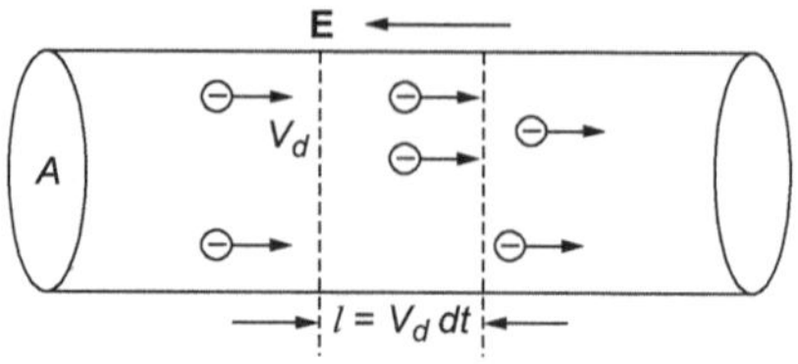

Fig. (3)

Suppose there are n charge carriers per unit volume and each charge carrier has a charge e. In time dt, the electron advances a distance l which is given by

$$l = v_d\, dt$$

The distance l is shown by dotted lines in Fig. (3).

We shall now calculate the number of electrons crossing the length l of the conductor in time dt. This will be equal to the number of electrons contained in a volume Al, *i.e.*, $Av_d\, dt$.

$\therefore$ Number of electrons = Volume × Number of electrons per unit volume

$$= Av_d\, dt \times n$$

Hence, charge crossing in time dt

= Number of electrons × charge on the electron

$$= Av_d\, dt\, ne$$

Further, current $$I = \frac{\text{Charge crossing in time } dt}{\text{Time } dt}$$

$$I = \frac{Av_d\, dt\, ne}{dt} = Av_d\, ne$$

So, $$\boxed{I = ne\, Av_d} \qquad \text{...(1)}$$

This is the relation between electric current and drift velocity.

We know that the current density is given by

$$J = \frac{I}{A} = \frac{ne\, Av_d}{A}$$

or $$\boxed{J = ne\, v_d} \qquad \text{...(2)}$$

This is relation between current density and drift velocity.

EXAMPLE 1. *Find the current crossing the portion of the* $y = 0$ *plane defined by* $-0.1 \le x \le 0.1$ m *and* $-0.002 \le z \le 0.002$ m.

If $$\mathbf{J} = (10)^2\, |x|\, \hat{\mathbf{a}}_y \ \ (\text{A/m}^2)$$

Solution. We know that $$I = \int_S \mathbf{J} \cdot d\mathbf{S} = \int_{-0.1}^{0.1}\int_{-0.002}^{0.002} (10)^2\, |x|\, \hat{\mathbf{a}}_y\, dx\, dz\, \hat{\mathbf{a}}_y$$

$$= (10)^2 \int_{-0.1}^{0.1} |x|\, dx \int_{-0.002}^{0.002} dz = (10)^2 \left[\left\{ \frac{x^2}{2} \right\}_{-0.1}^{0.1} \cdot \{z\}_{-0.002}^{0.002} \right]$$

$$= \mathbf{4\ mA}$$

EXAMPLE 2. *In a cylindrical conductor of radius* 2 mm, *the current density varies with distance from the axis according to*

$$I = 10^3 e^{-400\rho}\ (\text{A/m}^2)$$

Find the total current.

Solution. We know that

$$I = \int_S \mathbf{J}\cdot d\mathbf{S} = \int_S J\,dS = \int_0^{0.002}\int_0^{2\pi} 10^3 e^{-400\rho} d\rho\, d\phi$$

$$= 2\pi(10)^3 \int_0^{0.002} e^{-400\rho}\, d\rho$$

$$= 2\pi(10)^3 \left[\frac{e^{-400\rho}}{(-400)^2}(-400\rho - 1)\right]_0^{0.002} = \mathbf{7.51\ mA}$$

EXAMPLE 3. *Vector current density is given as*

$$\mathbf{J} = \left(\frac{4}{r^2}\cos\theta\,\hat{\mathbf{a}}_r + 20\,e^{-2r}\sin\theta\,\hat{\mathbf{a}}_\theta - r\sin\theta\cos\phi\,\hat{\mathbf{a}}_\phi\right)\text{A/m}^2$$

Find

(*a*) **J** at $P(r = 3, \theta = 0°, \phi = \pi)$

(*b*) *Total current passing through spherical cap* $r = 3$

$0° < \theta < 20°$, $0 < \phi < 2\pi$ *in direction of* $\hat{\mathbf{a}}_r$.

Solution. (*a*)

$$\mathbf{J} = \frac{4}{r^2}\cos\theta\,\hat{\mathbf{a}}_r + 20e^{-2r}\sin\theta\,\hat{\mathbf{a}}_\theta - r\sin\theta\cos\phi\,\hat{\mathbf{a}}_\phi$$

At $P(r = 3, \theta = 0°, \phi = \pi)$

$$\mathbf{J}_P = \frac{4}{(3)^2}\cos 0°\,\hat{\mathbf{a}}_r + 20e^{-2(3)}\cdot\sin 0°\,\hat{\mathbf{a}}_\theta - 3\sin 0°\cos\pi\,\hat{\mathbf{a}}_\phi$$

$$= \left(\frac{4}{9}\hat{\mathbf{a}}_r + 0 + 0\right)\text{A/m}^2$$

$$\mathbf{J}_P = \frac{4}{9}\hat{\mathbf{a}}_r\ \text{A/m}^2$$

(*b*)

$$I = \int_S \mathbf{J}\cdot d\mathbf{S}$$

$$d\mathbf{S} = r^2\sin\theta\,d\theta\,d\phi\,\hat{\mathbf{a}}_r$$

$$I = \int_0^{20°}\int_0^{2\pi}\left(\frac{4}{r^2}\cos\theta\cdot r^2\sin\theta\,d\theta\,d\phi\right)$$

At $r = 3$,

$$I = 4\int_0^{20°}\int_0^{2\pi}\sin\theta\cos\theta\,d\theta\,d\phi$$

$$= 2\int_0^{20°}\int_0^{2\pi}\sin 2\theta\,d\theta\,d\phi$$

$$= 2\times\left[\frac{-\cos 2\theta}{2}\right]_0^{20°}\times[\phi]_0^{2\pi} = \mathbf{1.47\ A}$$

3.3 CONTINUITY EQUATION

The current through any closed surface is given by

$$I = \int_S \mathbf{J} \cdot d\mathbf{S}$$

According to the law of conservation of charge, the *outward flow of positive charge must be balanced by a decrease of positive charge within closed surface.*

If q is charge inside closed surface, then rate of decrease in charge is given by

$$\text{Rate of charge decrement} = -\frac{dq}{dt} = I$$

where I is current flowing outward.

Therefore, $$I = \oint_S \mathbf{J} \cdot d\mathbf{S} = -\frac{dq}{dt}$$

As we have studied, $$q = \iiint_V \rho_V \, dV$$

So, $$\oint_S \mathbf{J} \cdot d\mathbf{S} = -\frac{d}{dt}\iiint_V \rho_V \, dV$$

$$\oint_S \mathbf{J} \cdot d\mathbf{S} = \iiint_V \left(-\frac{d\rho_V}{dt}\right) \cdot dV$$

Applying Stoke's theorem on left hand side, we get

$$\iiint_V (\vec{\nabla} \cdot \mathbf{J}) dV = \iiint_V \left(-\frac{d\rho_V}{dt}\right) \cdot dV$$

Therefore, $$\boxed{\vec{\nabla} \cdot \mathbf{J} = -\frac{d\rho_V}{dt}} \qquad \text{...(1)}$$

Eq. (1) is known as point form of continuity equation.

3.4 CONDUCTORS

On the basis of the response to applied electric field, the materials may be classified as conductors, insulators (or dielectrics) and semiconductors. In a conductor, a large number of free charge carriers (electrons) are available in conduction band. In the absence of an applied electric field, the free electrons are in random motion. The net flow of electrons through any cross-sectional area is zero. Therefore, the current flow through the conductor is zero. Now consider that an external electric field is applied to the conductor. The electrons drift along a direction opposite to the electric field. This process is known as *conduction*. Therefore, *the materials in which a large number of electrons participate in the process of conduction are known as conductors*.

On the basis of energy band gap, conductors, dielectrics and semiconductors can be classified as follows:

Fig. (4) represents energy band diagram of three types of materials at 0 K. In fig. (4*a*), there is no gap between valence band and conduction band. The materials are known as conductors.

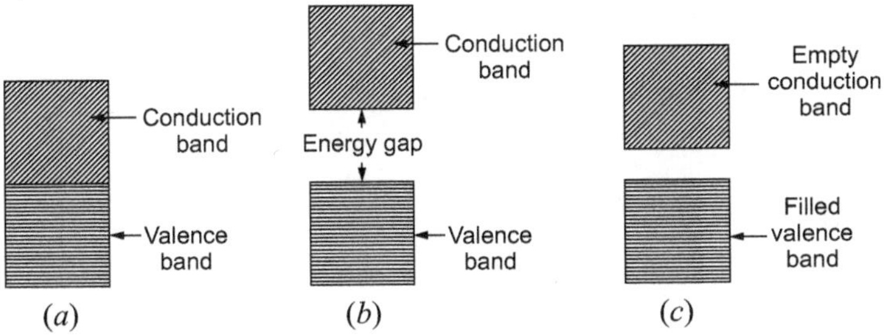

Fig. (4)

Fig. (4*b*) shows valence band and conduction band with some energy gap. Such materials are known as insulators. Fig. (4*c*) shows the structure of semiconductors.

The electrons with highest energy levels are located in valence band. If the valence band merges smoothly into conduction band, then additional kinetic energy may be given to valence electrons by an external field. This results to electron flow. This type of material is known as conductor.

Conductors in Electric Field

Consider a conductor with free electrons which move under the influence of electric field **E**. Now, electron with charge $q = -e$ will experience a force.

$$\mathbf{F} = -e\,\mathbf{E}$$

Drift velocity $\vec{v}_d$ of electron

$$\vec{v}_d = -\mu_e\,\mathbf{E}$$

where μ_e is mobility of electron and is a positive quantity.

$$\therefore \quad \mathbf{J} = \rho_e \vec{v}_d = -\rho_e \mu_e \mathbf{E}$$

where ρ_e free-electron charge density, a negative value. For metallic conductor.

$$\boxed{\mathbf{J} = \sigma\,\mathbf{E}} \qquad \text{where } \sigma = -\rho_e \mu_e. \qquad ...(1)$$

σ is known as conductivity.

Conductivity of electron is a positive quantity because ρ_e is negative quantity and μ_e is positive.

Therefore, in general $\boxed{\sigma = \rho_V \mu}$ ...(2)

As $\mathbf{J} = \sigma\,\mathbf{E}$,

the current density and field intensity have same direction irrespective of sign of charge carriers.

Equation (1) is also known as point form of Ohm's law.

Conductor Properties

The important properties of conductors are:

(1) No charge lies within an ideal conductor.

(2) Charge can exist on the surface of the conductor. The conductor now is known as charge conductor.

(3) Electric field intensity inside the conductor is zero.

(4) The tangential component of electric field intensity at the surface of the conductor is zero.

(5) Conductivity of an ideal conductor is infinity.

(6) Conductor surface is a equipotential surface.

EXAMPLE 1. *Find the magnitude of electric field intensity in material having conductivity* $\sigma = 6.17 \times 10^7$ mho/m and $\mu_e = 0.0056$ m^2/V.s. *if*

(*i*) *Drift velocity is* 2 mm/s,

(*ii*) *Current density is* 10^8 A/m^2.

Solution.

(*i*) The drift velocity $\vec{v}_d$ is given by

$$\vec{v}_d = -\mu_e \mathbf{E}$$

$$\therefore \qquad \mathbf{E} = -\frac{\vec{v}_d}{\mu_e}$$

$$\text{So,} \qquad |\vec{\mathbf{E}}| = \frac{|\vec{v}_d|}{\mu_e} = \frac{2 \times 10^{-3}}{0.0056} = \mathbf{0.3571\ V/m}$$

(*ii*) The current density $\vec{\mathbf{J}}$ is given by

$$\vec{\mathbf{J}} = \sigma \mathbf{E}$$

$$\therefore \qquad |\mathbf{E}| = \frac{|\mathbf{J}|}{\sigma} = \frac{10^8}{6.17 \times 10^7} = \mathbf{1.621\ V/m}.$$

EXAMPLE 2. *An aluminium conductor is* 30 m *long and has circular cross-section with a diameter of* 20.32 mm. *If there is a d.c. voltage of* 1.2 V *between ends, find*

(*a*) $|\vec{\mathbf{J}}|$, (*b*) *Current,*

(*c*) *Power dissipated.*

Given σ *for aluminium* $= 3.82 \times 10^7$ mho/m.

Solution.

(*a*) Length of conductor = 30 m

$$|\vec{\mathbf{E}}| = \frac{V}{d} = \frac{1.2}{30} = 0.04 \frac{\text{V}}{\text{m}}$$

$$|\mathbf{J}| = \sigma |\mathbf{E}| = 3.82 \times 10^7 \times 0.04 \text{ A/m}^2 = \mathbf{1.528 \times 10^6\ A/m^2}$$

(*b*)

$$I = \frac{V}{R}$$

$$\text{So,} \qquad R = \frac{1}{\sigma} \cdot \frac{L}{A} \qquad \left\{\because \frac{1}{\sigma} = \rho \text{ (Resistivity)}\right\}$$

$$R = \frac{1}{3.82 \times 10^7} \times \frac{30}{\pi\left(\frac{20.32 \times 10^{-3}}{2}\right)^2} \text{ ohm}$$

$$= \frac{30}{3.82 \times 10^7 \times \pi \times 103.22 \times 10^{-6}} \text{ ohm}$$

$$R = \mathbf{2.42 \times 10^{-3} \text{ ohm}}$$

$$\therefore \quad I = \frac{1.2}{2.42 \times 10^{-3}} = \mathbf{495.86 \text{ A}}$$

(*c*) Power dissipated = $I^2R = (495.86)^2 \times 2.42 \times 10^{-3}$ watt = **595.02 watt.**

EXAMPLE 3. *Find the total current in a circular conductor of radius* 4 mm *if the current density varies according to* $J = \left(\frac{10^4}{r}\right)$ A/m^2.

Solution. Current $I = \int_S \mathbf{J} \cdot d\mathbf{S} = \int_S J\, dS$

$$\therefore \quad I = \int_{r=0}^{0.004} \int_{\phi=0}^{2\pi} \frac{10^4}{r} r\, dr\, d\phi$$

$$= 10^4 \int_{r=0}^{0.004} dr \int_0^{2\pi} d\phi$$

$$= 10^4 [\{r\}_0^{0.004} \{\phi\}_0^{2\pi}] = 10^4 [0.004 \times 2\pi] = \mathbf{80\ \pi \text{ amp}}.$$

3.5 BOUNDARY CONDITIONS AT CONDUCTOR SURFACE IN FREE SPACE

Let us consider a boundary *AB* between conductor and free space as shown in Fig. (5). The aim of this article is to find the tangential and normal components of electric field and electric flux

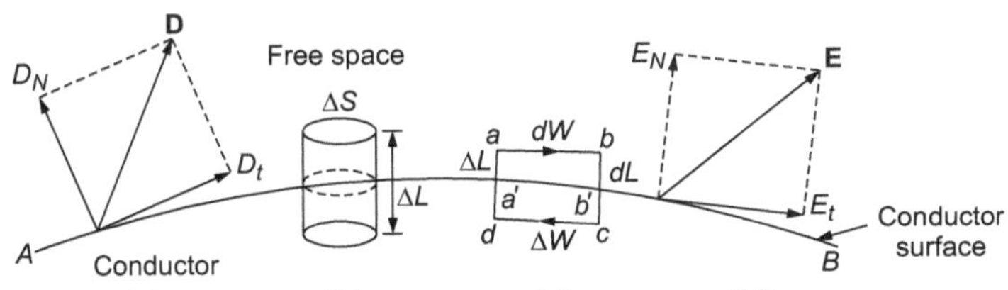

Fig. (5)

density **under static electric field condition.** These are known as boundary conditions.

The external electric field intensity **E** [shown in Fig. (5*d*)] is divided into two components, one tangential (E_t) and other normal (E_N) to the conductor surface. The tangential component (E_t) should be zero. If this component is not zero, then a tangential force will act on the surface charge. This force will disturb the static condition. Similarly, the electric flux density **D** [shown in Fig. (5*a*)] is divided into tangential component (D_t) and normal component D_N. For static condition, the tangential component should be zero. Therefore, for static conditions, the tangential components E_t and D_t should be zero.

$$D_t = \varepsilon E_t$$

Tangential Components

The above idea can be considered mathematically. For this purpose we consider a Gaussian surface *a b c d a* as shown in Fig. (5*c*). Let us move a unit positive charge along this closed path. No work is required, hence,

$$\oint \mathbf{E}\cdot d\mathbf{L} = 0 \text{ \{because network done in a close path zero for conservative forces\}.}$$

$$\oint \mathbf{E}\cdot d\mathbf{L} = \int_a^b \mathbf{E}\cdot d\mathbf{L} + \int_b^c \mathbf{E}\cdot d\mathbf{L} + \int_c^d \mathbf{E}\cdot d\mathbf{L} + \int_d^a \mathbf{E}\cdot d\mathbf{L} = 0$$

or $$\int_a^b \mathbf{E}\cdot d\mathbf{L} + \left[\int_b^{b'} \mathbf{E}\cdot d\mathbf{L} + \int_{b'}^c \mathbf{E}\cdot d\mathbf{L}\right] + \int_c^d \mathbf{E}\cdot d\mathbf{L} + \left[\int_d^{a'} \mathbf{E}\cdot d\mathbf{L} + \int_{a'}^a \mathbf{E}\cdot d\mathbf{L}\right] = 0$$

The parts of the Gaussian surface $b'c$, cd and da' lie inside the conductor. We know that inside the conductor, there does not exist the electric field and hence, $\mathbf{E} = 0$ for the above parts. Therefore,

$$E_t\,\Delta W + \left[(-E_N)\left(\frac{1}{2}\Delta L\right) + 0\right] + 0 + \left[0 + E_N\left(\frac{1}{2}\Delta L\right)\right] = 0$$

or $$E_t\,\Delta W = 0 \quad \text{or} \quad E_t = 0 \quad \text{because} \quad \Delta W \neq 0 \qquad \text{...(1)}$$

The tangential component of electric flus density

$$D_t = \varepsilon_0\,E_t = \varepsilon_0 \times 0 = 0 \qquad \text{...(2)}$$

The above eqns. (1) and (2) show that tangential components of electric field intensity and electric flux density at the boundary of conductor and free space are zero.

Normal Components

To derive the conditions for normal components of electric field intensity and electric flux density at the boundary, we consider a cylindrical Gaussian surface as shown in Fig. (5*b*). Applying Gauss's law, we have

$$\oint_S \mathbf{D}\cdot d\mathbf{S} = q$$

or $$\int_{\text{Top}} \mathbf{D}\cdot d\mathbf{S} + \int_{\text{Bottom}} \mathbf{D}\cdot d\mathbf{S} + \int_{\substack{\text{Side surface}\\ \text{in free space}}} \mathbf{D}\cdot d\mathbf{S} + \int_{\substack{\text{Side surface}\\ \text{in conductor}}} \mathbf{D}\cdot d\mathbf{S} = q \qquad \text{...(3)}$$

The bottom surface of the cylinder and same side surface are inside the conductor. As there is no charge inside the conductor, the second and fourth integrals in eq. (3) are zero. Now, we have

$$\int_{\text{Top}} \mathbf{D}\cdot d\mathbf{S} + \int_{\substack{\text{Side surface}\\ \text{in free space}}} \mathbf{D}\cdot d\mathbf{S} = q \qquad \text{...(4)}$$

As $dL << dW$, the area of side surface is negligible. Eq. (4) reduces to

$$\int_{\text{Top}} \mathbf{D}\cdot d\mathbf{S} = q \qquad \text{...(5)}$$

Then $$D_N\,\Delta S = q = \rho_S\,\Delta S$$

$$D_N\,\Delta S = \rho_S\,\Delta S$$

or $$\boxed{D_N = \rho_S} \qquad \text{...(6)}$$

The normal component of electric field intensity is given by

$$D_n = \varepsilon_0 E_N = \rho_S$$

or

$$\boxed{E_N = \frac{\rho_S}{\varepsilon_0}} \quad \text{...(7)}$$

Therefore, the electric flux leaves the conductor in direction normal to the surface. The value of electric flux density is numerically equal to the surface charge density.

EXAMPLE 1. *A potential field is given as* $V = 50\, e^{-5x}$ *sin 3y cos 4z volt. A point* $P\left(0.1, \frac{\pi}{12}, \frac{\pi}{24}\right)$ *is located at the conductor-free space boundary. Find (i) V, (ii)* **E**, *(iii),* E_N, *(iv)* E_t *and (v)* ρ_S *at this point.*

Solution.

(*i*) $V_P = 50\, e^{-5(0.1)} \sin 3\left(\frac{\pi}{12}\right) \cos 4\left(\frac{\pi}{24}\right) = \mathbf{18.57\ volt.}$

(*ii*) $\mathbf{E} = -\vec{\nabla} V = -\frac{\partial V}{\partial x}\hat{\mathbf{a}}_x - \frac{\partial V}{\partial y}\hat{\mathbf{a}}_y - \frac{\partial V}{\partial z}\hat{\mathbf{a}}_z$

$$= [250\, e^{-5x} \sin 3y \cos 4z\, \hat{\mathbf{a}}_x - 150\, e^{-5x} \cos 3y \cos 4z\, \hat{\mathbf{a}}_y + 200\, e^{-5x} \sin 3y \sin 4z\, \hat{\mathbf{a}}_z]$$

$$\therefore \qquad \mathbf{E}_p = [92.86\, \hat{\mathbf{a}}_x - 55.72\, \hat{\mathbf{a}}_y + 42.89\, \hat{\mathbf{a}}_z]$$

or

$$|\mathbf{E}_p| = \sqrt{[(92.86)^2 + (-55.72)^2 + (42.89)^2]}$$

$$= \mathbf{116.478\ V/m.}$$

(*iii*) At the conductor surface, only normal component exists.

$$\therefore \qquad |\mathbf{E}_N| = \mathbf{116.478\ V/m.}$$

(*iv*) $E_t = 0$.

(*v*) $\rho_S = |D_N| = \varepsilon_0 |\mathbf{E}| = (8.854 \times 10^{-12}) \times (116.478)\ \text{C/m}^2 = \mathbf{1.031\ nC/m^2}$.

3.6 DIELECTRICS

Dielectrics are the substances which do not contain free electrons or the number of such electrons is too low to constitute the electric current. In dielectrics, the electrons are tightly bound to the nucleus of the atom. Mica, glass, plastic, etc. are examples of dielectrics. When a non-conducting material is placed in an electric field, it modifies the electric field. Now, it acts as a source of storage of electrical charges. Therefore, when charge storage is the main function of a material then this is known as dielectric. So, to be a good dielectric, the material must be an insulator. In this way, any insulator is a dielectric.

Faraday, first realised the importance of dielectrics in electrical phenomena. He introduced a slab of dielectric medium between the plates of a parallel plate condenser and found that the charge on the capacitor with dielectric is greater than that without. Thus, the capacitance of the capacitor increases because the charge is larger. The dielectric helps in the following three ways:

(*i*) it helps in maintaining two large metal plates at very small separation.

(*ii*) it increases the potential difference which a capacitor can withstand without breakdown.

(*iii*) it increases the capacitance of a capacitor.

Difference Between Dielectrics and Conductors

Following are the differences between dielectrics and conductors:

(*i*) A dielectric or an insulator is a material which does not contain free electrons. The electrons are tightly bound to the nuclei of atoms. Conductors contain a fairly large number of free electrons. The free electrons roam through the conducting material.

(*ii*) The dielectric does not conduct electricity. On the application of an external electric field, the electrons may be able to move to and fro about their equilibrium positions but they do not leave the vicinity of their atoms. Of course, real dielectric do possess a feeble conductivity. On the other hand, conductors conduct electricity.

(*iii*) The charge given to a dielectric remains localised whereas charge resides on the outer surface in case of conductor.

(*iv*) For a particular field strength, the dielectric loses its insulation character. The field strength applied to the dielectric is now known as breakdown strength.

3.7 DIELECTRIC AN ATOMIC VIEW (POLAR AND NON-POLAR MOLECULES)

We know that atom consists of positive and negative charges in equal magnitudes. The positive charge of the nucleus may be supposed to be concentrated at a single point called as the centre of gravity of the positive charge. Similarly, the negative charge of electrons may be supposed to be concentrated at a single point called the centre of gravity of negative charge. When the two centres of gravity coincide, the molecule is known as ***non-polar molecule***. The non-polar molecules have symmetrical structure and zero dipole moment. Examples are: H_2, N_2, O_2, CO_2, benzene, etc. (see Fig. 6*a*).

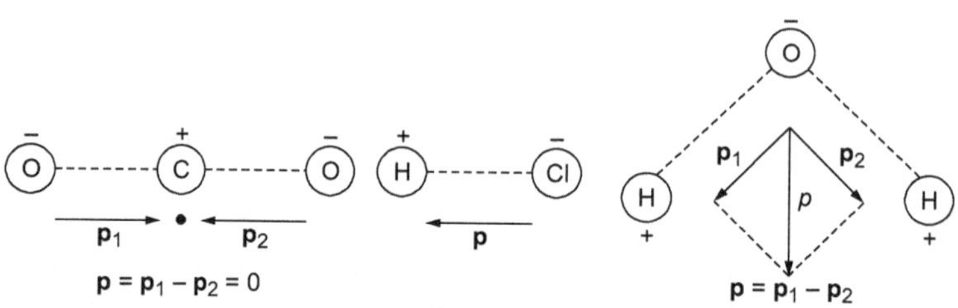

(*a*) Non-polar molecule (*b*) Polar molecule

Fig. (6)

On the other hand, when the two centres of gravity do not coincide (*i.e.*, they are slightly displaced), the molecule is called as ***polar molecule***. They have un-symmetrical structure and has a permanent dipole moment. Examples are: H_2O, HCl, CO, N_2O, NH_3, etc. (see Fig. 6*b*).

3.8 NON-POLAR DIELECTRIC IN AN ELECTRIC FIELD

When a dielectric slab is placed in an electric field, say between the plates of a charged condenser, the positive and negative charges are reoriented, *i.e.*, the centre of gravity of positive charges is pulled towards the negative plate of the condenser and vice-versa. Thus, the net effect of the applied field is to separate the positive charges from the negative charges. This effect is known as *polarisation* of dielectric. ***The dielectrics which are polarised only when they are placed in an electric field are called non-polar dielectric***. Fig. (7*a*) shows the random distribution of plus and minus charges in a non-polar dielectric. When this dielectric is placed in an electric field $\mathbf{E_0}$ (say between the plates of a parallel plate condenser), surface charges appear (positive charge on one surface and negative charge on the other surface) as shown in Fig. (7*b*).

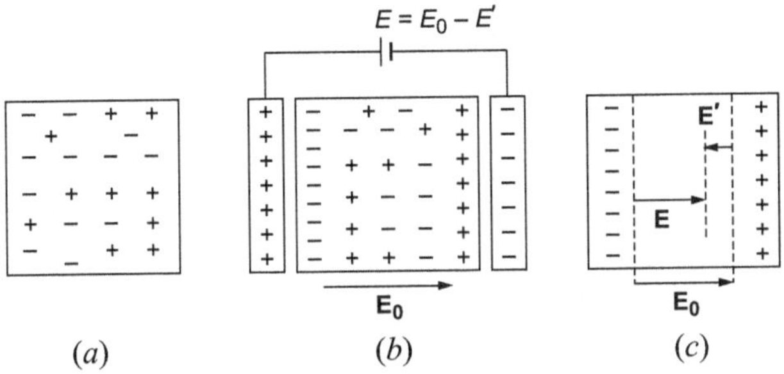

Fig. (7)

As shown in Fig. (7*c*), the induced surface charges appear in such a way that the electric field set up by them ($\mathbf{E}'$) opposes the external electric field $\mathbf{E_0}$. The resultant field $\mathbf{E}$ in the dielectric is the vector sum of $\mathbf{E}'$ and $\mathbf{E_0}$.

$$E = E_0 - E'$$

Thus, if the dielectric is placed in an electric field, induced surface charges appear which tend to weaken the original field within the dielectric.

3.9 POLAR DIELECTRIC IN ELECTRIC FIELD

We know that polar dielectrics have permanent dipole moments with their random orientations as shown in Fig. (8*a*). In presence of an electric field, the partial alignment of dipoles takes place as shown in Fig. (8*b*). As the molecules are always in thermal agitation, the alignment will

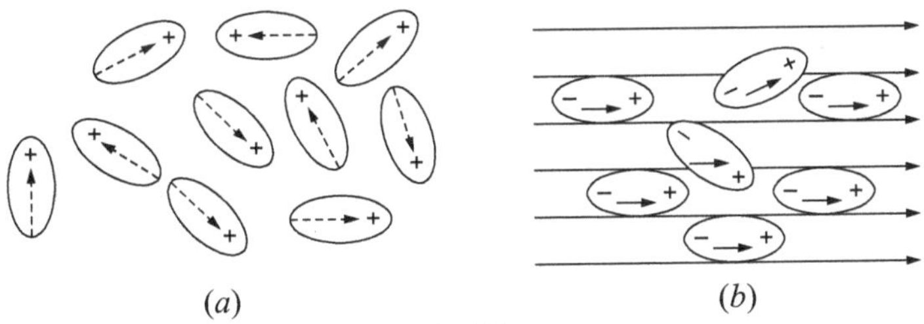

Fig. (8)

never be perfect. The alignment increases with the increase of electric field or with the decrease of temperature. The dipole moment of a polar molecule in an electric field will be $P_p + P_i$, where P_p is the permanent dipole moment and P_i is the induced dipole moment.

Thus, the non-polar molecules in an electric field become induced dipoles while polar molecules are re-oriented with dipole moment increased.

3.10 DIELECTRIC POLARISATION AND CHARGE DENSITY

We know that a conductor or insulator (di-electric) consists of a positive nuclei and negative electrons. In a conductor, the electrons are free to move while in dielectric they are bound to the nucleus.

Let a dielectric slab (say glass) is placed between the two metallic plates which can be charged to equal and opposite charges. Fig. (9*a*) shows the electronic structure of an atom when the two plates are not charged. When the plates are charged, *i.e.*, an electrostatic field is established between the two plates as shown in Fig. (9*b*), the electrons are slightly displaced towards the positively charged plate while the positively charged nucleus towards the negatively charged plate. The electronic displacement is a function of the magnitude of the charges on plates and the nature of dielectric. In this way, the dielectric is acted upon by the forces and is said to be polarised. Now, the distorted atom is called as *electric dipole*. We have studied that an electric dipole has an electric dipole moment. *The electric dipole moment per unit volume is called as dielectric polarisation* **P**.

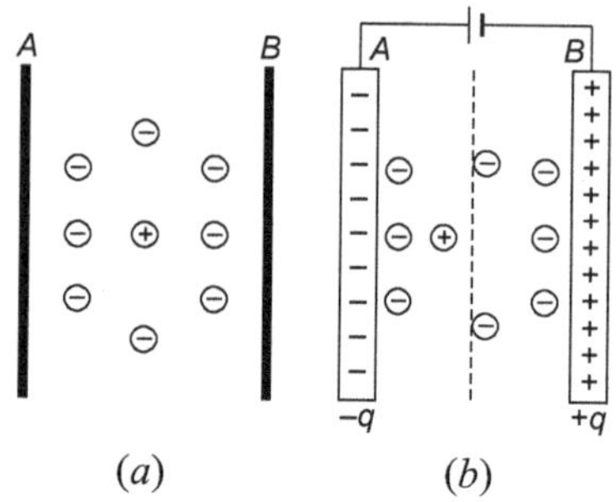

Fig. (9)

Suppose a dielectric slab of area of cross-section A and length l is placed in an electric field as shown in Fig. (10). Let the induced charges on faces, $ABCD$ and $EFGH$ be $-q'$ and $+q'$ respectively. The dipole moment **p** will be $q'l$. As volume of the slab is Al, hence, the electric dipole moment per unit volume, *i.e.*, dielectric polarisation **P** is given by

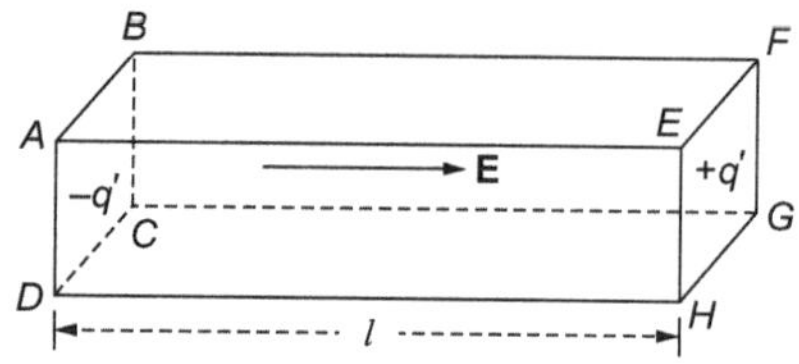

Fig. (10)

$$P = \frac{q'l}{Al} = \frac{q'}{A} \qquad \text{(charge per unit area, } i.e.\text{, charge density)}$$

Hence, dielectric polarisation is numerically equal to the surface charge density.

3.11 DIELECTRIC CONSTANT OR RELATIVE PERMITTIVITY

We define dielectric constant in the following ways:

1. We have studied that when a dielectric is placed between the plates of a condenser its capacity is increased. ***The ratio of the capacitance of a condenser with dielectric to the capacitance of the same condenser without dielectric is defined as dielectric constant. Thus,***

$$\boxed{k = \frac{C \text{ (Capacitance of the condenser with dielectric)}}{C_0 \text{ (Capacitance of the condenser without dielectric)}}} \qquad ...(1)$$

2. Instead of maintaining the two capacitors at the same potential difference, we can give the same charge to each capacitor. It has been observed that the potential difference V_d between the plates of the capacitor filled with dielectric is smaller than the potential difference V_0 without dielectric. ***The ratio of potential difference without dielectric to the potential difference with dielectric is defined as dielectric constant.*** Hence,

$$\boxed{\frac{V_0}{V_d} = k} \qquad ...(2)$$

3. According to Coulomb's law, the force of attraction of repulsion between two electric charges of magnitudes q_1 and q_2 separated at a distance r in free space is given by

$$F_0 = \frac{1}{4\pi\varepsilon_0} \times \frac{q_1 q_2}{r^2}$$

When the charges are placed in some other medium, then

$$F = \frac{1}{4\pi\varepsilon} \times \frac{q_1 q_2}{r^2}$$

where ε = permittivity of the medium.

$$\therefore \qquad \frac{F}{F_0} = \frac{\varepsilon_0}{\varepsilon} = \frac{1}{k}$$

where $$\boxed{k = \frac{\varepsilon}{\varepsilon_0}} \qquad ...(3)$$

So, $$\boxed{k = \frac{F_0}{F}} \qquad ...(4)$$

From eq. (3), ***the dielectric constant may be defined as the ratio of permittivity of the medium to the permittivity of free space.***

From eq. (4), ***the dielectric constant may be defined as the ratio of force between two charges in air or vacuum to the force between the same charges in dielectric medium.***

Dielectric constant k may also be represented by ε_r.

Dielectric constant is a ratio. Its value is one for vacuum and ∞ for metals.

3.12 ELECTRIC SUSCEPTIBILITY

When a dielectric is placed in electric field it is polarised. The polarisation vector **P** is proportional to the electric field **E**. Hence,

$$\mathbf{P} \propto \mathbf{E} \quad \text{or} \quad \mathbf{P} = \chi \mathbf{E}$$

where the constant of proportionality χ is known as electric susceptibility. ***The electric susceptibility may be defined as the ratio of polarisation vector to the electric intensity in the dielectric.***

3.13 GAUSS'S LAW IN DIELECTRICS

The Gauss's law states that the electric flus Φ_E through any closed surface is $\frac{1}{\varepsilon_0}$ times the charge enclosed by the surface.

Hence, $$\Phi_E = \oint \mathbf{E} \cdot d\mathbf{S} = \frac{q}{\varepsilon_0}$$

or $$\oint \mathbf{D} \cdot d\mathbf{S} = q$$

where **E** is the electric field vector and $d\mathbf{S}$ is the surface area.

The aim of this article is to consider the form of Gauss's law in case of a parallel plate condenser filled with a dielectric of dielectric constant k. Fig. (11) shows a parallel plate condenser (a) without and (b) with dielectric. The charge q on the plates is the same.

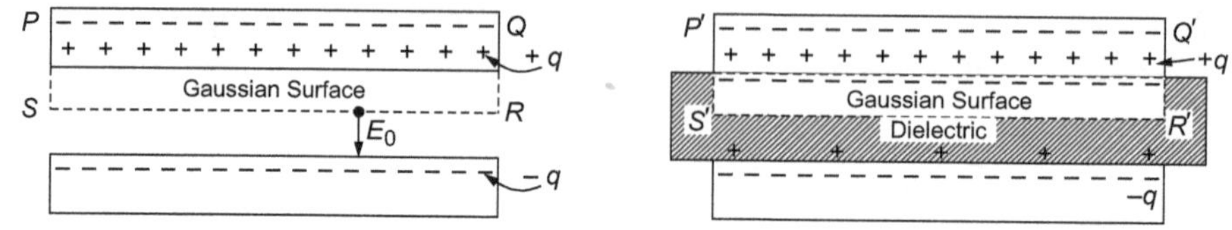

(a) Parallel plate condenser without dielectric. $PQRS$ is the Gaussian surface enclosing charge $+q$

(b) Parallel plate condenser with dielectric, $P'Q'R'S'$ is the Gaussian surface enclosing charge $(q - q')$

Fig. (11)

The Gaussian surfaces are also shown in the same figure. When no dielectric is present, then Gauss's law gives

$$\oint \mathbf{E}_0 \cdot d\mathbf{S} = \frac{q}{\varepsilon_0}$$

where q is charge enclosed.

or $$\mathbf{E}_0 \oint d\mathbf{S} = E_0 A = \frac{q}{\varepsilon_0} \quad \text{or} \quad E_0 = \left(\frac{q}{\varepsilon_0 A}\right) \qquad \text{...(1)}$$

where A is plate are.

When the dielectric is placed between the plates of the condenser (see Fig. 11b), the net charge within the Gaussian surface $P'Q'R'S'$ is $(q - q')$, where q' is the induced surface charge. Let E be the resultant field within dielectric. Then by Gauss's law

$$\oint \mathbf{E} \cdot d\mathbf{S} = \frac{(q - q')}{\varepsilon_0}$$

or $$EA = \frac{(q - q')}{\varepsilon_0}$$

$\therefore$ $$E = \frac{(q - q')}{\varepsilon_0 A} = \frac{q}{\varepsilon_0 A} - \frac{q'}{\varepsilon_0 A} \qquad \text{...(2)}$$

Here, it should be remembered that E is less than E_0 because induced charges produce their own field which opposes the external field E_0.

By the definition of dielectric constant k, $E = \dfrac{E_0}{k}$, hence, from eq. (2), we get

$$\frac{E_0}{k} = \frac{q_0}{\varepsilon_0 A} - \frac{q'}{\varepsilon_0 A}$$

Substituting the value of E_0 from eq. (1), we have

$$\frac{q}{k\varepsilon_0 A} = \frac{q}{\varepsilon_0 A} - \frac{q'}{\varepsilon_0 A}$$

or $$q' = q - \frac{q}{k} = q\left(1 - \frac{1}{k}\right) \qquad \text{...(3)}$$

Now, Gauss's law with dielectric present can be expressed as

$$\oint \mathbf{E}\cdot d\mathbf{S} = \frac{q}{\varepsilon_0} - \frac{q}{\varepsilon_0}\left(1-\frac{1}{k}\right)$$

or $$\oint \mathbf{E}\cdot d\mathbf{S} = \frac{q}{\varepsilon_0} - \frac{q}{\varepsilon_0} + \frac{q}{\varepsilon_0 k} = \frac{q}{\varepsilon_0 k}$$

or $$\boxed{k\oint \mathbf{E}\cdot d\mathbf{S} = \frac{q}{\varepsilon_0}}$$

or $$\boxed{k\oint \mathbf{D}\cdot d\mathbf{S} = q} \quad \text{...(4)}$$

This is Gauss's law in a dielectric.

3.14 THREE ELECTRIC VECTORS

The three electric vectors are: (*i*) Electric intensity, (*ii*) Dielectric polarisation and (*iii*) Electric displacement.

(*i*) **Electric Intensity E:** ***The electric intensity* E *at any point in the electric field is numerically equal to the force experienced by a unit positive charge placed at that point. The direction of* E *being the same as that of the field.***

(*ii*) **Dielectric polarisation P:** When a dielectric is polarised, the distorted atom is called an electric dipole. ***The electric dipole moment per unit volume is called as dielectric polarisation* P.**

(*iii*) **Electric displacement D:** When a dielectric slab is placed between the plates of a parallel plate condenser [see Fig. (11b)], the medium is polarised. Now induced surface charges appear. The charge is negative on the surface nearer the positive plate of the condenser and positive charge nearest the negative plate. Let q' be the induced surface charge. Now the charge q on the plate of the condenser and induced surface charge q' are related as

$$\frac{q}{k\varepsilon_0 A} = \frac{q}{\varepsilon_0 A} - \frac{q'}{\varepsilon_0 A} \quad \text{...(1)}$$

[See eq. (3) of article 3.13]

From eq. (1), we have

$$\frac{q}{\varepsilon_0 A} = \frac{q}{k\varepsilon_0 A} + \frac{q'}{\varepsilon_0 A}$$

or $$\frac{q}{A} = \varepsilon_0\left(\frac{q}{k\varepsilon_0 A}\right) + \frac{q'}{A} \quad \text{...(2)}$$

We know that

$$\frac{q}{k\varepsilon_0 A} = E \quad \text{and} \quad \frac{q'}{A} = P$$

$\therefore$ $$\frac{q}{A} = \varepsilon_0 E + P$$

We put the ratio $\left(\frac{q}{A}\right)$ as D. Hence, in vector form,

$$\boxed{\mathbf{D} = \varepsilon_0 \mathbf{E} + \mathbf{P}} \qquad ...(3)$$

where D is called as electric displacement.

In terms of dielectric polarisation **P**, a general form of the Gauss's law for **E** can be expressed as

$$\oint \mathbf{E} \cdot d\mathbf{S} = \frac{(q - q')}{\varepsilon_0} \quad \text{or} \quad \varepsilon_0 \oint \mathbf{E} \cdot d\mathbf{S} = (q - q')$$

or $$\varepsilon_0 \oint \mathbf{E} \cdot d\mathbf{S} = q - \oint \mathbf{P} \cdot d\mathbf{S}$$

or $$\oint (\varepsilon_0 \mathbf{E} + \mathbf{P}) \cdot d\mathbf{S} = q$$

or $$\oint \mathbf{D} \cdot d\mathbf{S} = q \qquad ...(4)$$

Hence, dielectric displacement* D *is defined as a vector quantity whose surface integral over any charge surface, the flux of* D*, is equal to the free charge only within the surface.

The following points should be noted:

(*i*) **D** is connected with free charge only. It is not altered by the introduction of the dielectric. The lines of **D** begin and end on free charges.

(*ii*) **P** is connected with polarisation charge only. The lines of **P** begin and end on polarisation charges.

(*iii*) **E** is connected with all charges that are actually present whether free or polarisation. **E** is reduced inside the dielectric, where there are fewer lines.

For these three points, see Fig. (12).

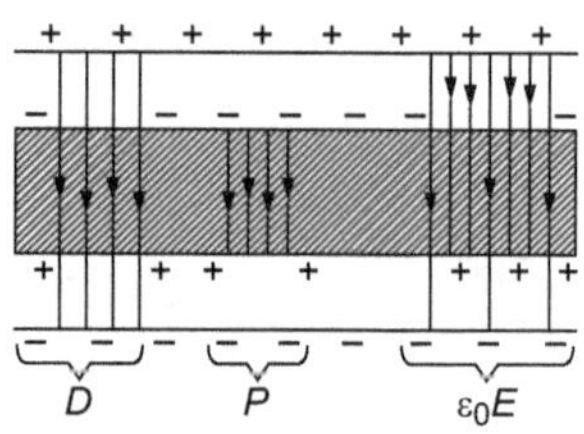

Fig. (12)

3.15 RELATIONS BETWEEN THREE VECTORS

The vectors **D** and **P** are separately connected to **E**.

We know that, $$E = \frac{q}{\varepsilon_0 k A} \quad \text{or} \quad \frac{q}{A} = \varepsilon_0 k E$$

$$\therefore \quad \boxed{D = k\varepsilon_0 E} \quad \because \quad \frac{q}{A} = D \qquad ...(5)$$

The product $k\varepsilon_0$ is called the *permittivity* ε of the medium and hence,

$$\boxed{\mathbf{D} = \varepsilon \mathbf{E}}$$

In vacuum $k = 1$ and hence, $\varepsilon = \varepsilon_0$. This is the reason that ε_0 is often referred as *permittivity of vacuum* or free space. The ratio $\frac{\varepsilon}{\varepsilon_0}$ is also termed as *relative permittivity* because k is a ratio.

Putting the value of D in eq. (3), we get

$$k\,\varepsilon_0\,\mathbf{E} = \varepsilon_0\,\mathbf{E} + \mathbf{P}$$

or $$\boxed{\mathbf{P} = \varepsilon_0(k-1)\mathbf{E}} \qquad ...(6)$$

The constant $\varepsilon_0(k-1)$ in eq. (6) is called as electric susceptibility χ of the dielectric. Hence,

$$\boxed{\mathbf{P} = \chi\,\mathbf{E}} \qquad ...(7)$$

Three quantities ε_0, k and χ are simply different ways of describing the same property of the dielectric. These are related by the following relation:

$$\chi = \varepsilon_0(k-1) = \varepsilon_0 k - \varepsilon_0 = \varepsilon - \varepsilon_0$$

$\therefore$ $$\varepsilon = \varepsilon_0 + \chi$$

Now, $$k = \frac{\varepsilon}{\varepsilon_0} = \frac{\varepsilon_0 + \chi}{\varepsilon_0} = 1 + \frac{\chi}{\varepsilon_0}$$

$\therefore$ $$\boxed{k = 1 + \frac{\chi}{\varepsilon_0}} \qquad ...(8)$$

3.16 BOUNDARY CONDITIONS AT THE DIELECTRIC SURFACE

The rules governing the behaviour of **E** and **D** at the boundary between two dielectrics are known as boundary conditions.

Following are the two boundary conditions:

(1) ***The normal component of electric displacement D is the same on both sides of the boundary of two media of different dielectrics or the normal component of displacement vector is continuous across the charge free boundary between two dielectrics.***

(2) ***The tangential components of the electrical intensities are the same (continuous) on both sides of the dielectrics.***

We shall derive these boundary conditions in the following way:

(1) Let AB represents a small portion of the boundary between two media of absolute permittivities ε_1 and ε_2 as shown in Fig. (13). The media are assumed to be homogeneous and isotropic. A medium with same properties at every point is homogeneous and a medium with same properties in all directions is isotropic. We now consider a small area dS on the boundary so that its curvature may be neglected. Suppose $\mathbf{D}_1$ and $\mathbf{D}_2$ be the electric induction vectors in the media on either side of dS respectively. θ_1 and θ_2 are the angle which $\mathbf{D}_1$ and $\mathbf{D}_2$ make with the normal to dS.

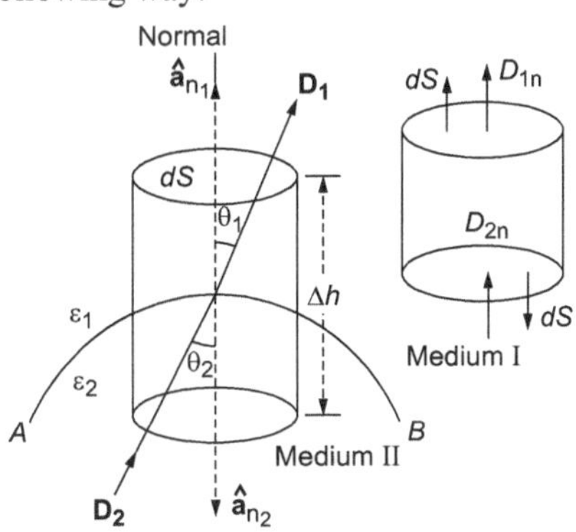

Fig. (13)

In order to find the boundary condition for **D**, let us construct a small pill box shaped surface which intersects the boundary with its end faces parallel to it. The height h of the pill box is assumed to be negligibly small in comparison with the diameter of the base. Applying Gauss's theorem to the pill box surface, we have

$$\oint \mathbf{D} \cdot d\mathbf{S} = q$$

or $$\int_{\text{Top}} \mathbf{D} \cdot d\mathbf{S} + \int_{\text{Bottom}} \mathbf{D} \cdot d\mathbf{S} + \int_{\text{Sides}} \mathbf{D} \cdot d\mathbf{S} = q \quad ...(1)$$

Let D_{1n} be the average normal component of displacement-vector D_1 to the face of the pill box in medium 1 and D_{2n} the average normal component of displacement vector D_2 to the face of the box in medium II. It should be remembered that D_{2n} is an inward normal. Because the height of the pill box is negligibly small in comparison with the diameter of the base ($\Delta n \to 0$) the only contribution of flux of **D** will come through the end faces. Thus, by Gauss's law

$$D_{1n}\, dS - D_{2n}\, dS = q \quad ...(2)$$

where q is the total charge enclosed by the surface. The second term of eq. (2) is negative because D_{2n} and dS oppositely directed.

From eq. (2), $$\boxed{D_{1n} - D_{2n} = \frac{q}{dS} = \rho_S} \quad ...(3)$$

where ρ_S is the charge per unit area on the boundary of two dielectrics. Eq. (3) shows that the normal component of the displacement vector **D** changes at the charged boundary between two dielectrics by an amount equal to the surface charge density.

If the boundary is free from charge, then $\rho_S = 0$. Now, eq. (3) reduces to

$$\boxed{D_{1n} = D_{2n}} \quad ...(4)$$

Thus, the normal component of displacement vector is continuous across the charge free boundary between two dielectrics.

(2) In order to consider the boundary condition for **E**, let us consider a rectangle *PQRS* of small width and its length parallel to the boundary separating two dielectrics as shown in Fig. (14). The rectangular lies with its longest sides parallel to the boundary.

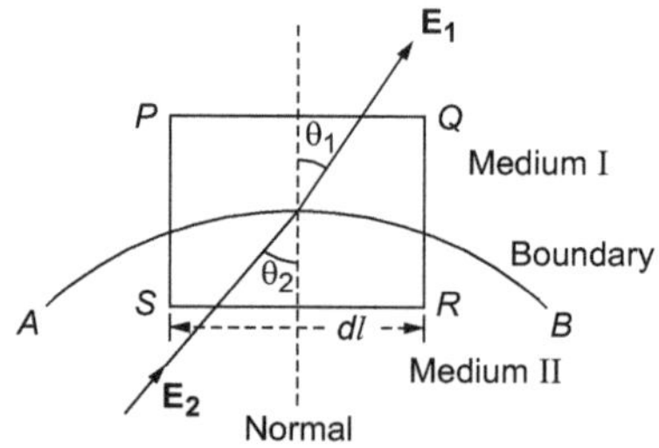

Fig. (14)

Let $\mathbf{E}_1$ and $\mathbf{E}_2$ be the electric intensities in two media. The directions of the electric field vectors make angles θ_1 and θ_2 with the normal to the boundary. Now the work done in taking a unit charge amount the rectangle *PQRS* must vanish, *i.e.*,

$$\oint \mathbf{E} \cdot d\mathbf{l} = 0$$

or $$E_1 \sin \theta_1\, dl - E_2 \sin \theta_2\, dl = 0 \quad ...(5)$$

Here, we have neglected the contribution of short sides *QR* and *SP* as they are taken very small.

From eq. (5), $E_{1t} = E_{2t}$ ...(6)

Thus, the tangential component of the field is continuous at the boundary of two dielectrics.

The boundary condition for **E** can also be derived in the following manner:

Consider a rectangular path (*a b c d*) half-way in each medium as shown in Fig. (15).

Let Δx be the length parallel to the boundary while a length Δy normal to the boundary. E_{1t} and E_{2t} are the electric field intensities tangential to the boundary in medium I and II respectively.

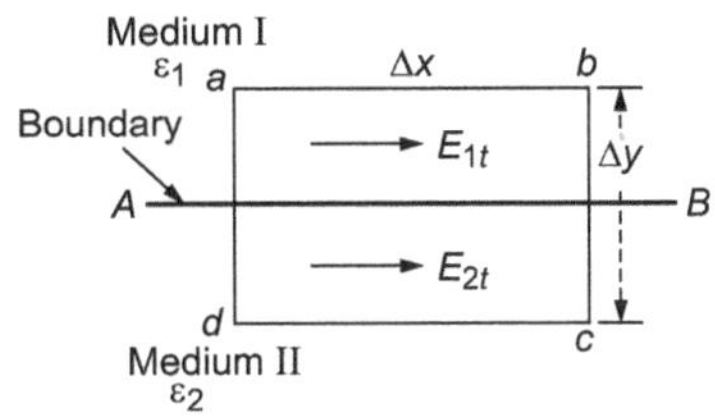

Fig. (15)

Consider the workdone per unit charge required in bringing a positive test charge around the closed path *a b c d*. The line integral **E** around the path should be zero, because for a closed path its value is zero. Therefore,

$$\oint \mathbf{E} \cdot d\mathbf{l} = 0$$

Now, $$\int_a^b \mathbf{E} \cdot d\mathbf{l} + \int_b^c \mathbf{E} \cdot d\mathbf{l} + \int_c^d \mathbf{E} \cdot d\mathbf{l} + \int_d^a \mathbf{E} \cdot d\mathbf{l} = 0$$

Let $\Delta y \to 0$ at the boundary

$$E_{1t}\,\Delta x + 0 - E_{2t}\,\Delta x + 0 = 0$$

or $$\boxed{E_{1t} = E_{2t}}$$

Thus, the tangential component of the field is continuous at the boundary of two dielectrics.

Solved EXAMPLES

EXAMPLE 1. *The areas of the plate of a parallel plate condenser is* 100 sq. cm. *The distance between plates is* 1 cm. *A potential difference of 100 volt is applied. A slab of thickness* 0.5 cm *and dielectric constant* 7 *is placed between plates. Calculate the values of E, D, P in air and dielectric.*

Solution. In air, Electric field $E = \dfrac{\text{Potential difference}}{\text{Distance}} = \dfrac{100}{1 \times 10^{-2}} = \mathbf{1 \times 10^4}$ **volt/metre.**

Electric displacement $D = \varepsilon_0 E = (8.9 \times 10^{-12})\,(1 \times 10^4)$
$= \mathbf{8.9 \times 10^{-8}}$ **coul/metre²**

Polarisation $P = 0$

In dielectric, Permittivity $\varepsilon = k\varepsilon_0 = 7 \times (8.9 \times 10^{-12})$
$= \mathbf{62.3 \times 10^{-12}}$

Electric displacement D = constant = $\mathbf{8.9 \times 10^{-8}}$ **coul/meter²**

Electric field in dielectric $= \dfrac{D}{\varepsilon} = \dfrac{8.9 \times 10^{-8}}{62.3 \times 10^{-12}} =$ **1429 volt/metre**

$$\text{Polarisation } P = D - \varepsilon_0 E = (8.9 \times 10^{-8}) - (8.9 \times 10^{-12})1429$$
$$= \mathbf{7.628 \times 10^{-8}\ coul/metre^2}$$

EXAMPLE 2. *The thickness of dielectric between parallel plates of a condenser is* 5 mm. *Dielectric constant is* 3. *Electric field in the dielectric is* 10^6 V/m. *Calculate the surface charge density on the condenser plate, surface charge density on the dielectric, polarisation, electric displacement and energy density.*

Solution. Electric displacement

$$D = \varepsilon E = k\,\varepsilon_0 E$$

$$\therefore \quad D = 3 \times (8.9 \times 10^{-12}) \times 10^6$$
$$= \mathbf{26.7 \times 10^{-6}\ coul/metre^2}$$

$\therefore$ Surface charge density on the plate of condenser

$$\sigma = 26.7 \times 10^{-6}\ \text{coul/metre}^2$$
$$= \mathbf{26.7\ \mu C/m^2}$$

Polarisation
$$P = \varepsilon_0(k-1)E = (8.9 \times 10^{-12})(3-1)(10)^6$$
$$= 17.8 \times 10^{-6}\ \text{coulomb/metre}^2$$
$$= \mathbf{17.8\ \mu C/m^2}$$

$\therefore$ Surface charge density on the dielectric = $\mathbf{17.8\ \mu C/m^2}$

$$\text{Energy density in the dielectric} = \frac{1}{2}\cdot \varepsilon E^2 = \frac{1}{2}\varepsilon_0 kE^2$$

$$= \frac{1}{2} \times (8.9 \times 10^{-12}) \times 3 \times (10^6)^2 = \mathbf{13.35\ J/m^3}$$

EXAMPLE 3. *The dielectric constant of water is* 78. *Calculate its electrical permittivity.*

Solution. We know that
$$\varepsilon = k\,\varepsilon_0$$
$$\varepsilon = 78 \times (8.85 \times 10^{-12}) = \mathbf{690.3 \times 10^{-12}\ F/m}$$

EXAMPLE 4. *The dielectric constant of medium is* 4. *Electric field in the dielectric is* 10^6 V/m. *Calculate electric displacement and polarisation. Take* $\varepsilon_0 = 9 \times 10^{-12}$ F/m.

Solution. We know that $D = k\,\varepsilon_0 E$ and $P = \varepsilon_0(k-1)E$

Given that, $k = 4$, $E = 10^6$ V/m and $\varepsilon_0 = 9 \times 10^{-12}$ F/m

$$\therefore \quad D = 4 \times (9 \times 10^{-12}) \times 10^6$$
$$= \mathbf{36 \times 10^{-6}\ C/m^2}$$

and
$$P = (9 \times 10^{-12}) \times (4-1) \times 10^6 = \mathbf{27 \times 10^{-6}\ C/m^2}$$

EXAMPLE 5. *A parallel plate capacitor consists of free space and two dielectrics of relative dielectric constants* $k = 3$ *and* $k = 4$ *respectively. If the plates of the capacitor have a charge density of* $\rho_S = 25 \times 10^{-6}$ C/m², *find the electric displacement, electric field intensity and polarisation in the three regions.*

Solution. The situation is shown in Fig. (16).

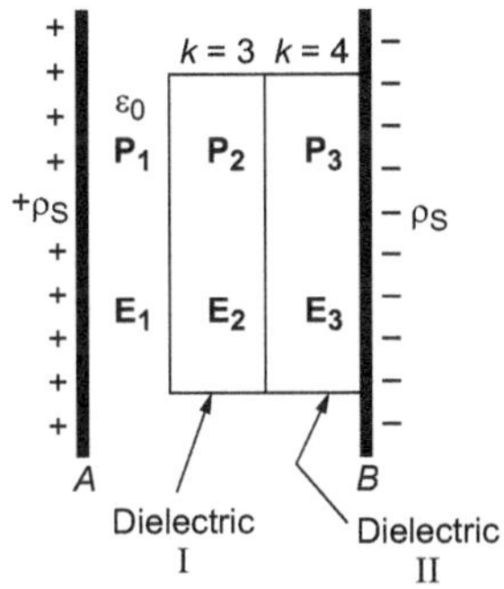

Fig. (16) Parallel plate capacitor with dielectrics

For free space,

Electric displacement, $D_1 = \rho_S = \mathbf{25 \times 10^{-6}\ C/m^2}$

Electric field intensity, $E_1 = \dfrac{\rho_S}{\varepsilon_0} = \mathbf{2.824 \times 10^6\ V/m}$

Polarisation, $P_1 = \mathbf{0}$

For dielectric I medium, $D_2 = \rho_S = \mathbf{25 \times 10^{-6}\ C/m^2}$

$$E_2 = \frac{D}{\varepsilon_0 k} = \frac{25\times10^{-6}}{3\,\varepsilon_0} = \mathbf{0.94 \times 10^6\ V/m}$$

$$P_2 = \varepsilon_0(k-1)E_2 = \mathbf{16.66 \times 10^{-6}\ C/m^2}$$

For dielectric II medium,

$$D_3 = \rho_S = \mathbf{25 \times 10^{-6}\ C/m^2}$$

$$E_3 = \frac{D}{\varepsilon} = \frac{D}{k\varepsilon_0} = \frac{25\times10^{-6}}{4\,\varepsilon_0} = \mathbf{0.706 \times 10^6\ V/m}$$

$$P_3 = \varepsilon_0(k-1)E_3 = \varepsilon_0(4-1)0.70 \times 10^6 = \mathbf{18.753 \times 10^{-6}\ C/m^2}$$

EXAMPLE 6. *The electric susceptibility of a material is* 36×10^{-12} C/N-m^2. *Calculate the value of dielectric constant and absolute permittivity of the material. Take* $\varepsilon_0 = 9 \times 10^{-12}$ F/m.

Solution. We know that $k = 1 + \dfrac{\chi}{\varepsilon_0}$ and $\varepsilon = k\varepsilon_0$

Given that $\chi = 36 \times 10^{-12}$ C/N-m^2 and $\varepsilon_0 = 9 \times 10^{-12}$ F/m

$\therefore$ $k = 1 + \dfrac{36\times10^{-12}}{9\times10^{-12}} = 1 + 4 = 5$

and $\varepsilon = 5 \times (9 \times 10^{-12}) = \mathbf{45 \times 10^{-12}\ F/m}$.

EXAMPLE 7. *The electric susceptibility of a material is* 35.4×10^{-12} coul^2/nt-m^2. *What are the values of dielectric coefficient and the permittivity of the material?*

Solution. The dielectric constant k of the material is related to the electric susceptibility χ by

$$k = 1+\frac{\chi}{\varepsilon_0} = 1+\frac{35.4\times10^{-12}}{8.85\times10^{-12}}$$
$$= 1 + 4 = \mathbf{5}$$

The permittivity is given by

$$\varepsilon = \varepsilon_0\, k = (8.85 \times 10^{-12}) \times 5$$
$$= \mathbf{44.3 \times 10^{-12}\ coul^2/nt\text{-}m^2}.$$

3.17 CAPACITOR AND CAPACITANCE

Capacitor: A capacitor is a device that stores electrical energy in form of electric field. It is a passive component. A capacitor consists of two electrical conductors in form of plates or surface separated by dielectric medium.

Capacitance: ***Thus, the capacitance is the ratio of the magnitude of charge q on either conductor to the potential V between them. When $V = 1$, then $C = q$. Hence, then capacity of a conductor may be defined as the charge required to raise its potential by unity.***

Unity of Capacitance: The unit of charge q is coulomb and the unit potential V is volt. Therefore, the unit of capacitance will be *coulomb/volt.* commonly referred. *Hence, the capacitance of a conductor is one farad if on giving a charge of* 1 *coulomb its potential is raised by* 1 *volt.* The farad is too large for practical purposes and hence, two other units are generally used.

$$1 \text{ mirco-farad } (\mu F) = 10^{-6} \text{ farad}$$

and 1 micro-micro farad or pico-farad ($\mu\,\mu$F or pF) = $\mathbf{10^{-12}}$ **farad**.

When n capacitors of capacitances $C_1, C_2, C_3, \ldots, C_n$ are **connected in series,** then the resultant capacitance C is given by

$$\frac{1}{C} = \frac{1}{C_1}+\frac{1}{C_2}+\frac{1}{C_3}+\ldots+\frac{1}{C_n}$$

So, the reciprocal of the effective capacitance of a number of capacitors connected in series is the sum of the reciprocal of their individual capacitances.

When n capacitors of capacitances $C_1, C_2, C_3, \ldots, C_n$ are **connected in parallel,** then the resultant capacitance C is given by

$$C = C_1 + C_2 + C_3 + \cdots + C_n$$

So, the effective capacitance of a number of capacitors connected in parallel is the sum of their individual capacitances.

3.18 CAPACITANCE OF PARALLEL PLATE CAPACITOR

As shown in Fig. (17), a parallel plate capacitor consists of two parallel conducting plates of area A and separated by a distance d. When these two plates are connected to the terminals of a battery, let a charge $+q$ appears on the one plate while a charge $-q$ on the other plate. If d is small compared with the dimensions of the plate, then electric field strength $\mathbf{E}$ between the plates

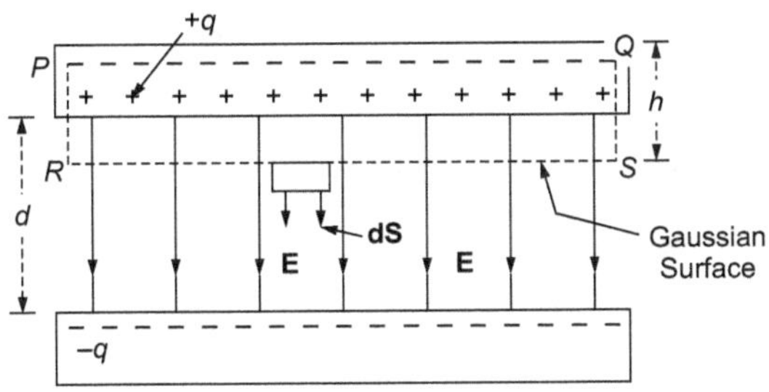

Fig. (17)

is uniform, *i.e.*, the lines of force are parallel and equally spaced. Here, we take the medium as vacuum or air between the plates.

In order to calculate the capacitance, we first calculate the value of electric field **E** between the plates by using Gauss's law. For this purpose, we consider a Gaussian surface *PQRS* in the form of a closed cylinder of height h and area of cross-section A as shown in Fig. (17). The flux through the ends *PR* and *QS* (cylindrical surface) of Gaussian surface is zero because **E** and $d\mathbf{S}$ are perpendicular to each other. The flux through the wall *PQ* of the Gaussian surface is also zero because **E** (electric field) inside a conductor carrying a static charge is zero. The only contribution to the flux is due to the surface *RS*. So, the flux through the entire Gaussian surface is given by

$$\Phi_E = \oint \mathbf{E} \cdot d\mathbf{S} = \oint E\, dS \cos\theta$$

or $$\Phi_E = E \oint dS = E A \qquad (\because \ E \text{ is constant})$$

Applying the Gauss's law

$$E A = \frac{q}{\varepsilon_0}$$

or $$E = \frac{q}{\varepsilon_0 A} \qquad \text{...(1)}$$

The potential difference V between the plates can be written as

$$V = E\, d = \frac{q\, d}{\varepsilon_0 A} \qquad \text{...(2)}$$

The capacity of a parallel plate condenser is, therefore,

$$C = \frac{q}{V} = \frac{q}{\left(\dfrac{q\, d}{\varepsilon_0 A}\right)}$$

$\therefore$ $$\boxed{C = \frac{\varepsilon_0 A}{d}} \qquad \text{...(3)}$$

Eq. (3) gives the capacity of a parallel plate capacitor.

3.19 CAPACITANCE OF A SPHERICAL CAPACITOR (Capacitance of Concentric Spheres)

A spherical capacitor consists of an inner metal sphere A of radius a surrounded by an outer metal sphere B of radius b as shown in Fig. (18). Suppose the inner sphere A is given a charge $+q$ while the outer sphere B is given a charge $-q$. These two charges produce an electric field **E** between the space of two spheres. The direction of **E** will be radially outwards as shown in the figure. Our aim is to calculate the capacitance of this spherical capacitor. For this purpose we consider a concentric spherical Gaussian surface of radius r. The flux Φ_E through the Gaussian surface is given by

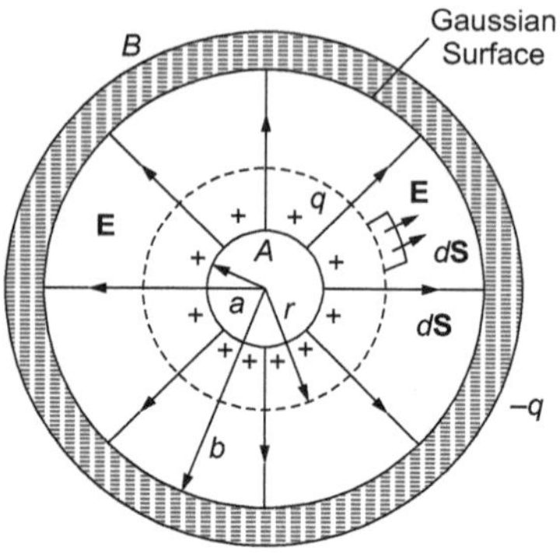

Fig. (18)

$$\Phi_E = \oint \mathbf{E} \cdot d\mathbf{S} = \oint \mathbf{E}\, dS \cos\theta$$

$$= E \oint dS = E(4\,\pi r^2)$$

According to Gauss's law

$$E(4\,\pi r^2) = \frac{q}{\varepsilon_0}$$

or
$$E = \frac{1}{4\,\pi\varepsilon_0} \cdot \frac{q}{r^2} \qquad ...(1)$$

We know that
$$E = -\frac{dV}{dr} \quad \text{or} \quad V = -\int E\, dr$$

i.e., the potential difference between the two sphere is equal to the work done in transferring a unit positive charge from outer sphere to inner sphere. Thus,

$$V = V_A - V_B = -\int_b^a E\, dr \qquad ...(2)$$

Substituting the value E from eq. (1) in eq. (2), we get

$$V = -\int_b^a \frac{1}{4\,\pi\varepsilon_0} \cdot \frac{q}{r^2} \cdot dr = -\frac{q}{4\,\pi\varepsilon_0} \int_b^a \frac{dr}{r^2}$$

$$= -\frac{q}{4\,\pi\varepsilon_0}\left[-\frac{1}{r}\right]_b^a = -\frac{q}{4\,\pi\varepsilon_0}\left[-\frac{1}{a}+\frac{1}{b}\right]$$

$$= -\frac{q}{4\,\pi\varepsilon_0}\left[\frac{1}{a}-\frac{1}{b}\right] = \frac{q}{4\,\pi\varepsilon_0}\,\frac{(b-a)}{ab} \qquad ...(3)$$

So,
$$\boxed{C = 4\,\pi\varepsilon_0 \frac{ab}{(b-a)}} \qquad \left(\because C = \frac{q}{V}\right) \qquad ...(4)$$

This is expression for the capacitance of a spherical condenser.

Capacitance of Isolated Conducting Sphere

When the outer surface is of infinite radius, *i.e.*, $b = \infty$, then the sphere left is known as isolated conducting sphere.

Now from eq. (4), we have

$$C = 4\pi\varepsilon_0 \frac{ab}{(b-a)} = \frac{4\pi\varepsilon_0}{\left(\frac{1}{a}-\frac{1}{b}\right)} = \frac{4\pi\varepsilon_0}{\left(\frac{1}{a}-\frac{1}{\infty}\right)}$$

or $$\boxed{C = (4\pi\varepsilon_0)a} \qquad ...(5)$$

Eq. (5) gives the capacitance of an isolated sphere.

3.20 CAPACITANCE OF CYLINDRICAL CAPACITOR

A cylindrical capacitor consists of two coaxial cylinders of radius a and b with length l as shown in Fig. (19). The cylinders are made of metallic material. The inner cylinder is given a charge $+q$ while the outer is given a charge $-q$. In order to calculate the capacitance of this condenser, we consider a coaxial cylindrical Gaussian surface of radius r.

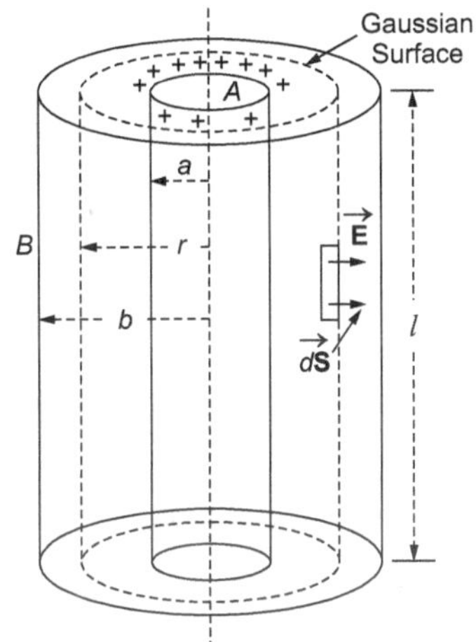

Fig. (19)

From Gauss's law $$\oint \mathbf{E}\cdot d\mathbf{S} = \frac{q}{\varepsilon_0}$$

$$E(2\pi rl) = \frac{q}{\varepsilon_0}$$

or $$E = \frac{q}{2\pi\varepsilon_0 rl} \qquad ...(1)$$

Now, the potential difference V between the two cylinders is equal to the workdone in moving a unit positive charge from the outer cylinder to the inner cylinder, *i.e.*,

$$V = -\int_b^a E\,dr \qquad ...(2)$$

Substituting the value of E from eq. (1) in eq. (2), we get

$$V = -\frac{q}{2\pi\varepsilon_0 l}\int_b^a \frac{dr}{r} = -\frac{q}{2\pi\varepsilon_0 l}[\log_e r]_b^a$$

$$= -\frac{q}{2\pi\varepsilon_0 l}(\log_e a - \log_e b) = \frac{q}{2\pi\varepsilon_0 l}\log_e\left(\frac{b}{a}\right) \qquad ...(3)$$

Hence, the capacitance of the cylindrical capacitance is

$$\boxed{C = \frac{2\pi\varepsilon_0 l}{\log_e\left(\frac{b}{a}\right)}} \qquad \left(\because C = \frac{q}{V}\right) \qquad ...(4)$$

3.21 PARALLEL PLATE CAPACITOR WITH DIELECTRIC SLAB

Fig. (20) shows a capacitor partially filled with a dielectric material.

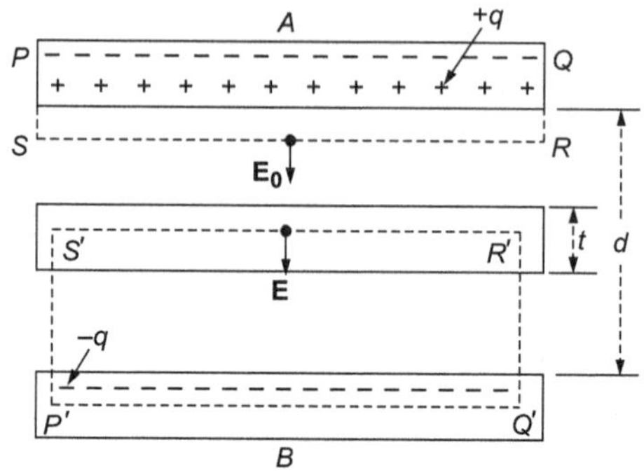

Fig. (20)

Let A and B be the plates of the capacitor with area A and separation d. The upper plate A is given a charge $+q$ while the lower plate B is given a charge $-q$. Now, a dielectric slab with thickness t and dielectric constant k is introduced between the plates. In order to consider the electric field $\mathbf{E}_0$ and $\mathbf{E}$ in air and dielectric respectively, we consider the Gaussian surfaces $PQRS$ and $P'Q'R'S'$ as shown in figure.

$$\int \mathbf{E}_0 \cdot d\mathbf{S} = \frac{q}{\varepsilon_0} \quad \text{or} \quad E_0 A = \frac{q}{\varepsilon_0}$$

$$\because \qquad E_0 = \frac{q}{\varepsilon_0 A} \qquad ...(1)$$

Similarly, for the Gaussian surface $P'Q'R'S'$, we have

$$\int k\mathbf{E} \cdot d\mathbf{S} = \frac{q}{\varepsilon_0} \quad \text{or} \quad kEA = \frac{q}{\varepsilon_0}$$

$$\therefore \qquad E = \frac{q}{k\varepsilon_0 A} \qquad ...(2)$$

The potential difference V between the two plates is the workdone in carrying a unit charge from one plate to another. Thus,

$$V = E_0(d-t) + Et \qquad ...(3)$$

(Here, the electric field E_0 is over a distance $(d-t)$ while the field E is over a distance t)

Substituting the value of E_0 and E from eqns. (1) and (2) in eqn. (3), we have

$$V = \frac{q}{\varepsilon_0 A}(d-t) + \frac{q}{k\varepsilon_0 A} \cdot t$$

$$= \frac{q}{\varepsilon_0 A}\left[(d-t) + \frac{1}{k}\right] = \frac{q}{\varepsilon_0 A}\left[d - t\left(1 - \frac{1}{k}\right)\right] \qquad ...(4)$$

The capacitance of the capacitor is given by

$$C = \frac{q}{V} = \frac{q}{\frac{q}{\varepsilon_0 A}\left[d - t\left(1 - \frac{1}{k}\right)\right]} = \frac{\varepsilon_0 A}{\left[d - t\left(1 - \frac{1}{k}\right)\right]}$$

$$\therefore \qquad \boxed{C = \frac{\varepsilon_0 Ak}{[kd - t(k-1)]}} \qquad \text{...(5)}$$

When the dielectric is completely filled, then $t = d$. In this case, the capacitance is given by

$$C = \frac{\varepsilon_0 Ak}{[kt - t(k-1)]} = \frac{\varepsilon_0 Ak}{t} \qquad \text{...(6)}$$

3.22 ENERGY STORED IN A CAPACITOR

The energy of a charged capacitor is stored inform of electric field. A charged capacitor stores an electric potential energy in it.

This energy is analogous to the potential energy stored in a compressed spring. The energy can be recovered back when the capacitor is allowed to discharge. When the charging of the capacitor is done by a battery, then the electrical energy stored in it is at the expense of chemical energy of the battery.

Consider a capacitor of capacitance C and carrying a charge q at any instant. Let the potential difference between the plates is V. Then

$$V = \frac{q}{C} \qquad \text{...(1)}$$

If an additional charge dq is to be given to this capacitor, then some work must be done against the potential difference.

So, workdone in increasing charge by dq is given by

$$dW = Vdq = \left(\frac{q}{C}\right)dq \qquad \text{...(2)}$$

$\therefore$ Total work to charge a capacitor to a charge q_0,

$$W = \int dW = \int_0^{q_0}\left(\frac{q}{C}\right)dq = \frac{q_0^2}{2C} \qquad \text{...(3)}$$

Now, the energy stored by a charged capacitor

$$W_E = U = W = \frac{1}{2}\frac{q^2}{C} = \frac{1}{2}CV^2 \qquad (\because\ q = CV) \quad \text{...(4)}$$

For a parallel plate capacitor of area A and plate separation d, the capacitance C is given by

$$C = \frac{\varepsilon_0 A}{d} \quad \text{and} \quad V = Ed$$

$$\therefore \text{ Energy stored} \qquad U = \frac{1}{2} + \frac{\varepsilon_0 A}{d} + E^2 d^2 = \frac{1}{2}\varepsilon_0 E^2\ Ad \text{ joules} \qquad \text{...(5)}$$

$$\text{Energy stored per unit volume} = \frac{1}{2}\varepsilon_0 E^2 \text{ joules/m}^2 \qquad \text{...(6)}$$

If the capacitor is placed in a medium of dielectric constant k, then the energy stored per unit volume is given by

$$\therefore \qquad W_E = U = \frac{1}{2}k\varepsilon_0 E^2 = \frac{1}{2}\varepsilon E^2$$

$$\therefore \qquad \boxed{W_E = U = \frac{\varepsilon E^2}{2}} \qquad \text{...(7)}$$

where ε is the permittivity of the medium.

EXAMPLE 1. *Calculate the capacitance of a spherical capacitor filled with material of dielectric constant* 2.5 *if the outer sphere of radius* 0.75 cm *is earthed. The diameter of inner sphere is* 0.5 cm.

Solution. The capacity of a spherical capacitor filled with dielectric and outer surface earthed is given by

$$C = 4\pi\varepsilon_0 k\left(\frac{ab}{b-a}\right)$$

Given that $k = 2.5$, $a = \frac{0.5}{2}$ cm $= 0.25 \times 10^{-2}$ m and $b = 0.75$ cm $= 0.75 \times 10^{-2}$ m

Substituting these values, we get

$$C = \frac{1}{9\times10^9}\times 2.5\times\left[\frac{(0.25\times10^{-2})(0.75\times10^{-2})}{(0.75\times10^{-2}-0.25\times10^{-2})}\right]$$

$$= \mathbf{1.04 \times 10^{-12}\ F}$$

EXAMPLE 2. *A parallel plate capacitor has circular plates of* 8 cm *radius and* 1 mm *separation. If the potential difference across the plates is* 100 V, *find the charge across the plates.* $\varepsilon_0 = 8.9 \times 10^{-12}$ C²/n-m².

Solution. The capacity of a parallel plate capacitor

$$C = \frac{A\varepsilon_0}{d}$$

∴ Charge on plate

$$q = \frac{A\varepsilon_0}{d}\times V \qquad (\because\ q = CV)$$

Given that,

$$r = 8 \times 10^{-2}\ \text{m}$$

$$\therefore \quad A = \pi r^2 = \pi(8 \times 10^{-2})^2 = \pi \times (64 \times 10^{-4})\text{m}^2$$

$$d = 1\ \text{mm} = 1 \times 10^{-3}\ \text{m} \quad \text{and} \quad V = 100\ \text{V}$$

$$\therefore \quad q = \frac{\pi(64\times10^{-4})(8.9\times10^{-12})(100)}{(10^{-3})}$$

or

$$q = \mathbf{1.79 \times 10^{-8}\ coulomb}.$$

EXAMPLE 3. *The capacity of a parallel condenser is* 0.2 μF *and potential difference between the plates is* 2 volt. *Calculate the energy stored by the charged condenser.*

Solution. Energy stored

$$U = \frac{1}{2}CV^2$$

Given that $C = 0.2$ μF $= 0.2 \times 0^{-6}$ F and $V = 2$ volt.

$$\therefore \quad U = \frac{1}{2}\times(0.2\times10^{-6})\times(2)^2 = \mathbf{4 \times 10^{-7}\ joules}.$$

EXAMPLE 4. *Calculate the energy stored in a spherical capacitor of* 10 cm *radius charged to a potential of* 200 volt.

Solution. The capacity of a spherical capacitor of radius 10 cm is given by

$$C = 4\pi\varepsilon_0 \times a = \frac{0.1}{(9 \times 10^9)} \text{ farad}$$

∴ Energy stored

$$U = \int \mathbf{J} \cdot d\mathbf{S} = \frac{1}{2} \times \frac{0.1}{(9 \times 10^9)} \times (200)^2$$

$$= \mathbf{0.2222 \times 10^{-6}\ joule.}$$

EXAMPLE 5. *Two coaxial cylindrical conductors are in one cable having radii* 5 cm *and* 10 cm *and length of each cylindrical conductor is* 10 km. *What is the capacity of cylindrical capacitor?*

Solution. The capacity of cylindrical capacitor is given by

$$C = \frac{2\pi\varepsilon_0 l}{\log_e\left(\frac{b}{a}\right)} = \frac{2\pi\varepsilon_0 l}{2.303 \log_{10}\left(\frac{b}{a}\right)}$$

Given that, $l = 10 \times 10^3$ m $= 10^4$ m, $a = 5$ cm $= 5 \times 10^{-2}$ m and $b = 10$ cm $= 10 \times 10^{-2}$ m

Again

$$4\pi\varepsilon_0 = \frac{1}{(9 \times 10^9)} \quad \text{or} \quad 2\pi\varepsilon_0 = \frac{1}{2 \times (9 \times 10^9)}$$

∴

$$C = \frac{1}{2 \times (9 \times 10^9)} \times \frac{10^4}{2.303 \log_{10}(2)}$$

$$= \frac{10^4}{2 \times (9 \times 10^9) \times 2.303 \times (0.3)}$$

$$= \mathbf{0.804 \times 10^{-6}\ F}$$

EXAMPLE 6. *Calculate the capacitance and energy stored in a parallel plate capacitor which consists of two metal plates of area* 60 cm^2 *separated by a dielectric of* 1.5 mm *thickness and* $\varepsilon_r = 3.5$ *if a voltage of* 1000 V *is applied across it.*

Solution. We know that the capacity and energy stored in a parallel plate capacitor are given by respectively as

$$C = \frac{kA\varepsilon_0}{d} \quad \text{and} \quad U = \frac{1}{2}CV^2$$

Given that, $A = 60 \times 10^{-4}$ m^2, $k = \varepsilon_r = 3.5$, $d = 1.5 \times 10^{-3}$ m and $V = 1000$ volt.

∴

$$C = \frac{(3.5)(60 \times 10^{-4})(8.85 \times 10^{-12})}{1.5 \times 10^{-3}} = \mathbf{1239 \times 10^{-6}\ F = 1239\ \mu F}$$

and

$$U = \frac{1}{2} \times (1239 \times 10^{-6}) \times (1000)^2 = \mathbf{6.145 \times 10^{-5}\ J.}$$

EXAMPLE 7. *(a) Two dielectric slabs of dielectric constants k_1 and k_2 have been filled in between the plates of a capacitor as shown in Fig. (22a and b). What will be the capacitor in each case?*

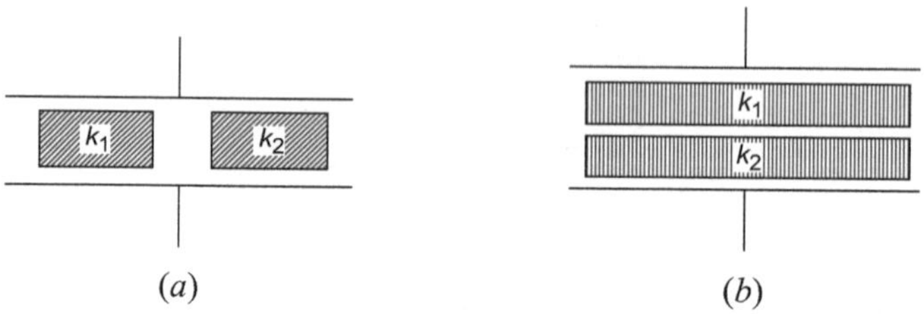

Fig. (22)

(b) A capacitor is filled with the dielectric of the dimensions but dielectric constant 2 and 3 respectively. Find the ratio of the capacitances in the two possible arrangements.

Solution.

(a) Let A be the area of each plate of the capacitor and d and the distance between the two plates. If the capacitance be C_1 and C_2 respectively, then

$$C_1 = \varepsilon_0 \frac{k_1 \cdot \frac{A}{2}}{d} \quad \text{and } C_2 = \varepsilon_0 \frac{k_2 \cdot \frac{A}{2}}{d}$$

Let C be the equivalent capacitance, then

$$C = C_1 + C_2$$

$$= \varepsilon_0 \frac{k_1 A}{2d} + \varepsilon_0 \frac{k_2 A}{2d} \qquad \{\because \text{ Two condensers are in parallel}\}$$

$$= \frac{\varepsilon_0 A}{2d}(k_1 + k_2) \qquad \text{...(1)}$$

The arrangement shown in Fig. (22*b*) is equivalent to two capacitors joined in series. Let their capacitances be C_1 and C_2 respectively. Then

$$C_1 = \varepsilon_0 \frac{k_1 A}{d/2} \quad \text{and} \quad C_2 = \varepsilon_0 \frac{k_2 A}{d/2}$$

Now,

$$\frac{1}{C} = \frac{1}{C_1} + \frac{1}{C_2} = \frac{d}{2\varepsilon_0 k_1 A} + \frac{d}{2\varepsilon_0 k_2 A} = \frac{d}{2\varepsilon_0 A}\left[\frac{k_2 + k_1}{k_1 k_2}\right]$$

$$C = \frac{2\varepsilon_0 A}{d}\left[\frac{k_1 k_2}{k_1 + k_2}\right] \qquad \text{...(2)}$$

(b) The ratio of capacitances in the two possible arrangements is given {from eqs. (1) and (2)},

$$= \frac{\frac{\varepsilon_0 A}{2d}(k_1 + k_2)}{\frac{2\varepsilon_0 A}{d}\left(\frac{k_1 k_2}{k_1 + k_2}\right)} = \frac{(k_1 + k_2)^2}{4 k_1 k_2} = \frac{(2+3)^2}{4 \times 2 \times 3} = \frac{25}{24} = \mathbf{25 : 24.}$$

EXAMPLE 8. *A parallel plate condenser consists of two metal plates of area A and separation d. A slab of thickness t and dielectric constant k is inserted between the plates with its faces parallel to the plates and having the same surface area as that of the plates. Find the capacitance of the system. If k = 2, for what value of t/d will the capacitance of the system be 3/2 times that of the air condenser alone? Calculate the energy in the two cases and account for the energy change.*

Solution. We know that the capacitance of a capacitor with a dielectric of dielectric constant k is given by $\frac{\varepsilon_0 kA}{d}$. Hence, the capacitance C_1 of the capacitor with dielectric constant k is given by

$$C_1 = \frac{\varepsilon_0 kA}{t}$$

where t is its thickness.

The capacitance C_2 of remaining capacitor is

$$C_2 = \frac{\varepsilon_0 A}{(d-t)}$$

Effective capacitance C of C_1 and C_2 is given by

$$C = \frac{C_1 C_2}{C_1 + C_2} = \frac{\varepsilon_0^2 A^2 k/t(d-t)}{\frac{\varepsilon_0 kA}{t} + \frac{\varepsilon_0 A}{(d-t)}}$$

On simplification, we have

$$C = \frac{\varepsilon_0 A}{d - \left(\frac{t}{2}\right)}$$

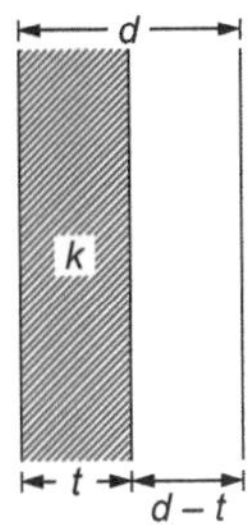

Fig. 21

The capacitance C_a of air condenser with thickness d is

$$C_a = \frac{\varepsilon_0 A}{d}$$

Now, $$C = \frac{3}{2} C_a$$

or $$\frac{\varepsilon_0 A}{d - \left(\frac{t}{2}\right)} = \frac{3}{2}\frac{\varepsilon_0 A}{d}$$

Solving we have $$\frac{t}{d} = \frac{2}{3}$$

If q = charge in the condenser (remains unchanged). Then initial energy E_i in the air condenser = $\left(\frac{q^2}{2\,C_a}\right)$ and final energy E_f after introducing dielectric = $\left(\frac{1 \cdot q^2}{2\,C}\right)$.

$$\therefore \quad \frac{F_i}{E_f} = \frac{3}{2}$$

When a dielectric is introduced, it decreases the potential energy of the condenser. This loss is used up to polarise the dielectric.

EXAMPLE 9. *A parallel plate capacitor shown in Fig. (23) has three dielectric layers. Area of each plate is* 20 cm^2. *Find (a) capacitance of the system and (b) percentage of total stored energy located in each of three regions.*

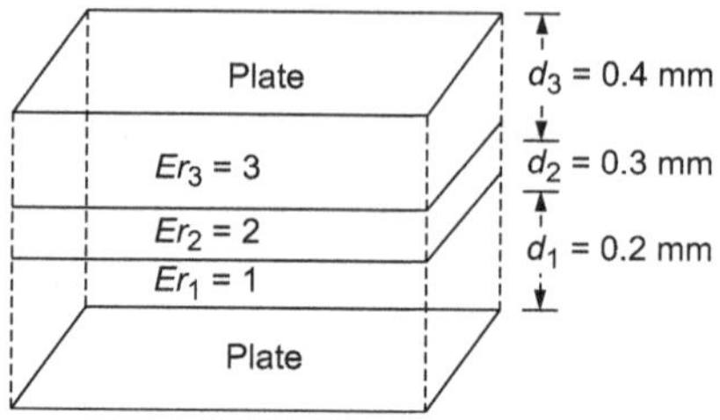

Fig. (23) Parallel plate capacitor with three dielectrics

Solution.

(*a*) The capacitor is a combination of three capacitors in series. Le the capacitance be C_1, C_2 and C_3 respectively. We have

$$C_1 = \frac{\varepsilon_0 \varepsilon_{r_1} A}{d} = \frac{\varepsilon_0 (1)(20 \times 10^{-4})}{0.2 \times 10^{-3}} = 10\ \varepsilon_0 \text{ farad}$$

$$C_2 = \frac{\varepsilon_0 (2)(20 \times 10^{-4})}{0.3 \times 10^{-3}} = 13.33\ \varepsilon_0 \text{ farad}$$

and
$$C_3 = \frac{\varepsilon_0 (3)(20 \times 10^{-4})}{0.3 \times 10^{-3}} = 15\ \varepsilon_0 \text{ farad}$$

The equivalent capacitance C is given by

$$\frac{1}{C} = \frac{1}{C_1} + \frac{1}{C_2} + \frac{1}{C_3}$$

$$\therefore \quad C = \frac{C_1 C_2 C_3}{C_1 C_2 + C_2 C_3 + C_3 C_1}$$

$$= \frac{(10\ \varepsilon_0)(13.33\ \varepsilon_0)(15\ \varepsilon_0)}{(10\ \varepsilon_0)(13.33\ \varepsilon_0) + (13.33\ \varepsilon_0)(15\ \varepsilon_0) + (15\ \varepsilon_0)(10\ \varepsilon_0)}$$

Substituting the value of ε_0 as 8.85×10^{-12} and solving, we get

$$C = \mathbf{36.636\ pF}$$

(*b*) We know that
$$U = \frac{1}{2} C V^2 = \frac{1}{2} \frac{q^2}{C}$$

$$\therefore \quad \frac{U_1}{U} = \frac{C}{C_1} = 0.41377 = \mathbf{41.377\%}$$

So, percentage of total shared energy located in region one is 41.377%.

Similarly,
$$\frac{U_2}{U} = \frac{C}{C_2} = 0.3103 = \mathbf{31.03\%}$$

So, percentage of total shared energy located in region second is 31.03%.

Further, $$\frac{U_3}{U} = \frac{C}{C_3} = 0.2785 = \mathbf{27.85\%}$$

So, percentage of total shared energy located in region third is 27.85%.

EXAMPLE 10. *Find the relative permittivity of a dielectric material present in a parallel-plate capacitor if (a)* $A = 0.12\ \text{m}^2$, $d = 80\ \mu\text{m}$, $V_0 = 12$ V *and capacitor contains* 1 μJ *of energy, (b) stored energy density is* 100 J/m³, $V_0 = 200$ V *and* $d = 45\ \mu\text{m}$.

Solution. (*a*) The energy stored U in a capacitor is given by

$$U = \frac{1}{2}CV_0^2 = \frac{1}{2}\left(\frac{\varepsilon_0\varepsilon_r A}{d}\right)V_0^2$$

$$\therefore \quad 1 \times 10^{-6}\ \text{J} = \frac{1}{2}\left(\frac{8.854\times10^{-12}\ \varepsilon_r \times 0.12}{80\times10^{-6}}\right)(12)^2$$

or $$\varepsilon_r = \frac{2\times(80\times10^{-6})\times(1\times10^{-6})}{(8.854\times10^{-12})\times(0.12)\times(12)^2} = \mathbf{1.045}.$$

(*b*) Given that stored energy density is 100 J/m³.

$$\therefore \quad 100 \times A \times d = \frac{1}{2}\left(\frac{\varepsilon_0\varepsilon_r A}{d}\right)V_0^2$$

or $$\varepsilon_r = \frac{100\times2\times(45\times10^{-6})^2}{(8.85\times10^{-12})\times(200)^2} = \mathbf{1.14}.$$

Exercises & Problems

EXERCISES

1. Define the current density and derive the expression for convection current density.
2. Explain the properties of a conductor.
3. Explain boundary conditions for conductor to conductor boundary.
4. What is a dielectric? Define dielectric constant.
5. Explain the polarization of a dielectric material.
6. Distinguish between polar and non-polar dielectric materials.
7. State and prove Gauss's law in dielectrics.
8. Explain the terms **E**, **D** and **P** in dielectric materials.
9. Obtain the different relations between **E**, **D** and **P**.
10. Discuss the boundary conditions or boundary relations at the boundary of two dielectrics in the following way:
    (*i*) electric field tangent to the boundary of two dielectrics.
    (*ii*) electric field normal to the boundary of two dielectrics.
11. Show that the normal component of electric displacement **D** and tangential component of electric field **E** are continuous at the charge free interface between two dielectrics.
12. Derive boundary relations for static electric field in the general form across a common boundary separated by two different perfect dielectric media.

13. What is an electrical capacitor? Define capacity of a condenser.

14. Obtain an expression for the capacitance of a parallel plate capacitor.

15. Find expression for the capacitance of a spherical concentric shells of radii r_1 and r_2 with $r_2 > r_1$.

16. A cylindrical capacitor consists of two coaxial cylinders of radii r_1 and r_2 and length l. Show that its capacitance is

$$C = \frac{2\pi\varepsilon_0 l}{\log_e\left(\frac{r_2}{r_1}\right)}$$

17. A dielectric slab of thickness t and dielectric constant k is introduced between the plates of a parallel plate capacitor of plate area A and separation d. Find its capacitance.

18. Show that energy stored in a capacitor is $\frac{1}{2} \cdot CV^2$ where C is capacitance and V is potential difference.

PRACTICE PROBLEMS

1. A certain current density

$$\mathbf{J} = \frac{100 \cos\theta}{r^2 + 1} \hat{\mathbf{a}}_r \text{ A/m}^2$$

How much current flows through spherical cap $r = 4, 0 < \theta < \frac{\pi}{6}, 0 < \phi < \pi$?

[**Ans.** 36.84 A]

[**Hint.** $I = \int_S \mathbf{J} \cdot d\mathbf{S}$

Here $J = \frac{100 \cos\theta}{r^2 + 1} \hat{\mathbf{a}}_r$ and $d\mathbf{S} = r^2 \sin\theta \, d\theta \, d\phi \, \hat{\mathbf{a}}_r$

$\therefore$ $I = \int_0^{\pi/6} \int_0^{\pi} \frac{100 \cos\theta}{r^2 + 1} r^2 \sin\theta \, d\theta \, d\phi$]

2. Find the current crossing the portion of the line $y = 0$ plane defined by $-0.1 \le x \le -0.1$ m and $-0.002 \le z \le 0.002$ m, if

$$\mathbf{J} = (10)^2 \, |x| \, \hat{\mathbf{a}}_y \text{ (A/m}^2\text{)}$$

[**Ans.** 4 mA]

[**Hint.** $I = \int \mathbf{J} \cdot d\mathbf{S} = \int_{-0.1}^{0.1} \int_{-0.002}^{0.002} (10)^2 \, |x| \, \hat{\mathbf{a}}_y \cdot dx \, dz \, \hat{\mathbf{a}}_y$]

3. A crystal is subjected to an electric field of 1000 V/m. The resulting polarisation is 4.3×10^{-8} C/m^2. Calculate the relative permittivity of the crystal.

[**Ans.** 5.86]

[**Hint.** $\mathbf{P} = \varepsilon_0 (k - 1)\mathbf{E}$ or $k = 1 + \frac{\mathbf{P}}{\varepsilon_0 \mathbf{E}}$

$\therefore$ $k = 1 + \frac{4.3 \times 10^{-8}}{(8.85 \times 10^{-12})(1000)}$]

4. A certain linear, homogeneous, isotropic dielectric material has a relative permittivity of 1.8 volt. If $V = -4000\,x$ volt in material, find **E**, **D** and **P**.

[**Ans.** $4000\,\hat{\mathbf{a}}_x$ V/m, $63.748 \times 10^{-9}\,\hat{\mathbf{a}}_x$ C/m^2, $28.332 \times 10^{-9}\,\hat{\mathbf{a}}_x$ C/m^2]

[**Hint.** $\mathbf{E} = -\vec{\nabla}\cdot V = -\dfrac{\partial V}{\partial x}\hat{\mathbf{a}}_x - \dfrac{\partial V}{\partial y}\hat{\mathbf{a}}_y - \dfrac{\partial V}{\partial z}\hat{\mathbf{a}}_z = 4000\,\hat{\mathbf{a}}_x$

$\mathbf{D} = k\varepsilon_0\,\mathbf{E} = 1.8 \times (8.54 \times 10^{-12}) \times 4000\,\hat{\mathbf{a}}_x$

and $\mathbf{P} = (k-1)\varepsilon_0\,\mathbf{E} = (1.8 - 1) \times (8.854 \times 10^{-12}) \times (4000\,\hat{\mathbf{a}}_x)$]

5. A dielectric slab of thickness 0.50 cm and dielectric constant 7 is placed between the plates of a parallel capacitor of plate area 100 cm^2 and separation 1 cm. A potential difference of 100 volt is applied with no dielectric present. The battery is then disconnected and dielectric slab is introduced. Find **E**, **D** and **P**.

[**Ans.** 1.43×10^3 V/m, 8.9×10^{-8} C/m^2 and 7.6×10^{-8} C/m^2]

[**Hint.** $E_0 = \dfrac{V_0}{d} = \dfrac{100}{10^{-2}} = 10^4$ V/m

$E = \dfrac{E_0}{R} = \dfrac{10^4}{7} = 1.43 \times 10^3$ V/m

$\mathbf{D} = k\varepsilon_0\,E = 7 \times (8.9 \times 10^{-12})(1.43 \times 10^3)$

$\mathbf{P} = (k-1)\varepsilon_0\,E = (7-1) \times (8.9 \times 10^{-12})(1.43 \times 10^3)$]

6. The capacity of a parallel plate condenser is 0.2 μF. Potential difference between the plates is 2 volt. Calculate the energy stored by charged condenser.

[**Ans.** 0.4×10^6 joule]

[**Hint.** $E = \dfrac{1}{2}CV^2 = \dfrac{1}{2}(0.2 \times 10^{-6}) \times (2)^2$]

7. Determine the capacitance to neutral per m of a single phase line taking into account the effect of the ground. The charge on the line is $+Q$, h is the height of the line above ground and r is the radius of the conductor.

[**Ans.** $8\,\pi\varepsilon h$]

[**Hints.** $F = \dfrac{Q \times (-Q)}{4\,\pi\varepsilon(2h)^2} = \dfrac{-Q^2}{4\,\pi\varepsilon(4h^2)}$

Workdone, $W = -\int_h^\infty \mathbf{F}\cdot d\mathbf{l} = -\dfrac{Q^2}{4\,\pi\varepsilon}\int_\alpha^\infty \dfrac{1}{4x^2}dx$ $(\because\ x = h)$

$$= \frac{Q^2}{4\,\pi\varepsilon \times 4d} \quad ...(1)$$

We know that $W = \dfrac{Q^2}{2C}$ ...(2)

$\therefore$ $\dfrac{Q^2}{2C} = \dfrac{Q^2}{4\,\pi\varepsilon \times 4d}$]

8. Find the capacitance of the system shown in Fig. (24), where the length of each conducting plate is L.

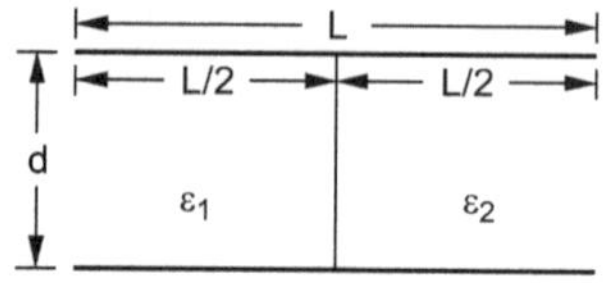

Fig. (24)

[**Ans.** $C = \dfrac{A}{2d}(\varepsilon_1 + \varepsilon_2)$]

[**Hint.** Fig. (24) represents two capacitors connected in parallel. Therefore,

$$C = C_1 + C_2$$

$$= \frac{\varepsilon_1\left(\frac{A}{2}\right)}{d} + \frac{\varepsilon_2\left(\frac{A}{2}\right)}{d} \text{]}$$

9. Calculate the capacitance of a parallel plate capacitor having a mica dielectric of dielectric constant $k = 6$, a plate area of 6.45×10^{-3} m^2 and a separation of 2.54×10^{-4} m.

[**Ans.** 1.349 nF]

[**Hint.** $C = \dfrac{\varepsilon_0 Ak}{d} = \dfrac{(8.9\times10^{-12})(6.45\times10^{-3})(6)}{2.54\times10^{-4}}$]

10. Two capacitors A and B are placed (*i*) in series, (*ii*) in parallel. C_A = 100 μF and C_B = 50 μF. Find the maximum energy stored in the circuit when 240 V, 50 Hz supply is applied.

[**Ans.** (*i*) 1.92 J, (*ii*) 8.65 J]

[**Hint.** In series $\dfrac{1}{C_1} = \dfrac{1}{C_A} + \dfrac{1}{C_B} = \dfrac{1}{100\times10^{-6}} + \dfrac{1}{50\times10^{-6}}$

In parallel, $C_2 = C_1 + C_2 = [(100 \times 10^{-6}) + (50 \times 10^{-6})]$

Energy stored $= \dfrac{1}{2}CV^2$ where $V = 240$ volt]

❑❑❑

4

Poisson's and Laplace's Equations and their Applications

4.1 POISSON'S AND LAPLACE'S EQUATIONS

We studied Gauss's law in chapter 2. This can be used to calculate the electric field intensity and electric flus density. The law is impractical when the charge distribution is not known. The electric field intensity can also be calculated by using $\mathbf{E} = -\vec{\Delta} V$. However, this method is also impracticable when the potential throughout the region is not known. Poisson and Laplace derived the two fundamental differential equations which are applicable in solving those problems which cannot be solved by Gauss's law. These equations provide a powerful method for calculating electric fields and potentials subjected to boundary conditions. The objective of the chapter in derive and apply these equation.

According to Gauss's theorem, the surface integral of flus density **D** over a closed surface is equal to the charge enclosed, *i.e.*,

$$\oint \mathbf{D} \cdot d\mathbf{S} = q \qquad ...(1)$$

If the charge is distributed in the volume V over the surface of which the integral is taken, then

$$\oint \mathbf{D} \cdot d\mathbf{S} = \int_V \rho_V \, dV \qquad ...(2)$$

where ρ_V is volume charge density in C/m^3.

For eqns. (1) and (2), we get

$$\oint \mathbf{D} \cdot d\mathbf{S} = \int_V \rho_V \, dV = q \qquad ...(3)$$

Using Gauss divergence theorem, we have

$$\oint \mathbf{D} \cdot d\mathbf{S} = \int_V \vec{\nabla} \cdot \mathbf{D} \, dV \qquad ...(4)$$

From eqns. (3) and (4), we get

$$\int_V \vec{\nabla} \cdot \mathbf{D} \, dV = \int_V \rho_V \, dV \qquad ...(5)$$

Eq. (5) is valid for all volumes. Therefore,

$$\boxed{\vec{\nabla} \cdot \mathbf{D} = \rho_V} \qquad ...(6)$$

This is Maxwell's first equation.

We know that $\mathbf{D} = \varepsilon\mathbf{E}$ and $\mathbf{E} = -\vec{\nabla}V$

$$\therefore \quad \mathbf{D} = \varepsilon(-\vec{\nabla}V) = -\varepsilon\vec{\nabla}V \qquad ...(7)$$

Substituting the value **D** from eq. (7) in eq. (6), we get

$$\vec{\nabla}\cdot(-\varepsilon\vec{\nabla}V) = \rho_V$$

or

$$-\varepsilon\nabla^2 V = \rho_V$$

or

$$\boxed{\nabla^2 V = -\frac{\rho V}{\varepsilon}} \qquad ...(8)$$

Eq. (8) is known as **Poisson's equation.**

If there is no charge present, then $\rho_V = 0$. Therefore,

$$\boxed{\nabla^2 V = 0} \qquad ...(9)$$

Eq. (9) is known as **Laplace's equation.**

4.2 LAPLACE'S EQUATION IN DIFFERENT COORDINATE SYSTEMS

Substituting the value of $\nabla^2 V$ in different coordinate systems, we have the following forms of Laplace's equation:

(1) Cartesian Coordinates

$$\nabla^2 V = \frac{\partial^2 V}{\partial x^2} + \frac{\partial^2 V}{\partial y^2} + \frac{\partial^2 V}{\partial z^2} = 0 \qquad ...(1)$$

(2) Cylindrical Coordinates

$$\nabla^2 V = \frac{1}{\rho}\frac{\partial}{\partial \rho}\left(\rho\frac{\partial V}{\partial \rho}\right) + \frac{1}{\rho^2}\frac{\partial^2 V}{\partial \phi^2} + \frac{\partial^2 V}{\partial z^2} = 0 \qquad ...(2)$$

(3) Spherical Coordinates

$$\nabla^2 V = \frac{1}{r^2}\frac{\partial}{\partial r}\left(r^2\frac{\partial V}{\partial r}\right) + \frac{1}{r^2\sin\theta}\frac{\partial}{\partial \theta}\left(\sin\theta\frac{\partial V}{\partial \theta}\right) + \frac{1}{r^2\sin^2\theta}\frac{\partial^2 V}{\partial \phi} = 0 \qquad ...(3)$$

4.3 APPLICATIONS

4.3.1 Capacity of Parallel Plate Capacitor using Laplace's Equation

Consider the two conducting plates A and B of a capacitor with area A and placed at $z = 0$ and $z = d$ as shown in Fig. (1). The separation between the plates of d. Let the potential difference between the two plates is V_0.

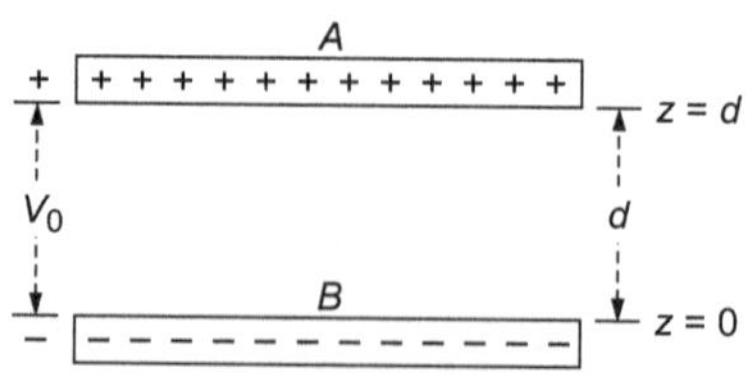

Fig. (1) Parallel plate capacitor

The aim of this article is to find the capacity of parallel plate capacitor using Laplace's equation. The Laplace's equation in Cartesian coordinates is

$$\frac{\partial^2 V}{\partial x^2} + \frac{\partial^2 V}{\partial y^2} + \frac{\partial^2 V}{\partial z^2} = 0$$

The variation of potential is along Z-axis and hence,

$$\frac{\partial^2 V}{\partial z^2} = 0 \qquad ...(1)$$

Integrating eq. (1), we get

$$\frac{\partial V}{\partial z} = A \qquad \text{where } A = \text{constant of integration}$$

or $$\partial V = A\partial z \qquad ...(2)$$

Again integrating eq. (2), we get

$$V = Az + B \qquad ...(3)$$

where B = another constant of integration.

The values of these constants (A and B) can be obtained by applying the given boundary conditions, *i.e.*, $z = 0$, $V = 0$ and $z = d$, $V = V = V_0$. Therefore,

$$0 = 0 + B \quad \text{or} \quad B = 0$$

Now, $$V = Az$$

Further, $$V_0 = Ad \quad \text{or} \quad A = \frac{V_0}{d}$$

Substituting the values A and B in eq. (3), we get

$$V = \frac{V_0}{d} z + 0 = \frac{V_0}{d} z \qquad ...(4)$$

(*i*) The electric field intensity **E** is given by

$$\mathbf{E} = -\vec{\nabla} V = -\left[\frac{\partial V}{\partial x}\hat{\mathbf{a}}_x + \frac{\partial V}{\partial y}\hat{\mathbf{a}}_y + \frac{\partial V}{\partial z}\hat{\mathbf{a}}_z\right]$$

or $$\mathbf{E} = -\frac{\partial V}{\partial z}\hat{\mathbf{a}}_z = -\frac{\partial}{\partial z}\left(\frac{V_0}{d} z\right)\hat{\mathbf{a}}_z = -\frac{V_0}{d}\hat{\mathbf{a}}_z \qquad ...(5)$$

(*ii*) The electric flux density,

$$\mathbf{D} = \varepsilon_0 \mathbf{E}$$

$\therefore$ $$\mathbf{D} = \varepsilon_0\left(-\frac{V_0}{d}\hat{\mathbf{a}}_z\right)$$

or $$D = -\frac{\varepsilon_0 V_0}{d}$$

(*iii*) Therefore, the surface charge density ρ_S is given by

$$\rho_S = D = -\frac{\varepsilon_0 V_0}{d} \qquad ...(6)$$

(*iv*) The value of charge q can be obtained in the following way:

$$q = \int_S \rho_S \, dS = -\int_S \left(\frac{\varepsilon_0 V_0}{d}\right) dS$$

or $$|q| = \frac{\varepsilon_0 V_0}{d} A \qquad \text{where } \int_S dS = A \quad ...(7)$$

(*v*) Now, the capacity C of the condenser is given by

$$C = \frac{\text{Charge}}{\text{Potential}} = \frac{(\varepsilon_0 V_0 A)/d}{V_0}$$

or $$\boxed{C = \frac{\varepsilon_0 A}{d}} \quad ...(8)$$

4.3.2 Capacity of Cylindrical Capacitor using Laplace's Equation

Consider a coaxial cable of inner radius a and outer radius b as shown in Fig. (2). The boundary conditions are: $V = V_0$ at $\rho = a$ and $V = 0$ at $\rho = b$.

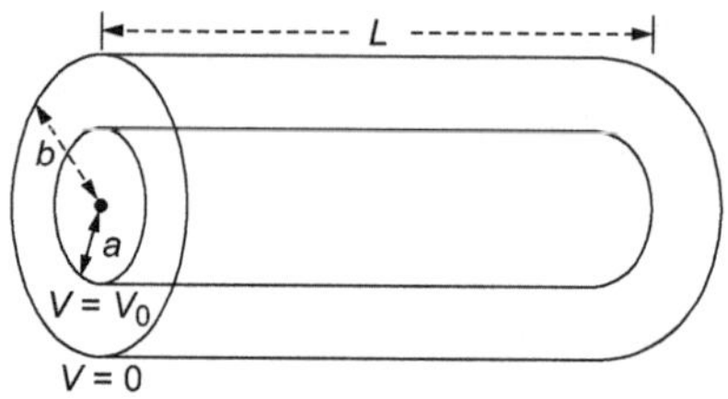

Fig. (2) Coaxial cable

The Laplace's equation in cylindrical coordinates is given by

$$\nabla^2 V = \frac{1}{\rho}\frac{\partial}{\partial \rho}\left(\rho \frac{\partial V}{\partial \rho}\right) + \frac{1}{\rho^2}\frac{\partial^2 V}{\partial \phi^2} + \frac{\partial^2 V}{\partial z^2} = 0$$

Here, the variation of potential is only with respect to ρ. Therefore, the Laplace's equation reduces to

$$\frac{1}{\rho}\frac{\partial}{\partial \rho}\left(\rho \frac{\partial V}{\partial \rho}\right) = 0 \quad ...(1)$$

Integrating eq. (1), we get $\rho \dfrac{\partial V}{\partial \rho} = A$ (A = constant of integration)

Now, $$\frac{\partial V}{\partial \rho} = \frac{A}{\rho}$$

or $$\partial V = \frac{A}{\rho} \partial \rho \quad ...(2)$$

Integrating eq. (2), we get

$$V = A \log_e \rho + B \quad (B = \text{constant of integration}) \quad ...(3)$$

Applying the boundary condition $V = V_0$ at $\rho = a$, we get

$$V_0 = A \log_e a + B \quad ...(4)$$

Similarly, applying boundary condition $V = 0$ at $\rho = b$, we get

$$0 = A \log_e b + B \quad ...(5)$$

From eq. (5),
$$B = -A \log_e b \quad ...(6)$$

Substituting this value in eq. (4), we get

$$V_0 = A \log_e a - A \log_e b = A \log_e\left(\frac{a}{b}\right)$$

$$\therefore \quad A = \frac{V_0}{\log_e\left(\frac{a}{b}\right)} \quad ...(7)$$

Substituting the values of A and B from eqns. (7) and (6) in eqn. (4), we get

$$V = \frac{V_0}{\log_e\left(\frac{a}{b}\right)}\log_e \rho - \frac{V_0}{\log_e\left(\frac{a}{b}\right)}\log_e b$$

or
$$V = \frac{V_0}{\log_e\left(\frac{a}{b}\right)}[\log_e \rho - \log_e b]$$

or
$$V = \frac{V_0}{\log_e\left(\frac{a}{b}\right)}\log_e\left(\frac{\rho}{b}\right) \quad ...(8)$$

The value of capacitance can be obtained by proceeding in the following steps:

(*i*) The electric field intensity is given by

$$\mathbf{E} = -\vec{\nabla} V = -\frac{\partial V}{\partial \rho}\hat{\mathbf{a}}_\rho = -\frac{\partial}{\partial \rho}\left[\frac{V_0 \log_e\left(\frac{\rho}{b}\right)}{\log_e\left(\frac{a}{b}\right)}\right]\hat{\mathbf{a}}_\rho$$

$$= -\frac{V_0}{\log_e\left(\frac{a}{b}\right)}\frac{\partial}{\partial \rho}\left[\log_e\left(\frac{\rho}{b}\right)\right]\hat{\mathbf{a}}_\rho = -\frac{V_0}{\log_e\left(\frac{a}{b}\right)}\left(\frac{b}{\rho}\times\frac{1}{b}\right)\hat{\mathbf{a}}_\rho$$

$$= -\frac{1}{\rho}\cdot\frac{V_0}{\log_e\left(\frac{a}{b}\right)}\hat{\mathbf{a}}_\rho$$

or
$$E = \frac{1}{\rho}\frac{V_0}{\log_e\left(\frac{b}{a}\right)} \quad ...(9)$$

(*ii*) $D = \varepsilon_0 E = \dfrac{\varepsilon_0 V_0}{\rho \log_e\left(\frac{b}{a}\right)}$ ...(10)

(iii) $\rho_S = D = \dfrac{\varepsilon_0 V_0}{\rho \log_e\left(\dfrac{b}{a}\right)}$ ...(11)

(iv) $q = \int_S \rho_S\, dS = \dfrac{\varepsilon_0 V_0}{\rho \log_e\left(\dfrac{b}{a}\right)} \times 2\pi a L = \dfrac{\varepsilon_0 V_0 \times 2\pi a L}{a \log_e\left(\dfrac{b}{a}\right)}$ $(\because \rho = a)$

$$= \frac{2\pi\varepsilon_0 V_0 L}{\log_e\left(\dfrac{b}{a}\right)} \quad ...(12)$$

(v) The capacity is given by

$$C = \frac{q}{V_0} = \frac{2\pi\varepsilon_0 V_0 L}{\log_e\left(\dfrac{b}{a}\right)} \times \frac{1}{V_0}$$

$$\therefore \quad \boxed{C = \frac{2\pi\varepsilon_0 L}{\log_e\left(\dfrac{b}{a}\right)}} \quad ...(13)$$

Eq. (13) gives the capacity of coaxial cable.

4.4 UNIQUENESS THEOREM

There are several methods which can be used to solve problems in electrostatic. ***Any solution of Laplace's or Poisson's equation which satisfies the same boundary conditions must be only solution irrespective of the method used.*** *This is known as uniqueness theorem.* The theorem is important because it leaves us free to use any method.

Let us suppose that a positive point charge q is situated at a distance d from a conducting plane. Our aim is to find the potential at every place on the conducting plane. Let V_1 and V_2 are the two solutions of Laplace's equation and both satisfy the given boundary condition. Now, we have

$$\nabla^2 V_1 = 0 \quad \text{and} \quad \nabla^2 V_2 = 0$$

$$\therefore \quad \nabla^2 (V_1 - V_2) = 0$$

Considering $V_d = (V_1 - V_2)$, we have

$$\nabla^2 V_d = \nabla^2 V_1 - \nabla^2 V_2 = 0$$

or $$V_d = 0 \quad \text{(on the boundary)} \quad ...(1)$$

According to Gauss divergence theorem,

$$\int_V \vec{\nabla} \cdot \mathbf{A}\, dV = \int_S \mathbf{A} \cdot d\mathbf{S} \quad ...(2)$$

Let $\mathbf{A} = V_d \vec{\nabla} V_d$, then we have the following vector identity.

$$(\vec{\nabla} \cdot \mathbf{A}) = \vec{\nabla}(V_d \vec{\nabla} V_d) = V_d \nabla^2 V_d + \vec{\nabla} V_d\, \vec{\nabla} V_d$$

$$= V_d\,(0) + \vec{\nabla} V_d\, \vec{\nabla} V_d \quad (\because \nabla^2 V_d = 0) \quad ...(3)$$

Substituting the value of $(\vec{\nabla}\cdot \mathbf{A})$ from eq. (3) in eq. (2), we get

$$\int_V (\vec{\nabla} V_d \cdot \vec{\nabla} V_d) dV = \int_S V_d \, \vec{\nabla} V_d d\mathbf{S}$$

or $$\int_V (\vec{\nabla} V_d)^2 dV = 0$$

or $$\vec{\nabla} V_d = 0 \qquad (\because \; dV \neq 0)$$

or $$V_d = V_2 - V_1 = \text{constant every where in volume}$$

$\therefore$ $$V_d = 0 \quad \text{or} \quad V_2 = V_1$$

Thus, V_1 and V_2 cannot be different solutions of the same problem.

Solved EXAMPLES

EXAMPLE 1. *Calculate magnitude for V and ρ_V at point P in free space if $V = 5\,\rho^2 \cos 2\phi$ at $P = \left[\rho = 3, \phi = \frac{\pi}{3} \text{ and } z = 2\right]$.*

Solution. The potential at point P is given by

$$V_p = 5(3)^2 \cos\left(2 \times \frac{\pi}{3}\right) = 5 \times 9 \times \left(-\frac{1}{2}\right) = \mathbf{-22.5 \text{ volt}}.$$

Further, $$\nabla^2 V = -\frac{\rho_V}{\varepsilon_0}$$

or $$\frac{1}{\rho}\frac{\partial}{\partial \rho}\left(\rho \frac{\partial V}{\partial \rho}\right) + \frac{1}{\rho^2}\frac{\partial^2 V}{\partial \phi^2} + \frac{\partial^2 V}{\partial z^2} = -\frac{\rho_V}{\varepsilon_0}$$

Solving we get $$\rho_V = 0.$$

EXAMPLE 2. *Find potential and volume charge density at P(0.5, 1.5, 1) in free space of potential field.*

$$V = (x^2 - y^2 - z^2) \text{ volt}$$

Solution. The potential at point $P(0.5, 1.5, 1)$ is given by

$$V = [(0.5)^2 - (1.5)^2 - (1)^2] = 0.25 - 2.25 - 1 = \mathbf{-3 \text{ volt}}$$

The volume charge density ρ_V can be calculated by applying Poisson's equation

$$\nabla^2 V = -\frac{\rho_V}{\varepsilon_0}$$

or $$\rho_V = -\varepsilon_0 (\nabla^2 V)$$

$\therefore$ $$\rho_V = -\varepsilon_0 \left[\frac{\partial^2 V}{\partial x^2} + \frac{\partial^2 V}{\partial y^2} + \frac{\partial^2 V}{\partial z^2}\right] = -\varepsilon_0 [2 - 2 - 2] = 2\,\varepsilon_0$$

$$= \mathbf{17.708 \times 10^{-12} \ C/m^3}.$$

EXAMPLE 3. *State whether following potential function satisfies Laplace's equation.*

$V = Cr\theta\phi$ *where C = constant.*

Solution. Given that $V = Cr\theta\phi$. Let us calculate the value of $\nabla^2 V$.

$$\nabla^2 V = \frac{1}{r^2}\frac{\partial}{\partial r}\left(r^2\frac{\partial V}{\partial r}\right) + \frac{1}{r^2\sin^2\theta}\frac{\partial}{\partial\theta}\left(\sin\theta\frac{\partial V}{\partial\theta}\right) + \frac{1}{r^2\sin^2\theta}\frac{\partial^2 V}{\partial\phi^2}$$

$$= \frac{1}{r^2}\frac{\partial}{\partial r}\left(r^2\frac{\partial}{\partial r}(Cr\theta\phi)\right) + \frac{1}{r^2\sin^2\theta}\frac{\partial}{\partial\theta}\left(\sin\theta\frac{\partial}{\partial\theta}(Cr\theta\phi)\right) + \frac{1}{r^2\sin^2\theta}\frac{\partial^2}{\partial\phi^2}(Cr\theta\phi)$$

$$= \frac{1}{r^2}\frac{\partial}{\partial r}(r^2(C\theta\phi)) + \frac{1}{r^2\sin\theta}\frac{\partial}{\partial\theta}(\sin\theta\,(Cr\phi)) + \frac{1}{r^2\sin^2\theta}(0)$$

$$= \frac{1}{r^2}(2rC\theta\phi) + \frac{1}{r^2\sin\theta}(\cos\theta Cr\phi) = \frac{2C\theta\phi}{r} + \frac{\cos\theta}{r\sin\theta}\cdot C\phi \neq 0$$

Therefore, the potential does not satisfy the Laplace's equation.

EXAMPLE 4. *Semi-infinite conducting planes $V = 0$ and $V = \frac{\pi}{6}$ are separated by an infinitesimal insulating gap. If $V(\phi = 0)$ and $V\left(\theta = \frac{\pi}{6}\right) = 100$ V. Calculate V and E in the region between the planes.*

Solution. Here, V depends only on ϕ. Therefore, Laplace's equation in cylindrical coordinates has the following form

$$\nabla^2 V = \frac{1}{\rho^2}\frac{d^2V}{d\phi^2} = 0$$

We exclude $\rho = 0$ due to insulating gap, hence

$$\frac{d^2V}{d\phi^2} = 0 \qquad ...(1)$$

Integrating twice this equation, we get

$$V = A\phi + B \qquad ...(2)$$

where A and B are two constants. The values of these constants can be calculated by applying the following boundary conditions:

when $\phi = 0, \quad V = 0$

and when $\phi = \phi_0, \quad V = V_0$

Applying these conditions in eq. (2), we get

$$0 = A \times 0 + B \quad \text{or} \quad B = 0$$

and $$V_0 = A\phi_0 + 0 \quad \text{or} \quad A = \frac{V_0}{\phi_0}$$

Substituting these values in eq. (2), we get

$$V = \frac{V_0}{\phi_0}\phi \qquad ...(3)$$

Now, $$\mathbf{E} = -\vec{\nabla}V = -\frac{1}{\rho}\frac{dV}{d\phi}\hat{\mathbf{a}}_\phi = -\frac{V_0}{\rho\phi_0}\hat{\mathbf{a}}_\phi \quad ...(4)$$

Substituting the given values in eqs. (3) and (4), we get

$$V = \frac{100}{(\pi/6)} \times \phi = \frac{600\,\phi}{\pi}$$

and $$\mathbf{E} = -\frac{600}{\pi\rho}\hat{\mathbf{a}}_\phi$$

EXAMPLE 5. *The potential field at any point in space containing a dielectric material of relative permittivity* 2.1 *is given by* $V = 5x^2y + 3yz^2 + 6xz$ *volt where x, y and z are in metre. Calculate the volume charge density at point P*(2, 5, 3) metre.

Solution. According to Poisson's equation

$$\nabla^2 V = -\frac{\rho_V}{\varepsilon} = -\frac{\rho_V}{\varepsilon_0\varepsilon_r}$$

Given that $V = 5x^2y + 2yz^2 + 6xz$

Let us calculate the value of $\nabla^2 V$ in cartesian coordinate system.

$$\nabla^2 V = \frac{\partial^2 V}{\partial x^2} + \frac{\partial^2 V}{\partial y^2} + \frac{\partial^2 V}{\partial z^2}$$

$$= \frac{\partial^2}{\partial x^2}[5x^2y + 3yz^2 + 6xz] + \frac{\partial^2}{\partial y^2}[5x^2y + 3yz^2 + 6xz] + \frac{\partial^2}{\partial z^2}[5x^2y + 3yz^2 + 6xz]$$

$$= \frac{\partial}{\partial x}(10xy + 6z) + \frac{\partial}{\partial y}(5x^2 + 3z^2) + \frac{\partial}{\partial z}(6yz + 6x)$$

$$= 10y + 0 + 6y = 16y = 16 \times 5 = 80$$

$$\therefore \quad 80 = -\frac{\rho_V}{\varepsilon_0\varepsilon_r}$$

or $$\rho_V = -80\,\varepsilon_0\varepsilon_r = -80 \times (8.854 \times 10^{-12}) \times 2.1 = \mathbf{-1.487 \times 10^{-9}\ C/m^3}.$$

EXAMPLE 6. *Conducting cylinders at* $\rho_1 = 1.6$ cm *and* $\rho_2 = 5$ cm *in free space are held at potential of* 80 V *and* – 40 V, *respectively. Calculate*

(*i*) *V and E at* $\rho = 2$ cm,

(*ii*) *The surface charge at which* $V = 0$,

(*iii*) *Capacitance per metre length between the conducting cylinders.*

Solution. For the given problem

$$\nabla^2 V = \frac{1}{\rho}\frac{\partial}{\partial\rho}\left(\rho\frac{\partial V}{\partial\rho}\right) = 0$$

or $$\frac{\partial}{\partial\rho}\left(\rho\frac{\partial V}{\partial\rho}\right) = 0 \qquad (\because\ \rho \neq 0)$$

Integrating once, we get $\rho\dfrac{\partial V}{\partial \rho} = A$ ...(1)

Integrating once again, $V = A \log_e \rho + B$ ...(2)

where A and B are constants of integrations. The values of A and B can be obtained by applying the given boundary conditions. We have

$$80 = A \log_e (0.016) + B$$

and

$$-40 = A \log_e (0.05) + B$$

Solving these equations, we get

$$A = -105.315 \quad \text{and} \quad B = -355.497$$

Substituting these values in eq. (2), we have

$$V = -105.315 \log_e \rho - 355.497 \qquad \text{...(3)}$$

(*i*) The value of V at $\rho = 2$ cm $= 0.02$ m is given by

$$V = -105.315 \log_e (0.02) - 355.497$$

$$= \mathbf{5265.75\ V/m}$$

$$\mathbf{E} = -\vec{\nabla} V = \frac{105.315}{\rho} = \frac{105.315}{0.02}\hat{\mathbf{a}}_r = \mathbf{5265.8\ \hat{a}_r}$$

(*ii*) When $V = 0$, then

$$0 = -105.315 \log_e (\rho) - 355.497$$

or

$$\rho = 0.034199 \text{ m} = \mathbf{3.420\ cm}$$

$$\mathbf{D} = \varepsilon_0 \mathbf{E} = \frac{105.315\varepsilon_0}{\rho}\hat{\mathbf{a}}_\rho$$

$$\rho_S = |\mathbf{D}_N| = \frac{105.315\,\varepsilon_0}{\rho}$$

(*iii*) The total charge q on the cylinder is given by

$$q = \rho_S \times 2\,\pi\rho h = \frac{105.315\,\varepsilon_0}{\rho} \times 2\,\pi\rho h$$

$$= 105.315\,\varepsilon_0\,(2\,\pi h)$$

Potential difference between the cylinders

$$V = 80 - (-40) = 120$$

$$\therefore \qquad C = \frac{q}{V} = \frac{105.315\,\varepsilon_0\,(2\,\pi h)}{120}$$

Capacitance per unit length

$$C = \frac{105.315 \times 8.85 \times 10^{-12} \times 2 \times 3.14}{120} = \mathbf{48.823\ pF/m}.$$

EXAMPLE 7. *Find the potential and electric field intensity for the region between two concentric right circular cylinders where* $V = 0$ *at* $r_a = 1$ mm *and* $V = 100$ volt *at* $r_b = 20$ mm.

Solution. The situation is shown in Fig. (3).

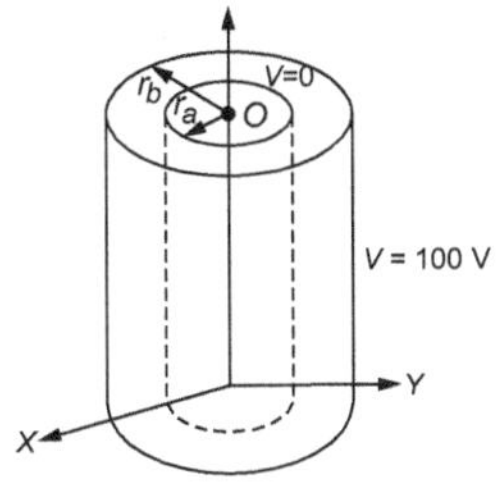

Fig. (3)

Here, V varies with respect to r only. Laplace's equation reduces to

$$\frac{1}{r}\frac{d}{dr}\left(r\frac{dV}{dr}\right) = 0 \quad \text{or} \quad \frac{d}{dr}\left(r\frac{dV}{dr}\right) = 0$$

Integrating once, we get $\quad r\dfrac{dV}{dr} = A \qquad (A = \text{constant})$

Further integrating, we get $\quad V = A \log_e r + B \qquad \text{...(1)}$

where B is another constant of integration.

The values of constants A and B can be obtained by applying the boundary conditions $V = 0$ at $r_a = 0.001$ m and $V = 100$ volt at $r_b = 0.020$ m. Therefore,

$$0 = A \log_e 0.001 + B$$

and

$$100 = A \log_e 0.020 + B$$

Solving, we get $\quad A = 33.36 \quad \text{and} \quad B = 230.49 \qquad \text{...(2)}$

Substituting the values of A and B in eq. (1), we get

$$V = 33.36 \log_e r + 230.49 \qquad \text{...(3)}$$

Now,

$$\mathbf{E} = -\vec{\nabla} V = -\frac{\partial V}{\partial r}\hat{\mathbf{a}}_r$$

or

$$\mathbf{E} = -\frac{33.36}{r}\hat{\mathbf{a}}_r \qquad \text{...(4)}$$

Exercises & Problems

EXERCISES

1. Express Poisson's and Laplace's equations in various coordinate systems.
2. Discuss the solution of Laplace's equation in one dimension.
3. Derive Poisson's equation and discuss its applications in electrostatics.
4. Explain uniqueness theorem.
5. Derive the capacitance of a parallel plate capacitor using Laplace's theorem.
6. Given $V = V_0$ at $\rho = a$
   and $V = 0$ at $\rho = b$. $(b > a)$
   Find **E**, **D** and ρ_S and capacitance of the system.
7. Use Laplace's equation to obtain the capacitance of a cylindrical capacitor.

PRACTICE PROBLEMS

1. State whether following potential functions satisfy Laplace's equation.
   (*i*) $V = Cxyz$ and (*ii*) $V = C\rho\phi z$
   [**Ans.** (*i*) Yes, (*ii*) No]

2. In a charge free region of space, a potential field is given as
   $$V(x, y) = 5x^3 + f(x) - 2y^2$$
   Find $f(x)$ if E_x and V both are zero at origin.
   [**Ans.** $f(x) = -5x^3 + 2x^2$]

3. A cylindrical capacitor has radii 1 and 2.5 cm. If the space between the plates is filled with an inhomogeneous dielectric with $\varepsilon_r = \dfrac{(10+\rho)}{\rho}$, where ρ is in centimeter, find the capacitance per metre of the capacitor.
   [**Ans.** 434.6 pF/m]

❑❑❑

UNIT - III

1. Steady Magnetic Field
2. Time Varying Fields and Maxwell's Equations

5

Steady Magnetic Field

In previous chapters, we studied static or stationary charges produce electrostatic fields. In 1820, Oersted discovered that a compass needle shows a deflection when brought near a current carrying wire. This shows that an electric charge in motion produces a magnetic field in the space around it. It should be noted that a charge (whether it is stationary or in motion) produces an electric field around it. However, when it is in motion, then in addition to electric field it produces a magnetic field. Therefore, the charges in motion with *constant velocity* produce *magnetostatic fields*. By using Biot-Savart law and Ampere's law we can calculate magnetic fields due to different current distributions. The objective of this chapter is to consider the different aspects of static magnetic field.

5.1 BIOT-SAVART'S LAW

A current carrying conductor produces a magnetic field around it. In 1820, Biot and Savart performed a series of experiments to study the magnetic field produced by various current carrying conductors. They obtained a relation by means of which the magnitude and direction of the field at any point of the space around a conductor in which current is passing can be obtained. The relation is called as Biot-Savart law.

As shown in Fig. (1), let AB be a conductor of any arbitrary shape in which a current I is flowing. Let P be a point at which field is to be determined. According to Biot and Savart,

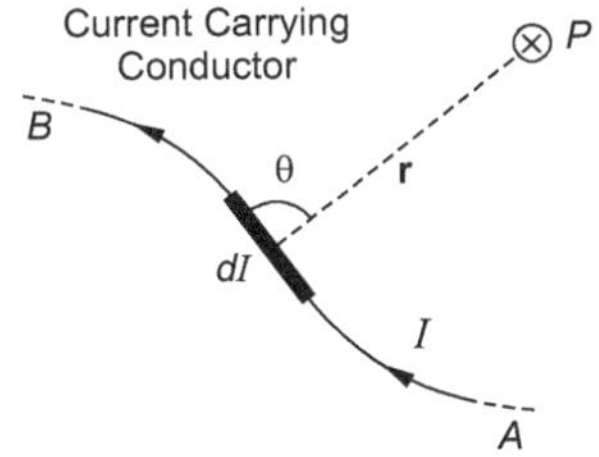

Fig. (1)

the field or *magnetic induction* or *magnetic flux density* **B** at any point can be computed by dividing the conductor into short current elements. The field distribution $d\mathbf{B}$ at that point due to each element is calculated. The total field **B** is then resultant of the infinitesimal field of all the elements of the conductor. Let $d\mathbf{l}$ be the length of one such element and **r** be the displacement vector of point P from this element. Biot and Savart observed that $d\mathbf{B}$ at point P due to small element $d\mathbf{l}$ of a current carrying conductor depends upon the following factors:

(*i*) It is directly proportional to the current I flowing through the conductor

$$dB \propto I$$

(*ii*) It is directly proportional to the length dl of the element considered

$$dB \propto dl$$

(*iii*) It is directly proportional to the sine of the angle θ between length of element and the line joining the element to the point P

$$dB \propto \sin\theta$$

(*iv*) It is inversely proportional to the square of the distance r of the point P from the element dl

$$dB \propto \frac{1}{r^2}$$

Combining all these factors

$$dB \propto \frac{Idl\sin\theta}{r^2}$$

When the conductor is place in vacuum or air, then

$$dB = \frac{\mu_0}{4\pi}\times\frac{Idl\sin\theta}{r^2} \qquad ...(1)$$

where $\left(\frac{\mu_0}{4\pi}\right)$ is proportionality constant μ_0 is the permeability of free space.

$\left(\frac{\mu_0}{4\pi}\right) = 10^7$ weber/(amp-metre).

In vector form

$$\boxed{d\mathbf{B} = \frac{\mu_0}{4\pi}\times\frac{Idl\times\hat{\mathbf{a}}_r}{r^2}} \qquad ...(2)$$

where $\hat{\mathbf{a}}_r$ is unit vector along the direction of $\mathbf{r}$.

For a medium

$$\boxed{d\mathbf{B} = \frac{\mu_0}{4\pi}\times\frac{Idl\times\hat{\mathbf{a}}_r}{r^2}} \qquad \text{where } \mu = \mu_0\mu_r \quad ...(3)$$

Here, μ = permeability of medium

μ_0 = permeability of free space = $4\pi \times 10^{-7}$

μ_r = relative permeability with respect to vacuum or free space

As $\hat{\mathbf{a}}_r = \frac{\mathbf{r}}{|\mathbf{r}|}$, we have the value of $d\mathbf{B}$ in free space as

$$\boxed{d\mathbf{B} = \frac{\mu_0}{4\pi}\times\frac{I\mathbf{dl}\times\mathbf{r}}{r^3}} \qquad ...(4)$$

The resultant field at P can be obtained by integrating eq. (4). Therefore,

$$\mathbf{B} = \int d\mathbf{B} \qquad ...(5)$$

The direction of **B** is given by right hand screw rule. With the screw placed in the direction of current flow, the direction of advance of the screw is the direction of $d\mathbf{B}$.

Different Forms

(*i*) Let us express the Biot-Savart law in terms of magnetic field intensity **H** (amphere/ metre). We know that

$$\mu_0 = \frac{\text{Magnetic flux density}}{\text{Magnetic field intensity}} = \frac{\mathbf{B}}{\mathbf{H}}$$

or $$\mathbf{B} = \mu_0 \mathbf{H}$$

$\therefore$ $$\boxed{d\mathbf{H} = \frac{1}{4\pi} \times \frac{Id\mathbf{l} \times \mathbf{r}}{r^3}} \qquad ...(6)$$

(*ii*) Let us express the Biot-Savart law in terms of current density. The current density **J** at a point on the element of length $d\mathbf{l}$ area of cross-section dA and volume dV is given by

$$Idl = \mathbf{J}dAdl = \mathbf{J}dV$$

$\therefore$ $$\boxed{\mathbf{B} = \frac{\mu_0}{4\pi} \int_V \frac{\mathbf{J} \times \mathbf{r}}{r^3} dV} \qquad ...(7)$$

5.2 MAGNETIC FIELD DUE TO LONG STRAIGHT CONDUCTOR CARRYING CURRENT

Consider an infinitely long conductor placed in vacuum and carrying a current I ampere as shown in Fig. (2). The aim is to derive an expression for magnetic induction **B** due to the conductor at a point P on X-axis using Biot and Savart law. Let the distance of P from the conductor be R. O is the foot of perpendicular from P to the conductor. Consider a small element AB of length $d\mathbf{l}$ of conductor at a distance l from O. Let **r** be the distance of the element from the point P. Suppose θ be the angle in clockwise direction which the direction of current makes with the line joining the elements to point P. The magnitude of the field $d\mathbf{B}$ due to element AB at P is given by

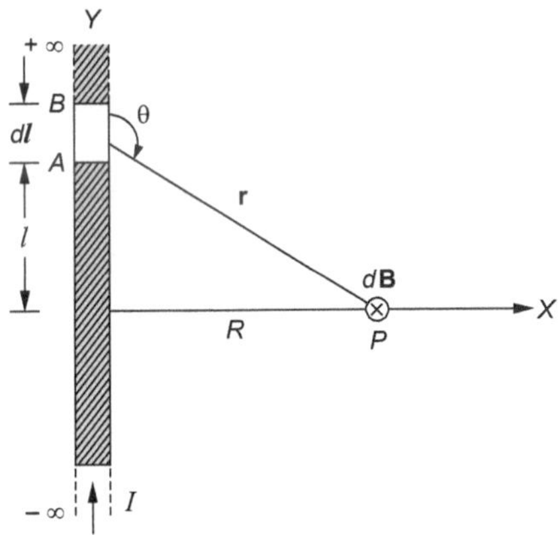

Fig. (2)

$$d\mathbf{B} = \frac{\mu_0}{4\pi} \times \frac{Id\mathbf{l} \times \mathbf{r}}{r^3}$$

In magnitude, $$dB = \frac{\mu_0}{4\pi} \times \frac{Idl \times \sin\theta}{r^2} \qquad ...(1)$$

The field due to the whole conductor is given by

$$B = \int dB = \frac{\mu_0 I}{4\pi} \int_{-\infty}^{+\infty} \frac{\sin\theta dl}{r^2} \qquad ...(2)$$

From figure, $r = (l^2 + R^2)^{1/2}$ and $\sin\theta = \sin(\pi - \theta) = \left(\frac{R}{r}\right)$

$$\therefore \qquad \sin\theta = \frac{R}{(l^2 + R^2)^{1/2}}$$

Substituting the values of r and $\sin\theta$ in eq. (2), we get

$$B = \frac{\mu_0 I}{4\pi}\int_{-\infty}^{+\infty} \frac{R\,dl}{(l^2 + R^2)^{3/2}} \qquad ...(3)$$

In order to evaluate this integral, we substitute

$$l = R\tan\alpha$$

$$\therefore \qquad dl = R\sec^2\alpha\, d\alpha$$

The limits of integration under this substitution become $-\frac{\pi}{2}$ to $\frac{\pi}{2}$.

$$\text{Hence,} \qquad B = \frac{\mu_0 I}{4\pi}\int_{-\pi/2}^{+\pi/2} \frac{R\cdot R\sec^2\alpha\, d\alpha}{(R^2\tan^2\alpha + R^2)^{3/2}} = \frac{\mu_0 I}{4\pi}\int_{-\pi/2}^{\pi/2} \frac{R^2\sec^2\alpha\, d\alpha}{R^3(1+\tan^2\alpha)^{3/2}}$$

$$= \frac{\mu_0 I}{4\pi}\int_{-\pi/2}^{\pi/2} \frac{d\alpha}{R\sec\alpha} = \frac{\mu_0 i}{4\pi R}\int_{-\pi/2}^{\pi/2} \cos\alpha\, d\alpha = \frac{\mu_0 I}{4\pi R}\cdot[\sin\alpha]_{-\pi/2}^{\pi/2}$$

$$= \frac{\mu_0 i}{4\pi R}[1+1]$$

$$\text{or} \qquad \boxed{B = \frac{\mu_0 I}{2\pi R}} \text{ weber/m}^2 \text{ or tesla} \qquad ...(4)$$

This is the expression for the magnetic field induction near a long straight conductor.

EXAMPLE 1. *Find the magnetic field intensity* **H** *at a point* (ρ, ϕ, z) *due to an infinitely long straight filament carrying a current I in Z-direction.*

Solution. The situation is shown in Fig. (3). In this figure, an infinitely long straight filament carrying a current I is place along Z-axis. $P(\rho, \phi, z)$ is a point where magnetic field intensity **H** is to be calculated.

Consider a current element

$$I\mathbf{dl} = Idz\hat{\mathbf{a}}_z$$

This is at a distance **R** as shown in Fig. (3).

According to Biot-Savart law, the magnetic field intensity $d\mathbf{H}$ at point P due to current carrying element is given by

$$d\mathbf{H} = \frac{I}{4\pi}\times\frac{dl\times\hat{\mathbf{a}}_R}{R^2} \qquad ...(1)$$

where $\hat{\mathbf{a}}_R$ is unit vector along the direction of **R**.

The value of unit vector $\hat{\mathbf{a}}_R$ is given by

$$\hat{\mathbf{a}}_R = \frac{\rho\hat{\mathbf{a}}_\rho - z\hat{\mathbf{a}}_z}{\sqrt{(\rho^2 + z^2)}} \qquad ...(2)$$

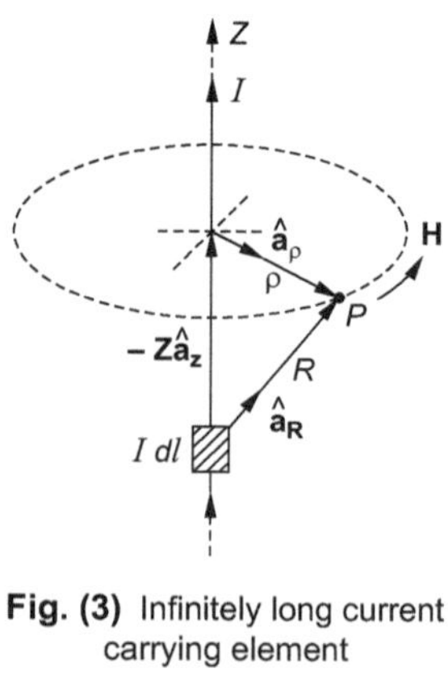

Fig. (3) Infinitely long current carrying element

Substituting the value of $\hat{\mathbf{a}}_R$ from eq. (2) and in eq. (1), we get

$$d\mathbf{H} = \frac{I}{4\pi} \times \frac{dz\hat{\mathbf{a}}_z \times (\rho\hat{\mathbf{a}}_\rho - z\hat{\mathbf{a}}_z)}{R^3}$$

or

$$d\mathbf{H} = \frac{1}{4\pi} \times \frac{Idz\rho\hat{\mathbf{a}}_\phi}{(\rho^2 + z^2)^{3/2}} \qquad ...(3)$$

In eq. (3) $\hat{\mathbf{a}}_\phi$ is a constant because the integration will be with respect to z. Therefore, total magnetic field intensity at P due to infinitely long straight filament is given by

$$\mathbf{H} = \frac{I\hat{\mathbf{a}}_\phi}{4\pi} \int_{z=-\infty}^{+\infty} \frac{\rho\, dz}{(\rho^2 + z^2)^{3/2}}$$

The integral can be solved as shown in article 5.2. The value comes out to be $(2/\rho)$. Therefore,

$$\mathbf{H} = \frac{I\hat{\mathbf{a}}_\phi}{4\pi} \times \frac{2}{\rho} = \frac{I\hat{\mathbf{a}}_\phi}{2\pi\rho} \qquad ...(4)$$

H is not a function of ϕ or z and varies inversely as the distance from the filament.

EXAMPLE 2. *A finite length of a current element AB is placed along Z-axis as shown in Fig. (4). A current I is flowing in it. Show that magnetic field intensity at a point P distant r is given by*

$$\mathbf{H} = \frac{I}{4\pi r}[\sin \alpha_2 - \sin \alpha_1]\hat{\mathbf{a}}_\phi$$

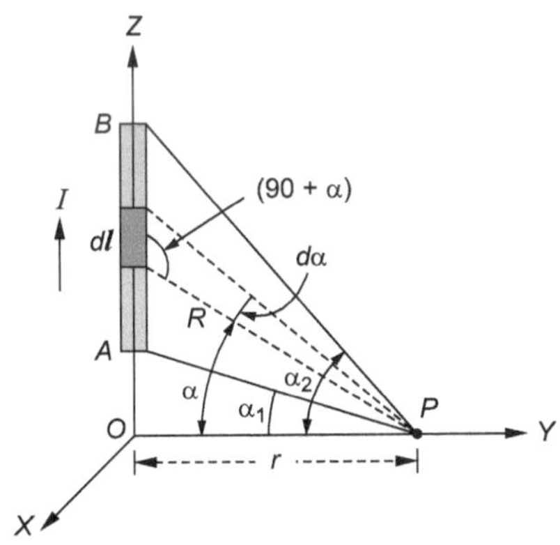

Fig. (4) Finite current element with current *I*

Solution. Consider a small element of $d\mathbf{l}$ of current element AB. The magnetic field intensity $d\mathbf{H}$ due to this element dl at point P is given by

$$d\mathbf{H} = \frac{Idl \times \hat{\mathbf{a}}_R}{4\pi R^2} \qquad ...(1)$$

where $\hat{\mathbf{a}}_R$ is the unit vector along the direction of **R**.

Now, $$d\mathbf{H} = \frac{Idl\sin(90+\alpha)}{4\pi R^2}\hat{\mathbf{a}}_\phi \quad \text{or} \quad d\mathbf{H} = \frac{Idl\cos\alpha}{4\pi R^2}\hat{\mathbf{a}}_\phi \qquad ...(2)$$

The value of dl can be obtained as follow:

From figure, $$Rd\alpha = dl\cos\phi \quad \text{or} \quad dl = \frac{Rd\alpha}{\cos\alpha} \qquad ...(3)$$

Substituting the value of dl from eq. (3) in eq. (2), we get

$$d\mathbf{H} = \frac{I}{4\pi R^2}\times\left(\frac{Rd\alpha}{\cos\alpha}\right)\cos\alpha\,\hat{\mathbf{a}}_\phi = \frac{Id\alpha}{4\pi R}\hat{\mathbf{a}}_\phi$$

$$= \frac{Id\alpha}{4\pi\left(\frac{r}{\cos\alpha}\right)}\hat{\mathbf{a}}_\phi \qquad \left(\because R = \frac{r}{\cos\alpha}\right)$$

$$= \frac{I}{4\pi r}\cos\alpha\,\hat{\mathbf{a}}_\phi \qquad ...(4)$$

To obtain the value of $\mathbf{H}$ at point P due to finite current element AB, we integrate eq. (3), in the limit α_1 and α_2. Hence,

$$\mathbf{H} = \frac{I}{4\pi r}\int_{\alpha_1}^{\alpha_2}\cos\alpha\,\hat{\mathbf{a}}_\phi$$

or $$\mathbf{H} = \frac{I}{4\pi r}[-\sin\alpha]_{\alpha_1}^{\alpha_2}\hat{\mathbf{a}}_\phi$$

$\therefore$ $$\mathbf{H} = \frac{I}{4\pi r}[\sin\alpha_1 - \sin\alpha_2]\hat{\mathbf{a}}_\phi \qquad ...(5)$$

Eq. (4) gives the total magnetic field intensity at P due to finite current element AB.

EXAMPLE 3. *Find the magnetic field intensity at point $P_2(4, 2, 0)$ due to current carrying conductor ($Id\mathbf{l} = 2\pi\,\hat{\mathbf{a}}_z$ μ Am) placed at point $P_1(0, 0, 2)$.*

Solution. We know that

$$d\mathbf{H} = \frac{Idl\times\hat{\mathbf{a}}_R}{4\pi R^2} = \frac{Idl\times\mathbf{R}}{4\pi R^3}$$

Given that $$\mathbf{R} = (4-0)\hat{\mathbf{a}}_x + (2-0)\hat{\mathbf{a}}_y + (0-2)\hat{\mathbf{a}}_z$$

$$|\mathbf{R}| = \sqrt{[(4)^2+(2)^2+(-2)^2]} = \sqrt{24} = 4.9 \text{ unit}$$

$\therefore$ $$d\mathbf{H} = \frac{2\pi\hat{\mathbf{a}}_z\times(4\hat{\mathbf{a}}_x + 2\hat{\mathbf{a}}_y - 2\hat{\mathbf{a}}_z)}{4\pi(4.9)^3}$$

$$= \frac{(4\hat{\mathbf{a}}_y - 2\hat{\mathbf{a}}_x)}{2\times(4.9)^3} = \frac{-2\hat{\mathbf{a}}_x + 4\hat{\mathbf{a}}_y}{(235.29)}\ \mu\text{A/m}$$

$$= (-8.5\,\hat{\mathbf{a}}_x + 17.01\,\hat{\mathbf{a}}_y)\ \text{nA/m}$$

5.3 MAGNETIC FIELD OF A CIRCULAR CARRYING LOOP

As shown in Fig. (5), consider a circular coil of radius a and carrying a current I. P is a point

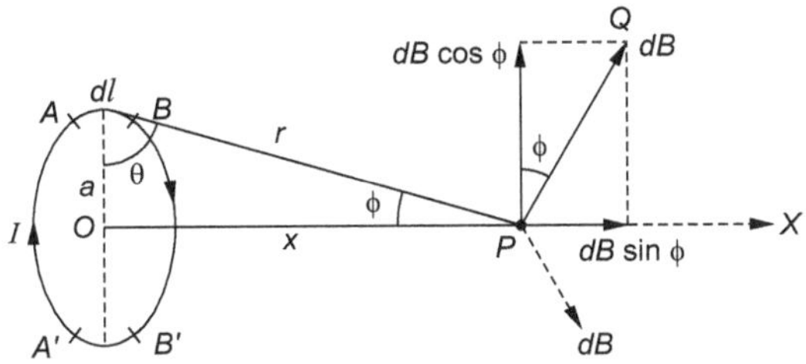

Fig. (5)

on the axis of the coil distant x from the centre. We are required to calculate the field at point P. Consider a small element AB of length $d\mathbf{l}$.

Let $\mathbf{r}$ be the distance of the element from the point P and θ be the angle which the direction of current makes with the line joining the element to the point O. The magnetic field $d\mathbf{B}$ at point P due to current element AB of length $d\mathbf{l}$ is given by

$$d\mathbf{B} = \frac{\mu_0}{4\pi} \times \frac{I d\mathbf{l} \times \mathbf{r}}{r^3} = \frac{\mu_0}{4\pi} \cdot \frac{I dl \sin 90^\circ}{r^2} = \frac{\mu_0}{4\pi} \cdot \frac{I dl}{r^2}$$

($\because$ angle between $d\mathbf{l}$ and $\mathbf{r} = 90^\circ$)

The vector $d\mathbf{B}$ at the point P, due to element $d\mathbf{l}$ would be perpendicular to $\mathbf{r}$. This can be resolved into two components: one of the magnitude $dB \cos \phi$ at right angle to axis and other of magnitude $dB \sin \phi$ along the axis of the coil. If we take another element $A'B'$ diametrically opposite to AB of the same length, it will also produce electric field $d\mathbf{B}$ at P. The direction of $d\mathbf{B}$ now will be opposite to the previous one. This can also be resolved into components $dB \cos \phi$ and $dB \sin \phi$. The components along the axis will add up while the components perpendicular to the axis will cancel. Similarly, if we divide the whole circular coil into a number of elements, the vertical components will cancel while the components along the axis will add up.

$\therefore$ Magnetic field along the axis $= B = \int dB \sin \phi$

$$\therefore \qquad B = \frac{\mu_0 I}{4\pi r^2} \int dl \sin \phi \quad \text{or} \quad B = \frac{\mu_0 I}{4\pi r^2} \int dl \cdot \left(\frac{a}{r}\right) \qquad \left[\because \sin \phi = \left(\frac{a}{r}\right)\right]$$

$$\text{or} \qquad B = \frac{\mu_0 I a}{4\pi r^3} \int dl$$

But $\int dl$ = circumference of the coil $= 2\,\pi a$

From Fig. (5), $\quad r = (a^2 + x^2)^{1/2}$

$$\therefore \qquad B = \frac{\mu_0 I a}{4\pi (a^2 + x^2)^{3/2}} \times 2\,\pi a = \frac{\mu_0 I a^2}{4\pi (a^2 + x^2)^{3/2}} \qquad \text{...(1)}$$

If there are N turns in coil, then

$$\boxed{B = \frac{\mu_0 N I a^2}{2(a^2 + x^2)^{3/2}}} \quad \text{weber/metre}^2 \quad \text{or} \quad \text{tesla} \qquad \text{...(2)}$$

The direction of B is along the axis of the coil.

(*i*) At the centre of the coil $x = 0$. Thus, at the centre of the coil

$$B = \frac{\mu_0 N I a^2}{2a^3} = \frac{\mu_0 N I}{2a} \quad ...(3)$$

(*ii*) At very far off from the loop $x >> a$ and $(a^2 + x^2)^{3/2} \approx x^3$

$$\therefore \quad B = \frac{\mu_0 N I a^3}{2x^2} \quad ...(4)$$

EXAMPLE 1. *Calculate the magnetic field due to a current carrying solenoid.*

Solution. Consider a long solenoid of length l metre and radius a metre [Fig. (7)]. Let N be the

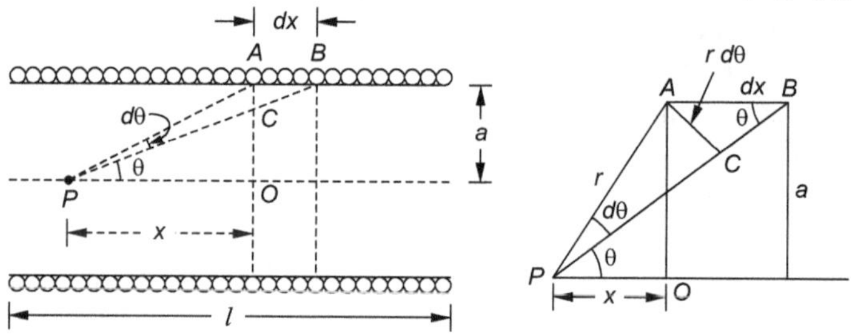

(*a*) (*b*) Enlarged view

Fig. (7)

total number of turns in the solenoid. Then the number of turns n per metre will be $\left(\frac{N}{l}\right)$.

Suppose the solenoid carries a current I ampere. Here, we shall calculate the field in the following cases:

(*i*) field at an inside point, (*ii*) field at an axial end point, and

(*iii*) field at the centre of solenoid of finite length.

EXAMPLE 2. *A circular loop of wire of radius a lying in XY plane with its centre at origin carries a current I in +* ϕ *direction. Find the value* **H** (0, 0, *z*) *and* **H** (0, 0, 0).

Solution. The situation is shown in Fig. (6)

The magnetic field intensity at point P due to a small element $d\mathbf{l}$ is given by

$$d\mathbf{H} = \frac{I\, d\mathbf{l} \times \hat{\mathbf{a}}_R}{4\pi R^2} \quad ...(1)$$

where $\hat{\mathbf{a}}_R = \frac{-a\hat{\mathbf{a}}_r + h\hat{\mathbf{a}}_z}{\sqrt{(a^2 + h^2)}}$ where $R = \sqrt{(a^2 + h^2)}$

Substituting the value in eq. (1), we get

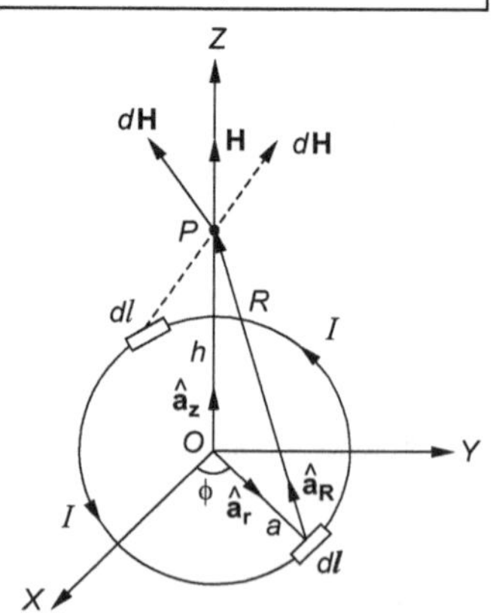

Fig. (6) Current carrying loop

$$d\mathbf{H} = \frac{I\, d\mathbf{l} \times (-a\,\hat{\mathbf{a}}_r + h\,\hat{\mathbf{a}}_z)}{4\pi R^2 \sqrt{(a^2+h^2)}}$$

$$= \frac{I\,(a\,d\phi\,\hat{\mathbf{a}}_\phi) \times (-a\,\hat{\mathbf{a}}_r + h\,\hat{\mathbf{a}}_z)}{4\pi\,(a^2+h^2)^{3/2}}$$

$$= \frac{I\,a\,d\phi\,(a\,\hat{\mathbf{a}}_z + h\,\hat{\mathbf{a}}_r)}{4\pi\,(a^2+h^2)^{3/2}} \qquad \text{...(2)}$$

Similarly, if we consider an element $d\mathbf{l}$ on other side, the magnetic field intensity due to this element will be in opposite direction. The horizontal components are cancelled and there remains the vertical components which are added up. Therefore, total field intensity at P will be

$$\mathbf{H} = \int_{\phi=0}^{2\pi} \frac{1\,a^2\,d\phi}{4\pi\,(a^2+h^2)^{3/2}}\,\hat{\mathbf{a}}_z \qquad \text{...(3)}$$

$$= \frac{I\,a^2\,\hat{\mathbf{a}}_z}{4\pi\,(a^2+h^2)^{3/2}} \int_0^{2\pi} d\phi = \frac{I\,a^2\,\hat{\mathbf{a}}_z}{4\pi\,(a^2+h^2)^{3/2}} \times 2\pi$$

$$\mathbf{H} = \frac{I\,a^2\,\hat{\mathbf{a}}_z}{2(a^2+h^2)^{3/2}}$$

At centre $\quad \mathbf{H} = \dfrac{I}{2a}\,\hat{\mathbf{a}}_z \qquad (\because \text{ at centre } h = 0)$

(i) Field at an Inside Point

Here, we shall calculate the magnetic induction **B** at a point P inside the solenoid on the axis. For this purpose we divide the solenoid into a number of narrow equidistant coils. Now, we consider one such coil of width dx. There will be $n\,dx$ turns. Let x be the distance of point P from the centre O of the coil.

The field at P due to elementary coil of width dx carrying a current I is given by

$$dB = \frac{\mu_0\,(n\,dx)\,I\,a^2}{2(a^2+x^2)^{3/2}} \text{ weber/metre}^2 \qquad \text{...(1)}$$

From Fig. (7*b*), considering triangle ABC, we have

$$\sin\theta = \frac{r\,d\theta}{dx} \quad \text{or} \quad dx = \frac{r\,d\theta}{\sin\theta}$$

From ΔAPO, $\quad a^2 + x^2 = r^2$

$\therefore \quad (a^2+x^2)^{3/2} = r^3$

Substituting these values in eq. (1), we get

$$dB = \frac{\mu_0 n \left(r\dfrac{d\theta}{\sin\theta}\right) I\,a^2}{2r^3}$$

or

$$dB = \frac{\mu_0 n\,I\,a^2 d\theta}{2r^2\sin\theta} = \frac{\mu_0 n\,I\,d\theta}{2\sin\theta}\left(\frac{a}{r}\right)^2$$

$$N = \frac{\mu_0 n\, I\, d\theta}{2\sin\theta}(\sin^2\theta) \qquad \left[\because \left(\frac{a}{r}\right)^2 = \sin^2\theta\right]$$

$$= \frac{\mu_0 n\, I\, d\theta \sin\theta}{2}(\sin^2\theta) \qquad ...(2)$$

The field induction B at P due to whole solenoid can be obtained by integrating the above eq. (2), between the limits θ_1 and θ_2. θ_1 and θ_2 are the semi-vertical angles subtended at P by first and last turns respectively of the solenoid. These angles are shown in Fig. (8).

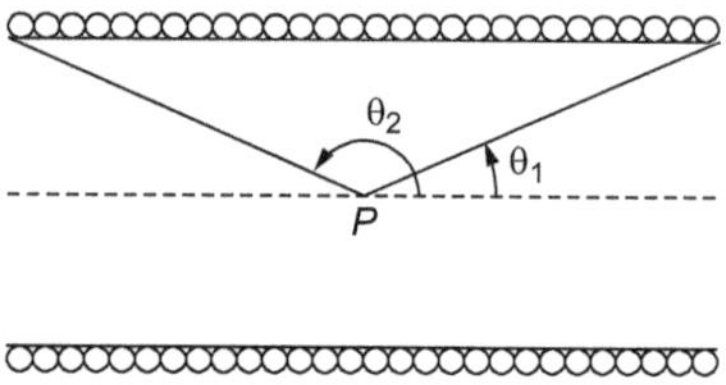

Fig. (8)

$$B = \int_{\theta_1}^{\theta_2} dB = \int_{\theta_1}^{\theta_2} \frac{\mu_0\, nI \sin\theta\, d\theta}{2}$$

$$= \frac{\mu_0\, nI}{2}[-\cos]_{\theta_1}^{\theta_2}$$

$$\therefore \qquad \boxed{B = \frac{\mu_0\, nI}{2}[\cos\theta_1 - \cos\theta_2]} \qquad ...(3)$$

At any axial point P when it is well inside a very long solenoid, $\theta_1 = 0$ and $\theta_2 = \pi$. Hence,

$$B = \frac{\mu_0\, nI}{2}[\cos 0 - \cos\pi] = \frac{\mu_0\, nI}{2}[1-(-1)] = \mu_0\, n\, I$$

$$\therefore \qquad \boxed{B = \mu_0\, nI} \qquad ...(4)$$

This can be taken as the field at the centre of a long solenoid.

(ii) Field at an Axial End Point

In this case, $\theta_1 = 0$ and $\theta_2 = 90°$

or $$B = \frac{\mu_0\, nI}{2}[\cos 0 - \cos 90] = \frac{\mu_0\, nI}{2}$$

$$\therefore \qquad \boxed{B = \frac{\mu_0\, nI}{2}} \qquad ...(5)$$

This shows that the field at either end is one half one of its magnitude at the centre.

(iii) Field at the Centre of a Solenoid of Finite Length

Consider that point P is at the centre, *i.e.*, it is a distance $\frac{l}{2}$ from either end.

In this case, $$\cos\theta_1 = \frac{\left(\frac{l}{2}\right)}{\left\{a^2 + \left(\frac{l}{2}\right)^2\right\}^{1/2}} = \frac{l}{(4a^2 + l^2)^{1/2}}$$

and $$\cos(\pi-\theta_2) = \frac{\left(\frac{l}{2}\right)}{\left\{a^2+\left(\frac{l}{2}\right)^2\right\}^{1/2}} = \frac{l}{(4a^2+l^2)^{1/2}}$$

$$\therefore \quad \cos\theta_2 = \frac{l}{(4a^2+l^2)^{1/2}}$$

Putting these values in eq. (3), we get

$$B = \frac{\mu_0\, nI}{2}\left[\frac{l}{(4a^2+l^2)^{1/2}}+\frac{l}{(4a^2+l^2)^{1/2}}\right]$$

$$= \frac{\mu_0\, nI}{2}\times\frac{2l}{(4a^2+l^2)^{1/2}} = \frac{\mu_0\, l\, nI}{(4a^2+l^2)^{1/2}}$$

$$\therefore \quad \boxed{B = \frac{\mu_0\, I\, n}{(4a^2+l^2)^{1/2}}} \qquad ...(6)$$

This expression gives the field at the centre of the solenoid of finite length.

5.4 AMPERE'S CIRCUITAL LAW

This law is analogous to Gauss' law in electrostatics. ***According to Ampere's law, the line integral of* H *along any closed path is exactly equal to the direct current enclosed by the path.***

$$\boxed{\oint \mathbf{H}\cdot d\mathbf{l} = I} \qquad ...(1)$$

In terms of magnetic flux density, the law can be expressed as

$$\boxed{\oint \mathbf{B}\cdot d\mathbf{l} = \mu_0\, I} \qquad ...(2)$$

where μ_0 is permeability constant.

Proof: Consider a long straight conductor carrying a current I as shown in Fig. (9). Let the conductor be perpendicular to the page directed outward. The magnitude of magnetic induction at a distance R from the conductor is given by

$$B = \frac{\mu_0\, I}{2\,\pi\, \mathrm{P}} \qquad \text{(Biot Savart law)} \quad ...(3)$$

The direction of magnetic induction **B** is tangent to the circle of radius R centred on the conductor. Regarding **B**, two points should be remembered:

(*i*) The magnitude of **B** is constant at all points on the circle.

(*ii*) This is parallel to the circuit element $d\mathbf{l}$.

Now, we shall calculate the line integral of **B** and $d\mathbf{l}$ which is given by

$$\oint \mathbf{B}\cdot dl = \oint B\, dl\cos 0° = B\oint dl = B(2\,\pi\, R) \qquad ...(4)$$

where $\oint dl = 2\,\pi\, R$ = circumference of the circle.

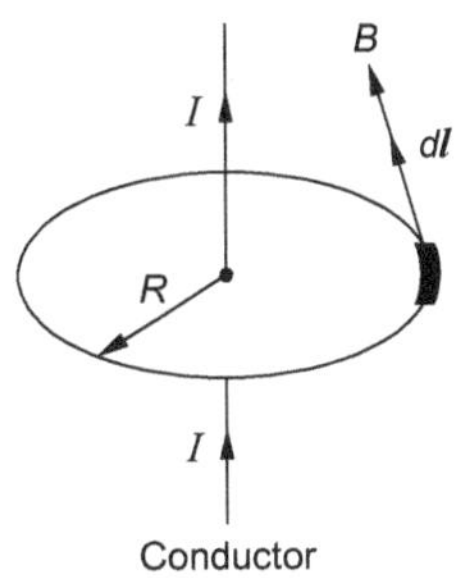

Fig. (9)

Substituting the value of B from eq. (3) in eq. (4), we get

$$\oint \mathbf{B} \cdot dl = \frac{\mu_0 I}{2 \pi R} \times 2 \pi R = \mu_0 I \quad ...(5)$$

Thus, the line integral $\oint \mathbf{B} \cdot dl$ is μ_0 times the current through the area bounded by the circle. This is Ampere's law. This law is true for any assembly of currents and for any closed surface.

Ampere's Law in Integral Form

The law can be applied inside or outside a conductor but the total current enclosed by the path should be known. The law is also applicable for line currents, sheet currents and volume currents. A path is taken in conducting medium as shown in Fig. (10). The general relation for this path is given by

$$\boxed{\oint \mathbf{H} \cdot dl = \oint_S \mathbf{J} \cdot d\mathbf{S} = I}$$

This is known as Ampere's law in integral form.

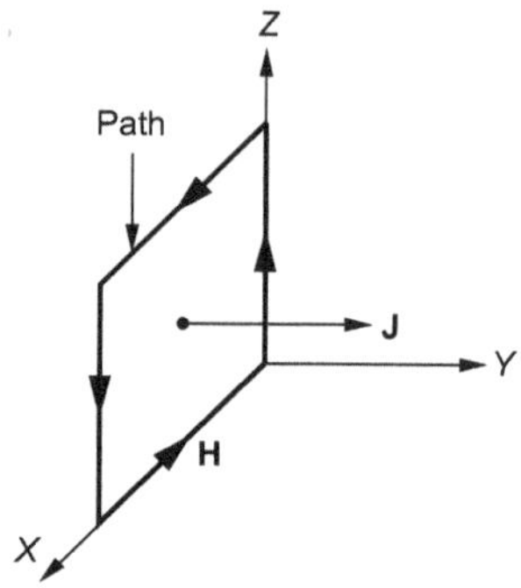

Fig. (10) Ampere's law inside a conductor

Ampere's Law in Differential Form

Let us consider a volume current distribution with current density vector $\mathbf{J}$ A/m^2 as a function of coordinates. A closed path ΔC is shown in Fig. (11). The surface ΔS is bounded by the closed path ΔC. The current enclosed by closed path ΔC is given by surface integral of current density $\mathbf{J}$ over the surface ΔS. Thus,

$$I = \int_{\Delta S} \mathbf{J} \cdot d\mathbf{S} \quad ...(6)$$

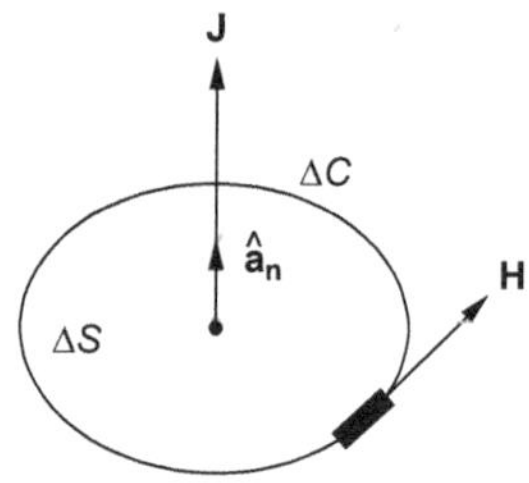

Fig. (11)

According to Ampere's law

$$\oint_{\Delta C} \mathbf{H} \cdot dl = \oint_{\Delta S} \mathbf{J} \cdot d\mathbf{S}$$

Dividing both sides by ΔS when $\Delta S \rightarrow 0$, we have

$$\lim_{\Delta s \rightarrow 0} \frac{\oint_{\Delta C} \mathbf{H} \cdot dl}{\Delta S} = \lim_{\Delta s \rightarrow 0} \frac{\oint_{\Delta S} \mathbf{J} \cdot d\mathbf{S}}{\Delta S}$$

$$= \frac{J\, \hat{\mathbf{a}}_n\, \Delta S}{\Delta S} = J \hat{\mathbf{a}}_n = \mathbf{J}$$

As per Stoke's theorem, the above equation becomes

$$\boxed{\begin{array}{l} curl\ \mathbf{H} = \mathbf{J} \\ \vec{\nabla} \times \mathbf{H} = \mathbf{J} \end{array}}$$

This is Ampere's circuital law in differential form

EXAMPLE 1. *A circular conductor of radius $r_0 = 1$ cm has internal field.*

$$\mathbf{H} = \frac{10^4}{r}\left[\frac{1}{a^2}\sin a\,r - \frac{r}{a}\cos a\,r\right]\hat{\mathbf{a}}_\phi\left(\frac{\text{A}}{\text{m}}\right)$$

where $a = \left(\frac{\pi}{2}r_0\right)$. *Find the total current in the conductor.*

Solution. According to Ampere's law, we have

$$I = \oint_{r=r_0}\mathbf{H}\cdot d\boldsymbol{l} = \int_0^{2\pi}\frac{10^4}{r}\left[\frac{4r_0^2}{\pi^2}\sin\frac{\pi}{2} - \frac{2r_0^2}{\pi}\cos\frac{\pi}{2}\right]r_0\,d\phi$$

$$= \frac{(4\times10^4)r_0^2}{\pi}\int_0^{2\pi}d\phi = \frac{(4\times10^4)r_0^2}{\pi}\times 2\pi$$

$$= \frac{(8\times10^4)(r_0^2)}{\pi} = \frac{(8\times10^4)(10^{-4})}{\pi} = \frac{8}{\pi}\text{ ampere}$$

EXAMPLE 2. *Find the magnetic field at a distance R from a long straight wire carrying a steady current using Ampere's law.*

Solution. See Fig. (12). By symmetry, the magnitude of **B** is constant around the circular loop of radius R, centred on the wire. Applying Ampere's law, we have

$$\oint \mathbf{B}\cdot d\boldsymbol{l} = \mu_0 I$$

Here, $$\oint B\,dl = \oint dl = B\,(2\,\pi\,R)$$

$$B(2\,\pi\,R) = \mu_0 I$$

or $$B = \frac{\mu_0\,I}{2\,\pi\,R}$$

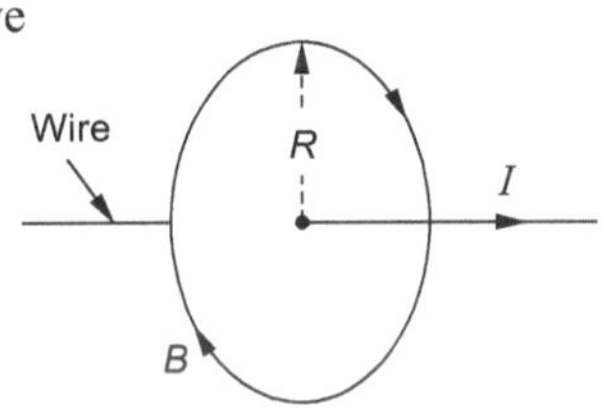

Fig. (12) Straight wire carrying a current I

EXAMPLE 3. *Find **H** for infinitely long coaxial transmission line where central conductor carries current I and outer conductor carries a current – I.*

Solution. The situation is shown Fig. (13).

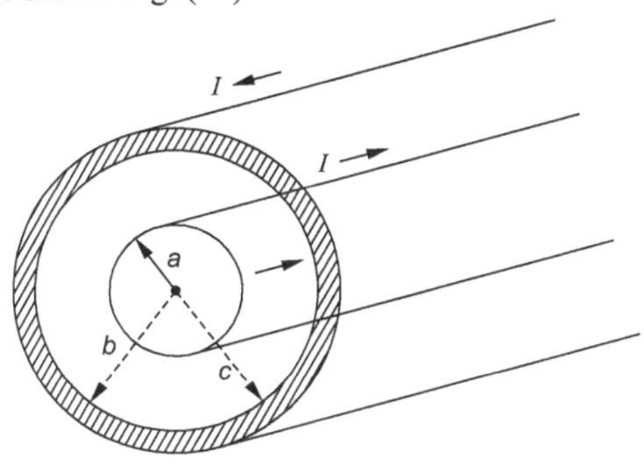

Fig. (13) Infinitely long coaxial transmission line

We consider the value of **H** under the following conditions.

(*i*) For $a < \rho < b$

In this case, $\oint \mathbf{H} \cdot dl = I$ or $H \oint \rho \, d\phi = I$

$\therefore$ $H \rho 2\pi = I$ or $H = \left(\frac{I}{2\pi\rho}\right)$

or $$\mathbf{H} = \frac{I}{2\pi\rho} \hat{\mathbf{a}}_\phi$$

(*ii*) For $0 < \rho < a$

In this case, I_{enclosed} = current density × area

$\therefore$ $$I_{\text{enclosed}} = \frac{I}{\pi a^2} \times \pi\rho^2 = \frac{I\rho^2}{a^2}$$

Now, $$2\pi\rho H_\phi = \frac{I\rho^2}{a^2}$$

$\therefore$ $$\mathbf{H} = \frac{I\rho}{2\pi a^2} \hat{\mathbf{a}}_\phi$$

(*iii*) For $\rho > c$

In this case, $I_{\text{enclosed}} = 0$

$\therefore$ $\mathbf{H} = 0$

(*iv*) $b < \rho < c$

In this case, $$2\pi\rho H_\phi = I - I\frac{(\rho^2 - b^2)}{(c^2 - b^2)} \quad \text{or} \quad 2\pi\rho H_\phi = I\left[\frac{(c^2-\rho^2)}{(c^2-b^2)}\right]$$

$\therefore$ $$\mathbf{H} = \frac{1}{2\pi\rho}\left[\frac{(c^2-\rho^2)}{(c^2-b^2)}\right]\hat{\mathbf{a}}_\phi$$

EXAMPLE 4. *Obtain an expression for* **B** *at a distance r from the axis of a long cylindrical wire of radius a and carrying a current* I_0.

Solution. Fig. (14) represents the section of a long cylindrical wire perpendicular to its length. Now, we draw a circle of radius r_1 (inside the cylindrical wire) and a circle of radius r_2 (outside the cylindrical wire). The magnetic field produced due to current flowing in wire at any point over the path will be tangential to it. The magnitude of **B** will be equal everywhere. Therefore,

$$\oint \mathbf{B} \cdot dl = \mu_0 I$$

or $$\oint \mathbf{B} \cdot dl = \oint \phi \, B \, dl = B \oint dl = B(2\pi r)$$

(*i*) For $r > a$, the current enclosed is I_0. Hence

$$B(2\pi r) = \mu_0 I_0$$

or $$B = \frac{\mu_0 I_0}{2\pi r}$$

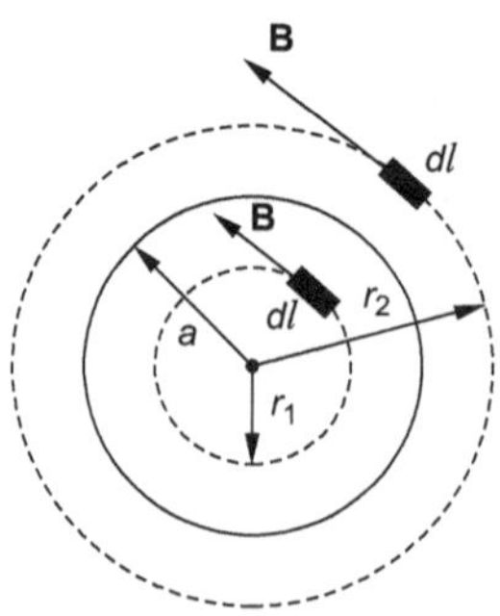

Fig. (14) Cylindrical wire carrying current I_0

(*ii*) For $r < a$, the current enclosed is given by

$$I = \text{current density} \times \text{area} = \frac{I_0}{\pi a^2} \times \pi r^2 = \frac{I_0 r^2}{a^2}$$

Now, $$B(2\pi r) = \mu_0 \left(\frac{I_0 r^2}{a^2}\right)$$

$\therefore$ $$B = \frac{\mu_0 I_0 r}{2\pi a^2}$$

(*iii*) For $r = a$, the current enclosed is I_0. Therefore,

$$B(2\pi a) = \mu_0 I_0$$

$\therefore$ $$B = \frac{\mu_0 I_0}{(2\pi a)}$$

EXAMPLE 5. *Calculate the magnetic induction of a long solenoid.*

Solution. Consider solenoid very long compared with its diameter and carrying a current I_0. Fig. (15) shows few turns of solenoid in which symbol ⊙ represents current in the wire coming

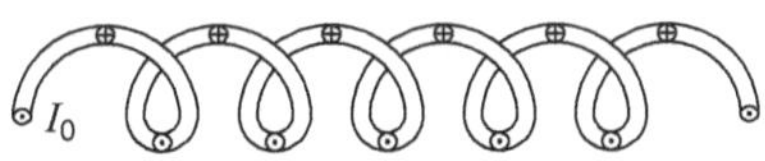

Fig. (15) Turns of rise of solenoid I_0

out of the paper while symbol ⊕ represents the current going into the paper. Following points should be remembered in case of a solenoid:

(*i*) The magnetic field outside the solenoid is very small as compared the field inside.

(*ii*) Magnetic field is uniform inside the solenoid except near the edges.

The field inside the solenoid is shown Fig. (16).

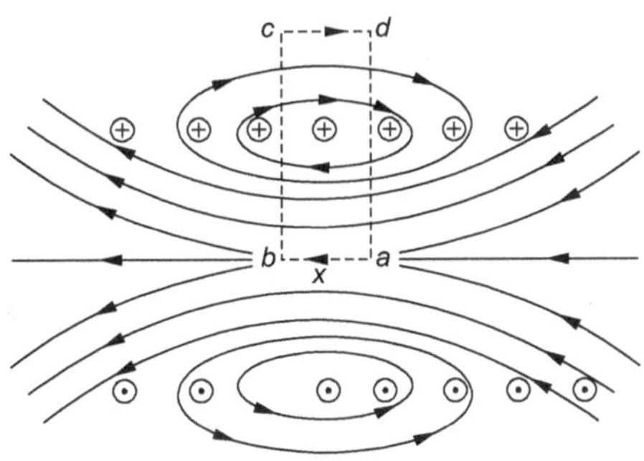

Fig. (16) Field inside the solenoid

In order to calculate **B**, we consider a rectangular path *a b c d* as shown in Fig. (16). The side *a b* of rectangle is parallel to the axis of solenoid. The sides *bc* and *da* are taken very long such that the field on line *cd* is negligible small. In case of a long solenoid, the field at *ab* will uniform. The direction of the field is perpendicular to sides *bc* and *ad*. Thus, the angle +between **B** and sides *bc* or *ad* is 90°.

Applying Ampere's law to the rectangular path *abcd*, we have

$$\oint_{abcd} = \mathbf{B} \cdot dl = \mu_0 I \quad ...(1)$$

where I is the net current enclosed by the rectangle.

Further $$\oint_{abcd} \mathbf{B} \cdot dl = \int_a^b \mathbf{B} \cdot dl + \int_b^c \mathbf{B} \cdot dl + \int_c^b \mathbf{B} \cdot dl + \int_d^a \mathbf{B} \cdot dl$$

As the sides *bc* and *da* are perpendicular to **B**, hence,

$$\int_b^c \mathbf{B} \cdot dl = \int_b^c B\, dl \cos 90° = 0$$

and $$\int_d^a \mathbf{B} \cdot dl = \int_d^a B\, dl \cos 90° = 0$$

Further $$\int_c^d \mathbf{B} \cdot dl = 0 \qquad (\because \mathbf{B} \text{ at } cd \text{ is zero})$$

$$\therefore \quad \oint_{abcd} \mathbf{B} \cdot dl = \int_a^b \mathbf{B} \cdot dl = \int_a^b B\, dl \cos 0 = \int_a^b B\, dl = B\,x \quad ...(2)$$

where x is the length *ab*.

Let n be the number of turns per unit length of the solenoid. The number of turns in length x is n. This also gives the number of turns passing through the rectangle *abcd*. As the current in each turn is I, and hence, the net current enclosed by the rectangle is $n\,x\,I_0$. Thus,

$$I = n\,x\,I_0 \quad ...(3)$$

From eqs. (1), (2) and (3), we have

$$B\,x = \mu_0\, n\, x\, I_0$$

or $$\boxed{B = \mu_0\, n I} \quad ...(4)$$

This is the field at a point well inside a long solenoid. Eq. (4) shows that the field is independent of length and diameter of solendoid. However, this expression is applied to long solenoid in vacuum and for points well inside the solenoid.

EXAMPLE 6. *Consider volume current density distribution in cylindrical coordinates as*

$$\mathbf{J}(\rho, \phi, z) = 0; \qquad 0 < \rho < a$$

$$= J_0 \left(\frac{\rho}{a}\right)\hat{\mathbf{a}}_z; \qquad 0 < \rho < b$$

$$= 0 \qquad b < \rho < \infty$$

Find the magnetic field intensity **H** *in various regions.*

Solution. The different regions are shown in Fig. (17).

(*i*) Consider the region $0 < \rho < a$.

For this region $\mathbf{J}(\rho, \phi, z) = 0$

Applying Ampere's law at a distance ρ, we have

$$\oint \mathbf{H} \cdot d\mathbf{l} = \int_S \mathbf{J} \cdot d\mathbf{S} = \int_{\rho=0}^{a} \int_{\phi=0}^{2\pi} J\,\hat{\mathbf{a}}_z \cdot \rho\, d\rho\, d\phi\, \hat{\mathbf{a}}_z$$

$$H \cdot 2\pi\rho = 0 \qquad \text{or} \qquad \mathbf{H} = 0$$

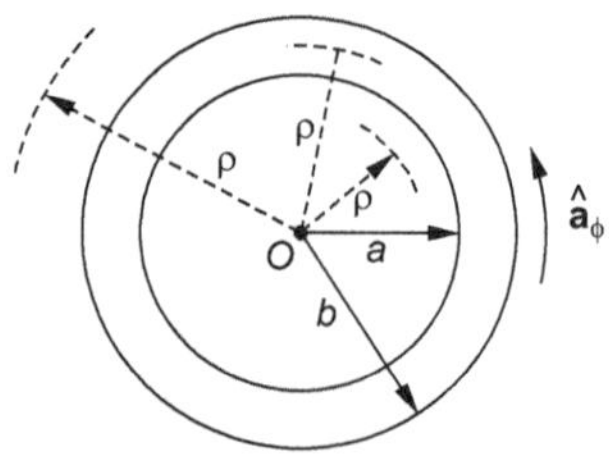

Fig. (17)

(*ii*) Consider the region, $a < \rho < b$

For this region, $\mathbf{J} = J_0\left(\frac{\rho}{a}\right)\hat{\mathbf{a}}_z$

Applying Ampere' law at any radius ρ, we get

$$\oint \mathbf{H}\cdot d\mathbf{l} = \int_a^{\rho}\int_0^{2\pi} J_0\left(\frac{\rho}{a}\right)\cdot \rho\, d\rho\, d\phi$$

$$H\cdot 2\pi\rho = \frac{J_0}{a}\left[\left(\frac{\rho^3}{3}\right)\right]_a^{\rho}(2\pi) = \frac{2\pi J_0}{3a}[\rho^3 - a^3]$$

$$\mathbf{H} = \frac{J_0}{3a\rho}(\rho^3 - a^3)\hat{\mathbf{a}}_\phi$$

At $\rho = b$,

$$\mathbf{H} = \frac{J_0}{3ab}(b^3 - a^3)\hat{\mathbf{a}}_\phi$$

(*iii*) Consider the region $b < \rho < \infty$

For this region, $\mathbf{J} = 0$. At any radius ρ, we have

Therefore,

$$\oint \mathbf{H}\cdot dl = \int_a^b\int_a^b \frac{J_0\rho}{a}\rho\, d\rho\, d\phi$$

$$\mathbf{H} = \frac{J_0}{3a\rho}(b^3 - a^3)\hat{\mathbf{a}}_\phi$$

5.5 MAGNETIC FORCE ON A MOVING CHANGE (LORENTZ FORCE)

If an electric charge q, moving with velocity **v** is placed in a magnetic field with magnetic flux density **B**, then it experiences a force **F**. The force is known as Lorentz force. The force is given by

$$\boxed{\mathbf{F} = q\,(\mathbf{v}\times\mathbf{B})} \quad ...(1)$$

We observed the following points from eq. (1).

(1) The force F is proportional to the magnitudes of charge q, its velocity v and magnetic field B.

(2) The force F is zero if the charge q is at rest.

(3) The force F is zero if the direction of **v** is either parallel or anti-parallel to **B** because $v\times B = vB\sin\theta$ [$\theta = 0$ when **v** and **B** are parallel] and [$\theta = \pi$, **v** and **B** are anti-parallel]

(4) The force F is maximum when **v** are perpendicular because

$$\mathbf{v}\times\mathbf{B} = vB\sin\theta = vB\sin 90° = vB$$

(5) The magnetic force is oppositively direction for charges of opposite sign.

(6) The force is perpendicular to plane defined by **v** and **B**. The direction of force is described by right-hand rule.

5.6 FORCE ON A CURRENT CARRYING CONDUCTOR

Consider the case of a current carrying conductor placed in magnetic field (perpendicular to paper) as shown in Fig. (18).

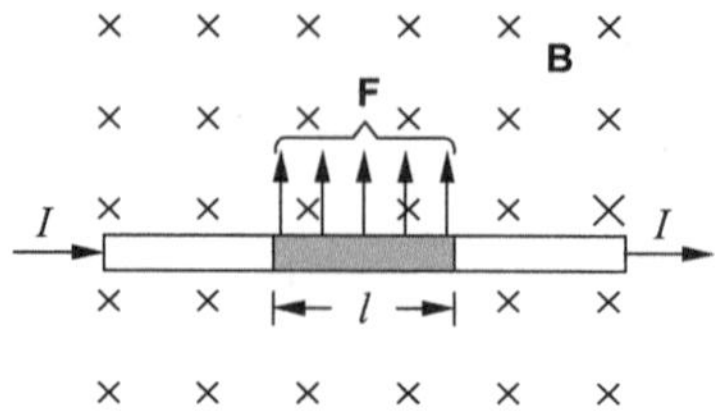

Fig. (18)

We know that current is an assembly of moving charges, therefore, the magnetic field will exert a side way force on the conductor carrying a current. If the charge is infinitely small, then

$$d\mathbf{F} = dq(\mathbf{v} \times \mathbf{B}) \qquad ...(1)$$

where **v** is displacement of charge per unit time.

$$\therefore \qquad d\mathbf{F} = dq\left(\frac{dl}{dt} \times \mathbf{B}\right) = \frac{dq}{dt}(dl \times \mathbf{B}) \qquad \left[\because \mathbf{v} = \frac{dl}{dt}\right]$$

$$\text{or} \qquad d\mathbf{F} = I(dl \times \mathbf{B}) \qquad ...(2)$$

When **B** is uniform over the length of wire, then integrating eq. (2), we have

$$\boxed{\mathbf{F} = I(\boldsymbol{l} \times \mathbf{B})} \qquad ...(3)$$

The force is perpendicular to both **I** and **B**.

Now, we consider the following cases:

(*i*) When the angle between magnetic induction **B** and length of vector ***l*** (direction of current flow) be θ, then the magnitude of force *F* is given by

$$F = I\,B\,l \sin\theta$$

The force will be perpendicular ***l*** and **B**.

(*ii*) When the angle between magnetic induction **B** and length of vector ***l*** be 90°, then the magnitude of force *F* will be maximum and is given by

$$F_{\max} = I\,B\,l$$

The force is again in a direction perpendicular to both ***l*** and **B**.

The direction of force can be obtained by ***Fleming's left hand rule***. According to this rule, stretch the first two fingers and thumb of your left hand in such a way that they are mutually perpendicular to each other. If the first finger points along the direction of field and second finger (central finger) in the direction of current then thumb points in the direction of force (*i.e.*, thrust).

5.7 FORCE ON A CLOSED CIRCUIT

The force a current carrying conductor placed in uniform magnetic field is given by

$$\mathbf{F} = I(dl \times \mathbf{B})$$

where *d***l** is the small length of the conductor.

For a closed loop, the force by uniform magnetic field is given by

$$\mathbf{F} = I\oint dl \times \mathbf{B} = -I\,\mathbf{B} \times \oint dl$$

The closed line integral $\oint dl = 0$

Therefore, force on a closed filamentary circuit in uniform magnetic field is zero,

5.8 FORCE BETWEEN TWO PARALLEL CONDUCTORS

Fig. (19) shows a portion of two infinitely long straight parallel conductors a and b separated at a distance d and carrying I_1 and I_2 respectively. Here each conductor lies in the magnetic field set up by other conductor and hence each will experience a force.

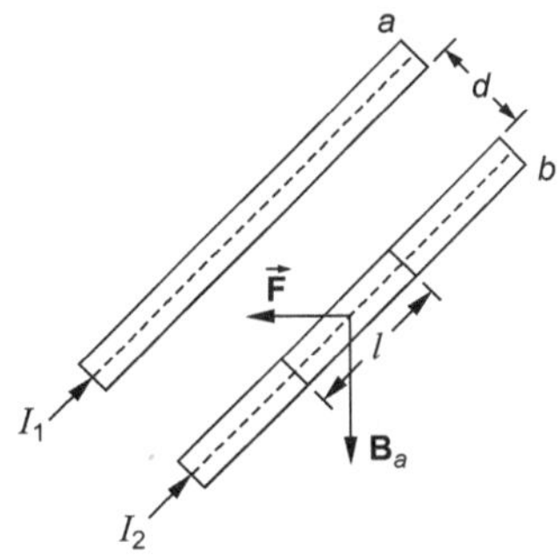

Fig. (19)

Let $\mathbf{B}_a$ be the field of induction produced by conductor a carrying a current I_1, near all points of conductor b. The magnitude of $\mathbf{B}_a$ is given by

$$\mathbf{B}_a = \frac{\mu_0}{4\pi} \cdot \frac{2I_1}{d} \quad ...(1)$$

The force on a length l of the conductor b is

$$F_b = I_2 B_a l = \frac{\mu_0}{4\pi} \cdot l \cdot \frac{2I_1 I_2}{d} \quad ...(2)$$

The force per unit length is given by

$$\frac{F_b}{l} = \frac{\mu_0}{4\pi} \frac{2I_1 I_2}{d}$$

or

$$\boxed{\frac{F_b}{l} = \frac{\mu_0 2I_1 I_2}{2\pi d}} \quad ...(3)$$

The direction of $\mathbf{B}_a$ at wire b will be downward and direction of $\mathbf{F}$ will be towards left.

Similarly, we can start with wire b, compute its field of induction $\mathbf{B}_b$ at wire a and then find the force on wire a. The direction of force will be towards right.

Let $\mathbf{B}_b$ be the field of induction produced by conductor b carrying a current I_2 near all points of conductor a. The magnitude of $\mathbf{B}_b$ is given by

$$\mathbf{B}_b = \frac{\mu_0}{4\pi} \times \frac{2I_2}{d}$$

The force on a length l of conductor a is

$$F_a = I_1 B_b l = \frac{\mu_0}{4\pi} \frac{2I_1 I_2}{d} \times l$$

or

$$\frac{F_a}{l} = \frac{\mu_0}{4\pi} \frac{2I_1 I_2}{d}$$

So,

$$\boxed{\frac{F_a}{l} = \frac{\mu_0 I_1 I_2}{2\pi d}} \quad ...(3)$$

The two forces will be equal and opposite. Hence, the ***conductors attract each other***. If the direction of either current is reversed, the forces will be in opposite directions. Thus, the parallel conductors carrying current in opposite directions ***repel each other***.

EXAMPLE 1. *Calculate the vector force on a current element I_2 and dl_2 by another current element I_1 dl_1. The .separation between two current elements is $\mathbf{R}_{12}$.*

Solution. Let $d\mathbf{B}_1$ be the field induction produced by current element $I_1 d\mathbf{l}_1$ near the current element $I_1 d\mathbf{l}_2$. Therefore,

$$d\mathbf{B}_1 = \frac{\mu_0}{4\pi} \times \frac{I_1\, dl_1 \times \hat{\mathbf{a}}_{R_{12}}}{(R_{12})^2} \qquad \text{...(1)}$$

where $\hat{\mathbf{a}}_{R_{12}}$ is unit vector in the direction of $\mathbf{R}_{12}$.

The force $d\mathbf{F}_2$ on current element $I_2\, d\mathbf{l}_2$ is given by

$$d\mathbf{F}_2 = (I_2\, d\mathbf{l}_2) \times d\mathbf{B}_1 \qquad \text{...(2)}$$

Substituting the value of $d\mathbf{B}_1$ from eq. (1) in eq. (2), we get

$$d\mathbf{F}_2 = (I_2\, d\mathbf{l}_2) \times \left[\frac{\mu_0}{4\pi} \times \frac{I_1\, dl_1 \times \hat{\mathbf{a}}_{R_{12}}}{(R_{12})^2}\right]$$

or

$$dF_2 = \left[\frac{\mu_0 I_1\, I_2}{4\pi\, (R_{12})^2}\, dl_2 \times (dl_1 \times \hat{\mathbf{a}}_{R_{12}})\right]$$

Solved EXAMPLES

EXAMPLE 1. *A point charge $q = 18$ nC has velocity of 5×10^6 m/s in the direction $\hat{\mathbf{a}}_v = 0.04\,\hat{\mathbf{a}}_x - 0.05\,\hat{\mathbf{a}}_y + 0.2\,\hat{\mathbf{a}}_z$. Calculate the magnitude of force exerted on charge by the field*

(*a*) $\mathbf{B} = -3\,\hat{\mathbf{a}}_x + 4\hat{\mathbf{a}}_y + 6\hat{\mathbf{a}}_z$ mT (*b*) $\mathbf{E} = -3\,\hat{\mathbf{a}}_x + 4\hat{\mathbf{a}}_y + 6\hat{\mathbf{a}}_z$ kV/m

(*c*) **B** *and* **E** *acting together.*

Solution.

(*a*) The force due to magnetic field is given by

$$\begin{aligned}
\mathbf{F} &= q\,(\mathbf{v} \times \mathbf{B}) \\
&= (18 \times 10^{-9})\,(5 \times 10^6 \times 10^{-3}) \\
&\quad \left[\frac{0.04\hat{\mathbf{a}}_x - 0.05\hat{\mathbf{a}}_y + 0.2\hat{\mathbf{a}}_z}{\sqrt{(0.04)^2 + (0.05)^2 + (0.2)^2}} \times -3\hat{\mathbf{a}}_x + 4\hat{\mathbf{a}}_y + 6\hat{\mathbf{a}}_z\right] \\
&= 90 \times 10^{-6}[0.19\,\hat{\mathbf{a}}_x - 0.238\hat{\mathbf{a}}_y + 0.952\,\hat{\mathbf{a}}_z] \times (-3\hat{\mathbf{a}}_x + 4\hat{\mathbf{a}}_y + 6\hat{\mathbf{a}}_z) \\
&= 90 \times 10^{-6}\,[-5.238\hat{\mathbf{a}}_x - 3.996\hat{\mathbf{a}}_y + 1.475\hat{\mathbf{a}}_z]\ \text{N} \\
&= [-471.42\hat{\mathbf{a}}_x - 359.64\hat{\mathbf{a}}_y + 132.75\hat{\mathbf{a}}_z]\ \mu\text{N}
\end{aligned}$$

or
$$|\,\mathbf{F}\,| = \sqrt{[(471.42)^2 + (359.64)^2 + (132.75)^2]}\ \mu\text{N}$$

$\therefore$
$$|\,\mathbf{F}\,| = \mathbf{607.618\ \mu N}$$

(*b*) The force due to electric field is given by

$$\mathbf{F} = q\,\mathbf{E} = (18 \times 10^{-9})\,[-3\hat{\mathbf{a}}_x + 4\hat{\mathbf{a}}_y + 6\hat{\mathbf{a}}_z]\ \mu\text{N}$$

$$= 18(-3\hat{\mathbf{a}}_x + 4\hat{\mathbf{a}}_y + 6\hat{\mathbf{a}}_z)\ \mu\text{N} = (-54\hat{\mathbf{a}}_x + 72\hat{\mathbf{a}}_y + 108\hat{\mathbf{a}}_z)\ \mu\text{N}$$

or $$|\mathbf{F}| = \sqrt{[(54)^2 + (72)^2 + (108)^2]}\ \mu\text{N}$$

$\therefore$ $$|\mathbf{F}| = \mathbf{140.58\ \mu N}$$

(*c*) Combined effect

$$\mathbf{F} = q\,E + q\,(\mathbf{v} \times \mathbf{B})$$

$$= (-54\hat{\mathbf{a}}_x + 72\hat{\mathbf{a}}_y + 108\hat{\mathbf{a}}_z) + (-417.42\hat{\mathbf{a}}_x - 359.64\hat{\mathbf{a}}_y + 132.75\hat{\mathbf{a}}_z)$$

$$= (-528.42\hat{\mathbf{a}}_x - 287.64\hat{\mathbf{a}}_y + 240.75\hat{\mathbf{a}}_z)\ \mu\text{N}$$

or $$|\mathbf{F}| = \mathbf{645.57\ \mu N}$$

EXAMPLE 2. *A test charge of* 1.6×10^{-19} C *is moving with a velocity of* $(2\hat{\mathbf{a}}_x + 3\hat{\mathbf{a}}_y)$ **m/s** *in an electric field intensity of* $(3\hat{\mathbf{a}}_x + 6\hat{\mathbf{a}}_y + \hat{\mathbf{a}}_z)$ *and a magnetic field of* $2\hat{\mathbf{a}}_x + 3\hat{\mathbf{a}}_z$ tesla. *Determine the magnitude and direction of Lorentz force acting on the test charge.*

Solution. The Lorentz force on test charge is given by

$$\mathbf{F} = q\,[\mathbf{E} + \mathbf{v} \times \mathbf{B}] \qquad \text{...(1)}$$

Let us first calculate the value of $\mathbf{v} \times \mathbf{B}$. We have

$$\mathbf{v} \times \mathbf{B} = \begin{vmatrix} \hat{\mathbf{a}}_x & \hat{\mathbf{a}}_y & \hat{\mathbf{a}}_z \\ 2 & 3 & 0 \\ 2 & 0 & 3 \end{vmatrix}$$

$\therefore$ $$\mathbf{v} \times \mathbf{B} = \hat{\mathbf{a}}_x(9-0) + \hat{\mathbf{a}}_y(-6) + \hat{\mathbf{a}}_z(-6) = 9\hat{\mathbf{a}}_x - 6\hat{\mathbf{a}}_y - 6\hat{\mathbf{a}}_z \qquad \text{...(2)}$$

Substituting the value of **E** and **v** and **B** in eq. (1), we get

$$\mathbf{F} = 1.6 \times 10^{-19}[(3\hat{\mathbf{a}}_x - 6\hat{\mathbf{a}}_y - \hat{\mathbf{a}}_z) + (9\hat{\mathbf{a}}_x - 6\hat{\mathbf{a}}_y - 6\hat{\mathbf{a}}_z)]$$

$$= 1.6 \times 10^{-19}[12\hat{\mathbf{a}}_x - 5\hat{\mathbf{a}}_z]$$

$\therefore$ $$|\mathbf{F}| = 1.6 \times 10^{-19}[\sqrt{(12)^2 + (5)^2}] = \mathbf{20.8 \times 10^{-19}\ N}$$

EXAMPLE 3. *The field* $\mathbf{B} = (-2\hat{\mathbf{a}}_x + 3\hat{\mathbf{a}}_y + 4\hat{\mathbf{a}}_z)$ mT *is present in free space. Find vector force exerted on straight wire carrying* 12 **A** *in* $\hat{\mathbf{a}}_{AB}$ *direction, given* $A(1, 1, 1)$ *and* $B(2, 1, 1)$.

Solution. We know that

$$d\mathbf{F} = I\,(d\mathbf{l} \times \mathbf{B}) \qquad \text{...(1)}$$

Now, $$d\mathbf{l} \times \mathbf{B} = \begin{vmatrix} \hat{\mathbf{a}}_x & \hat{\mathbf{a}}_y & \hat{\mathbf{a}}_z \\ dx & dy & dz \\ -2 & 3 & 4 \end{vmatrix}$$

$$= \hat{\mathbf{a}}_x(4\,dy - 3\,dz) - \hat{\mathbf{a}}_y(4\,dx + 2\,dz) + \hat{\mathbf{a}}_z(3dx + 2\,dy)$$

Let us calculate the value of $d\mathbf{l} \times \mathbf{B}$ at A (1, 1, 1) and B (2, 1, 1). We have

$$= \hat{\mathbf{a}}_x \left[4\int_1^1 dy - 3\int_1^1 dz\right] - \hat{\mathbf{a}}_x \left[4\int_1^1 dx + 2\int_1^1 dz\right] + \hat{\mathbf{a}}_z \left[3\int_1^1 dx + 2\int_1^1 dy\right]$$

$$= \hat{\mathbf{a}}_x [(0-0)] - \hat{\mathbf{a}}_y [(4+0)] + \hat{\mathbf{a}}_z [(3+0)] = -4\hat{\mathbf{a}}_y + 3\hat{\mathbf{a}}_z$$

From eq. (1), $\mathbf{F} = 12 \times 10^{-3} (-4\hat{\mathbf{a}}_y + 3\hat{\mathbf{a}}_z) \text{N} = 12(-4\hat{\mathbf{a}}_y + 3\hat{\mathbf{a}}_z) \text{ mN}$

$\therefore$ $\mathbf{F} = (-48\hat{\mathbf{a}}_y + 36\hat{\mathbf{a}}_z) \text{ mN}$

EXAMPLE 4. *A current* 20 A *flows through each of the parallel long wires which are* 4 cm *apart. Compute the force exerted per unit length of each wire.*

Solution. The force per unit length is given by

$$\frac{F}{l} = \frac{\mu_0 I_1 I_2}{2\pi d} = \frac{\mu_0 I^2}{2\pi d} = \frac{4\pi \times 10^{-7} I^2}{2\pi d}$$

or
$$\frac{F}{l} = \frac{2 \times 10^{-7} I^2}{d} = \frac{2 \times 10^{-7} (20)^2}{4 \times 10^{-2}} = \mathbf{2 \times 10^{-3} \text{ N}}$$

This is a force of attraction if the currents in the conductors are in the same direction but the force is repulsive when the currents are in opposite directions.

EXAMPLE 5. *Two different current elements* $I_1 d\mathbf{l}_1 = 3 \times 10^{-6} \hat{\mathbf{a}}_y$ A.m. *at* $P_1(1, 0, 0)$ *and* $I_2 d\mathbf{l}_2 = 3 \times 10^{-6} (-0.5\hat{\mathbf{a}}_x + 0.4\hat{\mathbf{a}}_y + 0.3\hat{\mathbf{a}}_z)$ A.m. *at* $P_2(2, 2, 2)$ *are located in free space. Calculate the vector force exerted on* $I_2 d\mathbf{l}_2$ by $I_1 d\mathbf{l}_1$.

Solution. The unit vector along R_{12}, *i.e.*, $\hat{\mathbf{a}}_{R_{12}}$ is given by

$$\hat{\mathbf{a}}_{R_{12}} = \frac{(2-1)\hat{\mathbf{a}}_x + (2-0)\hat{\mathbf{a}}_y + (2-0)\hat{\mathbf{a}}_z}{\sqrt{[(1)^2 + (2)^2 + (2)^2]}} = \frac{1 \cdot \hat{\mathbf{a}}_x + 2 \cdot \hat{\mathbf{a}}_y + 2 \cdot \hat{\mathbf{a}}_z}{3}$$

$$= 0.33\hat{\mathbf{a}}_x + 0.67\hat{\mathbf{a}}_y + 0.67\hat{\mathbf{a}}_z$$

$$R_{12} = 3$$

The magnetic induction $d\mathbf{B}_1$ produced by current element $I_1\ d\mathbf{l}_1$ near $I_2\ d\mathbf{l}_2$ is

$$d\mathbf{B}_1 = \frac{\mu_0}{4\pi} \times \frac{I_1\ dl_1 \times \hat{\mathbf{a}}_{R_{12}}}{(R_{12})^2}$$

$$= \frac{4\pi \times 10^{-7}}{4\pi} \times \frac{(3 \times 10^{-6})\hat{\mathbf{a}}_y \times (0.33\hat{\mathbf{a}}_x + 0.67\hat{\mathbf{a}}_y + 0.67\hat{\mathbf{a}}_z)}{(3)^2}$$

$$= \frac{10^{-13}}{3} [(-0.33)\hat{\mathbf{a}}_z + 0 + (0.67)\hat{\mathbf{a}}_x]$$

$$= \frac{10^{-13}}{3} [0.67\hat{\mathbf{a}}_x - 0.33\hat{\mathbf{a}}_z]$$

$$\therefore \quad d\mathbf{F}_2 = I_2\, d\boldsymbol{l}_2 \times d\mathbf{B}_1$$

$$= 3 \times 10^{-6} (-0.5\,\hat{\mathbf{a}}_x + 0.4\,\hat{\mathbf{a}}_y + 0.3\,\hat{\mathbf{a}}_z) \times \frac{10^{-13}}{3} [0.67\,\hat{\mathbf{a}}_x - 0.33\,\hat{\mathbf{a}}_z]$$

$$= 10^{-19} [(-0.5)(-0.33) - \hat{\mathbf{a}}_y + (0.4)(0.67)(-\hat{\mathbf{a}}_z)$$

$$+ (0.4)(-0.33)\,\hat{\mathbf{a}}_x + (0.3)(0.67)\,\hat{\mathbf{a}}_y]$$

$$= 10^{-19}[-0.132\,\hat{\mathbf{a}}_x + 0.036\,\hat{\mathbf{a}}_y - 0.268\,\hat{\mathbf{a}}_z]$$

$$= 10^{-20}[-1.32\,\hat{\mathbf{a}}_x + 0.36\,\hat{\mathbf{a}}_y - 2.67\,\hat{\mathbf{a}}_z]\,\text{N}$$

EXAMPLE 6. *For the above problem, calculate vector force on current element $I_1\, d\mathbf{l}_1$ by current element $I_2\, d\mathbf{l}_2$.*

Solution. In this case $\mathbf{R}_{21} = -\hat{\mathbf{a}}_x - 2\,\hat{\mathbf{a}}_y - 2\,\hat{\mathbf{a}}_z$

$$|\mathbf{R}_{21}| = 3 \text{ and } \hat{\mathbf{R}}_{21} = (-\hat{\mathbf{a}}_x - 2\hat{\mathbf{a}}_y - 2\hat{\mathbf{a}}_z)/3$$

The magnetic induction $d\mathbf{B}_2$ production by current element $I_2\, dl_2$ near current element $I_1\, dl_1$ is given by

$$d\mathbf{B}_2 = \frac{\mu_0}{4\pi} \times \frac{I_2\, dl_2 \times \hat{\mathbf{a}}_{R_{21}}}{(R_{21})^2}$$

$$\therefore \quad d\mathbf{B}_2 = \frac{4\pi \times 10^{-7}}{4\pi} \cdot \frac{3 \times 10^{-6} (-0.5\,\hat{\mathbf{a}}_x + 0.4\,\hat{\mathbf{a}}_y + 0.3\,\hat{\mathbf{a}}_z) \times (-\hat{\mathbf{a}}_x - 2\hat{\mathbf{a}}_y - 2\hat{\mathbf{a}}_z)}{(3)^2 \times 3}$$

$$= \frac{10^{-13}}{(3)^2} [\{(-0.5)(-2)\,\hat{\mathbf{a}}_z + (-0.5)(-2)(-\hat{\mathbf{a}}_y)\} \times \{(0.4)(-1)(-\hat{\mathbf{a}}_z)$$

$$+ (0.4)(-2)(\hat{\mathbf{a}}_x)\} + \{(0.3)(-1)(\hat{\mathbf{a}}_y) + (0.3)(-2)(-\hat{\mathbf{a}}_x)\}]$$

$$= \frac{10^{-13}}{(3)^2} [\{\hat{\mathbf{a}}_z - \hat{\mathbf{a}}_y\} + \{0.4\,\hat{\mathbf{a}}_z - 0.8\,\hat{\mathbf{a}}_x\} + \{-0.3\,\hat{\mathbf{a}}_y + 0.6\,\hat{\mathbf{a}}_x\}]$$

$$= \frac{10^{-13}}{(3)^2} [-0.2\,\hat{\mathbf{a}}_x - 1.3\,\hat{\mathbf{a}}_y + 1.4\,\hat{\mathbf{a}}_z]$$

$$\therefore \quad d\mathbf{F}_2 = I_1\, d\boldsymbol{l}_1 \times d\mathbf{B}_2 = 3 \times 10^{-6}\,\hat{\mathbf{a}}_y \times \frac{10^{13}}{9} [-0.2\hat{\mathbf{a}}_x - 1.3\hat{\mathbf{a}}_y + 1.4\hat{\mathbf{a}}_z]$$

$$= \frac{10^{-19}}{3} [-0.2(-\hat{\mathbf{a}}_z) + 1.4\,\hat{\mathbf{a}}_x]$$

$$= \frac{10^{-19}}{3} [1.4\hat{\mathbf{a}}_x + 0.2\hat{\mathbf{a}}_z]$$

$$= (4.66\hat{\mathbf{a}}_x + 0.667\hat{\mathbf{a}}_z) \times 10^{-20}\,\text{N}$$

EXAMPLE 7. *A loop of wire is constructed of three straight segments connecting* (0, 0, 0) *to* (0.6, 0, 0) *to* (0.4, 1, 0.7) *to* (0, 0, 0). *A current of* 4 mA *is in the direction* $\hat{\mathbf{a}}_x$ *in first segment. The uniform magnetic field is*

$$\mathbf{B} = 0.2\,\hat{\mathbf{a}}_x - 0.1\,\hat{\mathbf{a}}_y + 0.2\,\hat{\mathbf{a}}_z \text{ tesla}$$

Find the force on segment extending from (0, 0, 0) *to* (0.6, 0, 0). *What will be total force on the loop?*

Solution. The current loop is shown in fig (20).

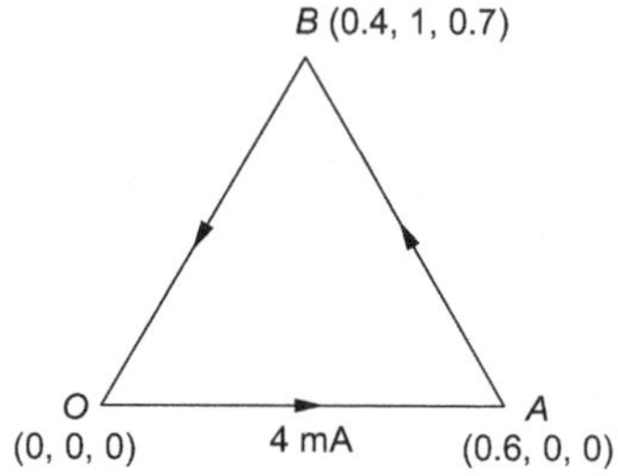

Fig. (20) Showing current loop

We know that the force on a current carrying conductor is given by

$$d\mathbf{F} = I\, d\boldsymbol{l} \times \mathbf{B}$$

Consider the force on a small segment dx of OA. The force is given by

$$d\mathbf{F} = I(dx\,\hat{\mathbf{a}}_x) \times \mathbf{B}$$

$$= (4 \times 10^{-3})dx\,\hat{\mathbf{a}}_x \times (0.2\,\hat{\mathbf{a}}_x - 0.1\,\hat{\mathbf{a}}_y + 0.2\,\hat{\mathbf{a}}_z)$$

$$= (-\,0.4\,\hat{\mathbf{a}}_z - 0.8\,\hat{\mathbf{a}}_y) \times 10^{-3}\, dx \text{ N}$$

$$= (-\,0.4\,\hat{\mathbf{a}}_z - 0.8\,\hat{\mathbf{a}}_y)\, dx \text{ mN}$$

The force on the segment extending from (0, 0, 0) to (0.6, 0, 0) is given by

$$\mathbf{F} = \int_0^{0.6} d\mathbf{F} = \hat{\mathbf{a}}_z \int_0^{0.6} dx - 0.8\,\hat{\mathbf{a}}_y \int_0^{0.6} dx$$

$$= -\,0.4\,\hat{\mathbf{a}}_z\,(0.6) - 0.8\,\hat{\mathbf{a}}_y\,(0.6) = (-\,0.48\,\hat{\mathbf{a}}_y - 0.24\,\hat{\mathbf{a}}_z) \text{ mN}$$

The total force on the loop will be zero because the closed path is subjected to uniform magnetic flux density.

EXAMPLE 8. *Fig.* (21) *shows a long wire carrying a current of* 30 amp. *The rectangular loop carries a current of* 20 amp. *Calculate the resultant force acting on the loop. Assume that* $a = 1$ cm, $b = 80$ cm *and* $l = 30$ cm.

Solution. Consider the left vertical branch of the loop. The current in the wire and vertical left branch of loop is in the same direction. So, the force is attractive and is given by

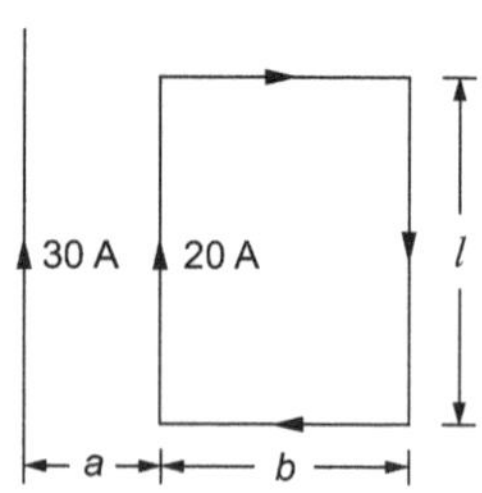

Fig. (21)

$$F_1 = \frac{\mu_0 I_1 I_2 l}{2\pi a} \quad ...(1)$$

The currents between long wire and right branch of loop are in opposite directions. Hence, the force is repulsive. This is given by

$$F = \frac{\mu_0 I_1 I_2 l}{2\pi (a+b)} \quad ...(2)$$

There is no contribution to force due to horizontal branches of the loop.

∴ Resultant force F is given by

$$F = \frac{\mu_0 I_1 I_2 l}{2\pi a} - \frac{\mu_0 I_1 I_2 l}{2\pi(a+b)} = \frac{\mu_0 I_1 I_2 l}{2\pi}\left[\frac{1}{a} - \frac{1}{(a+b)}\right]$$

$$= \frac{\mu_0 I_1 I_2 l}{2\pi} \times \frac{b}{a(a+b)} = \frac{(4\pi \times 10^{-7})(30)(20)(0.3)(0.08)}{2\pi \times (0.1)(0.01+0.08)}$$

$$= \mathbf{3.2 \times 10^{-3}\ N} \text{ directed towards long wire}$$

5.9 NATURE OF MAGNETIC MATERIALS

We have studied the magnetic field due to a current carrying loop. Now we shall investigate the characteristics of a material made of very large number of atoms and molecules. *The response of a material at atomic and molecular level to a magnetic field constitutes magnetic properties*. Here, we shall consider the following basic magnetic parameters which help to understand the response of a magnetic field to magnetic materials.

(1) The Magnetic Dipole (p_m)

We know that matter is made up of atoms. Each atom has a positively charged nucleus at its centre. Around the nucleus, the electrons revolve in various circular orbits. It is called the orbital motion of electrons. The electron also rotates about its own axis. This is called spin motion. The orbital motion of the electron constitutes the current. *Thus, the orbiting electron behaves like a current loop*. The magnetic field produced by the current loop is similar to electric field produced by a small electric dipole. Therefore, a small current loop is called as magnetic dipole. ***The magnetic dipole moment $\mathbf{p}_m$ is defined as the product of area of the plane loop and the magnitude of circulating current***. The direction of the dipole moment is perpendicular to the circulating current as shown in Fig. (22).

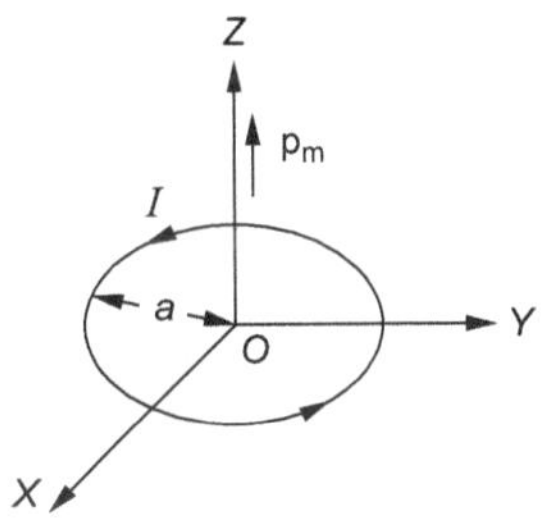

Fig. (22) Showing the direction of dipole moment

The field lines of a magnetic dipole are shown in Fig. (23).

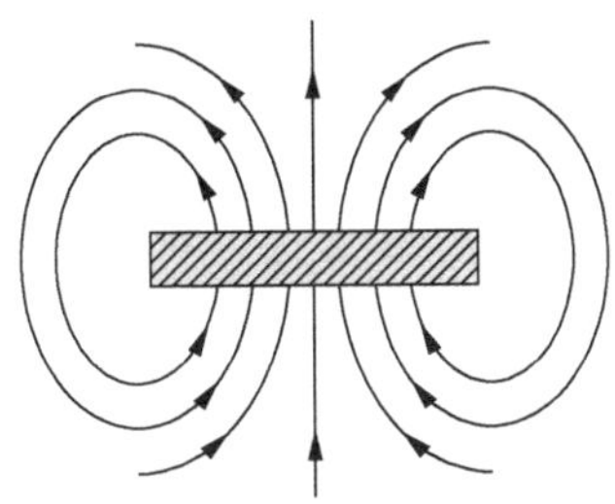

Fig. (23) Showing field lines of a magnetic dipole

The magnetic dipole moment p_m = current in loop × loop area

$$\therefore \quad \boxed{\mathbf{p}_m = \pi a^2 I\, \hat{\mathbf{a}}_z} \text{ A.m}^2$$

(2) Magnetisation and Intensity of Magnetisation (M)

In an unmagnetised substance, the dipoles are oriented at random as shown in Fig. (24a). Hence, the resultant magnetic dipole moment is zero. When a magnetisable substance is placed in the

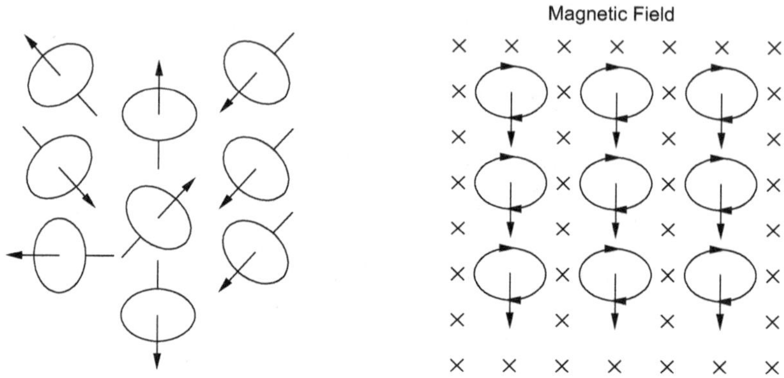

(*a*) Random orientation of magnetic dipoles (*b*) Magnetisation of dipoles

Fig. (24)

magnetic field, the dipoles are aligned parallel to the field as shown in Fig. (24b). Now, the resultant magnetic field is no more zero. As the substance produces its own magnetic field due to alignment of atomic dipoles, we say that the substance is magnetised.

Consider a block of substance having a large number of atomic magnetic dipoles aligned in one direction and distributed uniformly. Such a substance is called uniformly magnetised substance. ***The magnetic moment per unit volume is called magnetisation or intensity of magnetisation***. This is represented by vector **M** and its unit is ampere/metre (A/m).

If N is the number of atoms per unit volume of the substance and p_i is the magnetic dipole moment corresponding to atomic current in the ith atom, then the macroscopic intensity of magnetisation in the substance is

$$\mathbf{M} = \frac{1}{V} \lim_{\Delta V \to 0} \sum_{i=1}^{n\,\Delta V} \mathbf{p}_i$$

where $\sum_{i=1}^{n\,\Delta V} \mathbf{p}_i$ is the vector sum of magnetic moments in volume ΔV of the substance.

The magnetisation **M** of a material may also be expressed in terms of its elementary magnetic dipole moment $\mathbf{p}_m$ as dipole moment per unit volume, *i.e.*,

$$\boxed{\mathbf{M} = N\,\mathbf{p}_m}$$

where N is the number of dipoles per unit volume.

(3) Magnetic Induction (B)

When a substance is magnetised, it has two types of magnetic lines of force, (*i*) due to magnetised field and (*ii*) due to magnetisation of the substance itself. The lines of force in this case are known as lines of magnetisation. These are shown in Fig. 25(a). Fig. 26(b) shows the resultant field. It is obvious from Fig. 25(b), that lines of force are increased inside the substance whereas

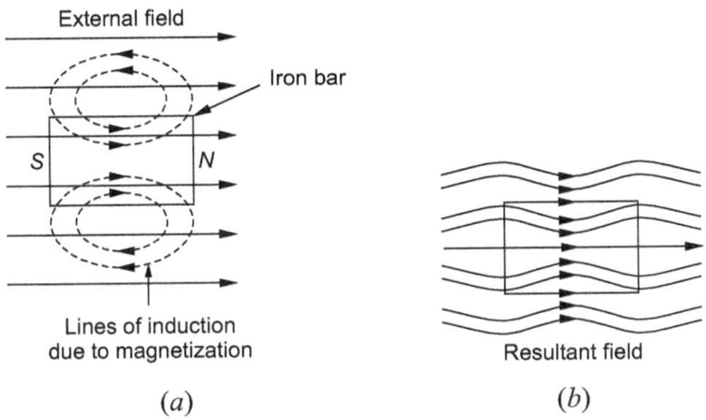

Fig. (25)

they are much reduced at certain places outside the bar. The resultant inside field in the substance is referred as magnetic induction **B**.

***Thus, the magnetic induction* B *is defined as the total number of lines of force per unit area due both to the magnetising field and to the induced magnetism in the substance.* Unit of B is weber per metre^2.**

(4) Magnetic Permeability (μ)

The magnetic permeability of any medium is defined as the ratio of number of lines of magnetic induction per unit area (magnetic flux density) in that medium to the of number of lines per unit area when the medium is replaced by vacuum (magnetic field intensity), *i.e.*,

$$\boxed{\mu = \frac{\mathbf{B}}{\mathbf{H}}}$$

μ is called as permeability of the medium. Its unit is henry/metre. For vacuum $\mu = \mu_0 = 4\pi \times 10^{-7}$ henry/metre.

The relative permeability is defined as the ratio of permeability of the medium to the permeability of free space

$$\mu_r = \left(\frac{\mu}{\mu_0}\right)$$

This is dimensionless quantity.

(5) Magnetic Susceptibility (χ_m)

The magnetic susceptibility of a material is a measure of the ease with which the material may be magnetised. ***It is defined as the ratio of the intensity of magnetisation* M *to the magnetic field intensity* H.**

$$\boxed{\chi_m = \left(\frac{\mathbf{M}}{\mathbf{H}}\right)}$$

5.10 RELATION BETWEEN B, H AND M

The magnetic flux density **B** in a material due to magnetising force may be regarded as the sum of the magnetic flux density in vacuum **B** produced by that magnetising force and the magnetic flux density $\mathbf{B}_m$ due to the magnetisation of material. Thus,

$$\mathbf{B} = \mathbf{B}_0 + \mathbf{B}_m \qquad ...(1)$$

But $\mathbf{B}_0 = \mu_0 \mathbf{H}$ = lines of force crossing unit area in a normal direction due to external field in vacuum.

$\mathbf{B}_m = \mu_0\mathbf{M}$ = lines of force crossing unit area normal due to intensity of magnetisation **M**.

The two sets of lines of force inside the substance are called as lines of induction.

$$\therefore \qquad \mathbf{B} = \mu_0\mathbf{H} + \mu_0\mathbf{M}$$

or

$$\boxed{\mathbf{B} = \mu_0\,[\mathbf{H} + \mathbf{M}]} \qquad ...(2)$$

5.11 RELATION BETWEEN RELATIVE PERMEABILITY AND SUSCEPTIBILITY

We know that

$$\mathbf{B} = \mu_0\mathbf{H} + \mu_0\mathbf{M} \qquad ...(1)$$

Dividing by **H**, we get

$$\frac{\mathbf{B}}{\mathbf{H}} = \mu_0 + \mu_0\,\frac{\mathbf{M}}{\mathbf{H}} \qquad ...(2)$$

Here, $\dfrac{\mathbf{B}}{\mathbf{H}} = \mu$, permeability of the material.

$\dfrac{\mathbf{M}}{\mathbf{H}} = \chi_m$, susceptibility of the material

$$\therefore \qquad \mu = \mu_0 + \mu_0\,\chi_m$$

or

$$\frac{\mu}{\mu_0} = 1 + \chi_m$$

or

$$\boxed{\mu_r = 1 + \chi_m} \qquad ...(3)$$

or

$$\boxed{\chi_m = (\mu_r - 1)} \qquad ...(4)$$

5.12 RELATION BETWEEN MAGNETISING CURRENT DENSITY $\mathbf{J}_m$ AND INTENSITY OF MAGNETISATION (M)

The intensity of magnetisation in a substance is given by

$$\mathbf{M} = \frac{1}{V}\lim_{\Delta V \to 0}\sum_{i=1}^{n\Delta V}\mathbf{p}_i \qquad ...(1)$$

The magnetisation creates a current I_m which is bound to the atom. Applying Ampere's circuital law to the atom, we can write

$$I_m = \oint \mathbf{H}\cdot d\mathbf{l} = \int_S \mathbf{J}_m \cdot d\mathbf{S} \qquad ...(2)$$

Using Stoke's theorem, we have

$$\int_S \vec{\nabla} \times \mathbf{M} \cdot d\mathbf{S} = \int_S \mathbf{J}_m \cdot d\mathbf{S} \qquad ...(3)$$

Thus, the magnetisation can be related to magnetisation current. The integrals of eq. (3) must be valid for any arbitrary surface, we obtain

$$\boxed{\vec{\nabla} \times \mathbf{M} = \mathbf{J}_m} \qquad ...(4)$$

If I be the convectional current enclosed by a closed path, I_m be the bound or magntisation current and I_t be the total current, then

$$\vec{\nabla} \times \mathbf{H} = \mathbf{J}$$

$$\vec{\nabla} \times \frac{\mathbf{B}}{\mu_0} = J_T$$

and

$$\vec{\nabla} \times \mathbf{M} = J_m$$

EXAMPLE 1. *Find magnetisation in a magnetic material where*

(*i*) $\mu = 1.8 \times 10^{-5}$ h/m, $|\mathbf{H}| = 120$ A/m

(*ii*) $\mu_r = 22$.

There are 8.3×10^{28} atoms/m^3 *and each atom has a dipole moment of* 4.5×10^{-27} A.m^2.

(*iii*) $B = 300\ \mu$T *and* $\chi_m = 15$.

Solution.

(*i*) Given that $\mu = 1.8 \times 10^{-5}$ h/m and $|\mathbf{H}| = 120$ A/m

We know that, $\chi_m = \frac{M}{H}$ or $M = \chi_m H$

Further, $\chi_m = (\mu_r - 1)$

$\therefore$ $M = (\mu_r - 1)H$

Here, $\mu_r = \frac{\mu}{\mu_0} = \frac{1.8 \times 10^{-5}}{4\pi \times 10^{-7}} = 14.33$

$\therefore$ $M = (14.33 - 1) \times 120 =$ **1599.74 A/m**

(*ii*) Magnetisation M is defined as magnetic dipole moment per unit volume

or $\mathbf{M} = N\mathbf{p}_i$

$\therefore$ $M = (8.3 \times 10^{28}) \times (4.5 \times 10^{-27}) =$ **373.5 A/m**

(*iii*) We know that

$$B = \mu_0 \mu H \quad \text{or} \quad H = \frac{B}{(\mu_0 \mu_r)}$$

$\therefore$ $H = \frac{300 \times 10^{-6}}{4\pi \times 10^{-7} \times 16}$ $\qquad (\because \mu_r = 1 + \chi_m = 1 + 15 = 16)$

Now, $M = \chi_m H = 15 \times (14.92) =$ **223.925 A/m**

EXAMPLE 2. *A specimen of iron is uniformly magnetised by a magntising field of* 500 *A/m. If the magnetic induction in the specimen is* 0.2 Wb/m^2, *find the relative permeability and the susceptibility.*

Solution. We know that

$$B = \mu H = \mu \mu_r H$$

$$\mu_r = \frac{B}{\mu_0 H} = \frac{0.2}{4\pi \times 10^{-7} \times 500} = \mathbf{318.5}$$

Now,

$$\chi = (\mu_r - 1) = 318.5 - 1 = \mathbf{317.5}$$

EXAMPLE 3. *Magnetisation in a magnetic material for which* $\chi_m = 8$ *is given in a certain region as* $H = 150\, z^2\, \hat{\mathbf{a}}_x$ A/m *at* $z = 4$ cm. *Find the magnitude of (i)* $\mathbf{J}_T$, *(ii)* $\mathbf{J}$ *and (iii)* $\mathbf{J}_m$.

Solution.

(*i*) $\mathbf{J}_T = \vec{\nabla} \times \left(\dfrac{\mathbf{B}}{\mu_0}\right) = \vec{\nabla} \times (\mu_0\, \mu_r\, \mathbf{H})/\mu_0$

$\therefore$ $$\mathbf{J}_T = \vec{\nabla} \times \mu_r\, \mathbf{H}$$

where $$\mu_r = \chi_m + 1 = 9$$

or $$J_T = \vec{\nabla} \times (9 \times 150\, z^2\, \hat{\mathbf{a}}_x) \text{ A/m}^2$$

$$= \begin{vmatrix} \hat{\mathbf{a}}_x & \hat{\mathbf{a}}_y & \hat{\mathbf{a}}_z \\ \dfrac{\partial}{\partial x} & \dfrac{\partial}{\partial y} & \dfrac{\partial}{\partial z} \\ 1350 z^2 & 0 & 0 \end{vmatrix}$$

$$= \hat{\mathbf{a}}_x\, (0 - 0) + \hat{\mathbf{a}}_y\, (2700\, z - 0) + \hat{\mathbf{a}}_z\, (0 - 0)$$

$\therefore$ $$\mathbf{J}_T = 2700\, z\, \hat{\mathbf{a}}_y$$

Magnitude of $\mathbf{J}_T$ at $z = 4$ cm $= 0.4$ m is given by

$$|\,\mathbf{J}_T\,| = 2700 \times 0.04 = \mathbf{108\ A/m^2}$$

(*ii*) $$\mathbf{J} = \vec{\nabla} \times \mathbf{H} = \begin{vmatrix} \hat{\mathbf{a}}_x & \hat{\mathbf{a}}_y & \hat{\mathbf{a}}_z \\ \dfrac{\partial}{\partial x} & \dfrac{\partial}{\partial y} & \dfrac{\partial}{\partial z} \\ 150 z^2 & 0 & 0 \end{vmatrix} = \hat{\mathbf{a}}_x\, (0 - 0) + \hat{\mathbf{a}}_y\, (300\, z - 0) + \hat{\mathbf{a}}_z\, (0 - 0) = 300\, z\, \hat{\mathbf{a}}_y$$

At $z = 4$, the magnitude of $|\,\mathbf{J}\,|$ is given by

$$|\,\mathbf{J}\,| = 300 \times 0.04 = \mathbf{12\ A/m^2}$$

(*iii*) $$\mathbf{J}_m = \mathbf{J}_T - \mathbf{J} = 108\, \hat{\mathbf{a}}_y - 12\, \hat{\mathbf{a}}_y = 96\, \hat{\mathbf{a}}_y$$

$$|\,\mathbf{J}_m\,| = \mathbf{96\ A/m^2}$$

5.13 CLASSIFICATION OF MAGNETIC MATERIALS

On the basis of magnetic behaviour, the materials can be classified as *diamagnetic, paramagnetic* and *ferromagnetic*. The magnetic behaviour may be due to electron orbital motion, electron spin and nuclear spin. Let us discuss the classification one by one.

(1) Diamagnetic Materials

The electron moving in an orbit is equivalent to a small current loop and acts like a dipole. When an external magnetic field is applied, the orbital magnetic dipole has a direction opposite to the applied magnetic field. Therefore, ***the substances which when placed in a magnetic field acquire feeble magnetism opposite to the direction of the field are known as diamagnetic substances***. Examples are: Bismuth (Bi), Zinc (Zn), Copper (Cu), Silver (Ag), Gold (Au), Salt (NaCl), Water (H_2O), Mercury (Hg), Hydrogen (H_2), etc. The permeability μ for a diamagnetic substance is less than one and susceptibility is negative. When a diamagnetic substance is less than one and susceptibility is negative. When a diamagnetic substance is placed in a non-uniform field, then it tends to move towards the weaker part from the stronger part of the field.

(2) Paramagnetic Substances

In such materials, the magnetic moments created by the orbiting and spinning electrons do not completely cancel each other and thus, leaving the atom with a small net magnetic moment. In other words, we can say that individual dipoles do not interact with each other and are randomly distributed in the absence of external applied field. This causes paramagnetism. When an external magnetic field is applied, it tends to align the dipoles in the direction of applied field. Therefore, ***the substances which when placed in a magnetic field, acquire feebble magnetism in the direction of magnetic field are known as paramagnetic substances***. Examples are : Aluminium (A1), Platinum (Pt), Manganese (Mn). Copper chloride ($CuCl_2$), Oxygen (O_2) solutions of salts of iron, etc. The permeability μ is greater than one and susceptibility is very small and positive. In non-uniform magnetic field, the paramagnetic substances are attracted towards the stronger part of the magnetic field from the weaker part of the field.

(3) Ferromagnetic Substances

The ferromagnetic material consists of *domains*. Each domain has all the magnetic moments aligned in the same direction and separated by other domains by domain wall as shown in Fig. (26). The net magnetisation of randomly oriented domains is zero.

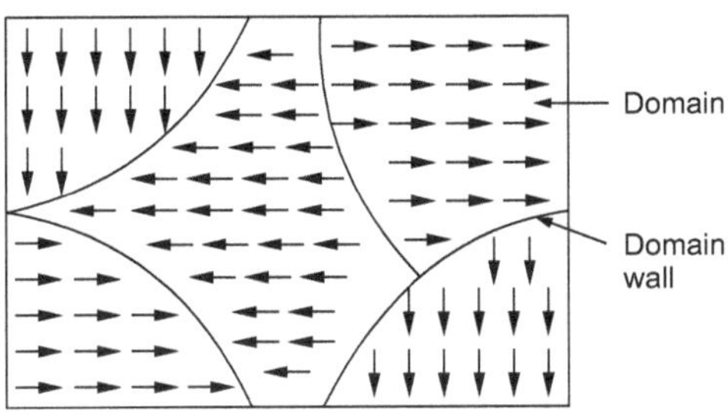

Fig. (26) Domain structure of a ferromagnetic specimen. The magnetic moments of all the atoms in one domain are pointed in the same direction.

Under the influence of an external magnetic field, the domains that have their magnetic fields aligned with applied field grow, *i.e.*, there is a displacement of domain wall. Thus, the magnetic moment of the substance or magnetism increases. Therefore, ***the substances which when placed in external magnetic field, become strongly magnetised in the direction of field are known as ferro-magnetic substances***. The examples are : Iron (Fe). Nickel (Ni), Cobalt (Co), Magnetite (Fe_3O_4), etc.

They possess all the properties of paramagnetic substances with much greater intensity. The susceptibility of a ferromagnetic substance decreases steadily with rise in temperature according with Curie's law.

$$\chi \propto \frac{1}{T}$$

where T is absolute temperature. Above certain temperature, called Curie temperature, ferromagnetic substance change to paramagnetic substances.

5.14 BOUNDARY RELATIONS IN MAGNETIC FIELDS

In a single medium, the magnetic field is continuous. At the boundary between two magnetic materials, the magnetic field may change abruptly both in magnitude and direction. The aim of this article is to consider that how the field vectors **B** and **H** change in passing an interface between two media. The two media may be two materials with different magnetic properties or material and vacuum. Consider two magnetically isotropic homogeneous linear materials [medium 1 and medium 2] with permeabilities μ_1 and μ_2 respectively in contact as shown in Fig. (27).

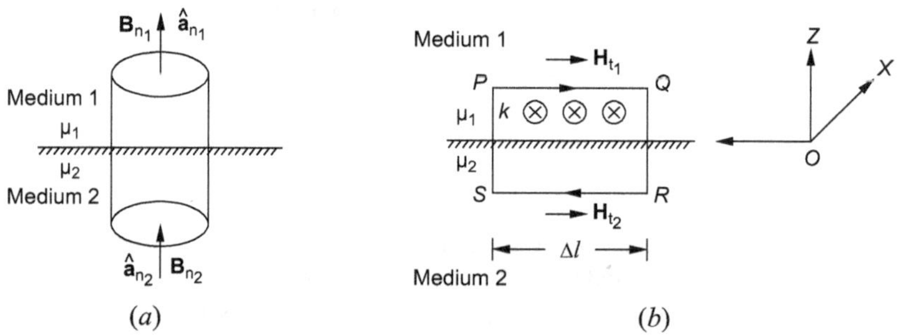

Fig. (27) Boundary between two media 1 and 2 with field components

(1) Condition for Normal Component of B

Consider a cylindrical surface which lies half in both the media of permeability μ_1 and μ_2. The height of the cylinder is assumed to be small so that the magnetic flux through the curved surface is negligible.

Applying Gauss's law to magnetic field, we have

$$\oint \mathbf{B} \cdot d\mathbf{S} = 0$$

Let $\mathbf{B}_{n_1}$ and $\mathbf{B}_{n_2}$ be the normal components of magnetic flux densities respectively.

Then

$$\mathbf{B}_{n_1} \cdot \hat{\mathbf{a}}_{n_1} \Delta S - \mathbf{B}_{n_2} \cdot \hat{\mathbf{a}}_{n_2} \Delta S = 0$$

or $$B_{n_1} \Delta S - B_{n_2} \Delta S = 0$$

or $$\boxed{B_{n_1} = B_{n_2}} \quad ...(1)$$

Thus, the normal components of magnetic flux densities are equal at the boundary between the two magnetic media, *i.e.*, ***normal component of* B *is continuous across the boundary between two media***.

Let us express eq. (1) in terms of field intensities, *i.e.*,

$$\mu_1 H_{n_1} = \mu_2 H_{n_2}$$

or $$\boxed{H_{n_2} = \frac{\mu_1}{\mu_2} H_{n_1}} \quad ...(2)$$

This shows that the normal component of **H** is discontinuous by the ratio $\frac{\mu_1}{\mu_2}$.

(2) Condition for Tangential Component of H

Let us consider a closed rectangular path *PQRS* with its long sides *PQ* and *RS* parallel to the interface and close to it.

Applying Ampere's law at the boundary

$$\oint \mathbf{H} \cdot dl = I = k\,\Delta l$$

At boundary, there exists a sheet current of density k.

For a closed path, $H_{t_1} \Delta l - H_{t_2} \Delta l = k\,\Delta l$

or $$\boxed{H_{t_1} - H_{t_2} = k} \quad ...(3)$$

If the surface current density is zero, then

$$\boxed{H_{t_1} = H_{t_2}} \quad ...(4)$$

The direction of vectors in eq. (4) are specified by the relation

$$\hat{\mathbf{a}}_n (\mathbf{H}_1 - \mathbf{H}_2) = \mathbf{k} \quad ...(5)$$

For tangential components,

$$\mathbf{H}_{t_1} - \mathbf{H}_{t_2} = \mathbf{k} \times \hat{\mathbf{a}}_n \quad ...(6)$$

For tangential **B**, we have

$$\boxed{\frac{B_{t_1}}{\mu_1} - \frac{B_{t_2}}{\mu_2} = k} \quad ...(7)$$

If H_1 (or B_1) and H_2 (or B_2) make angles θ_1 and θ_2 with the normal to the interface between two media as shown in Fig. (28), then we have

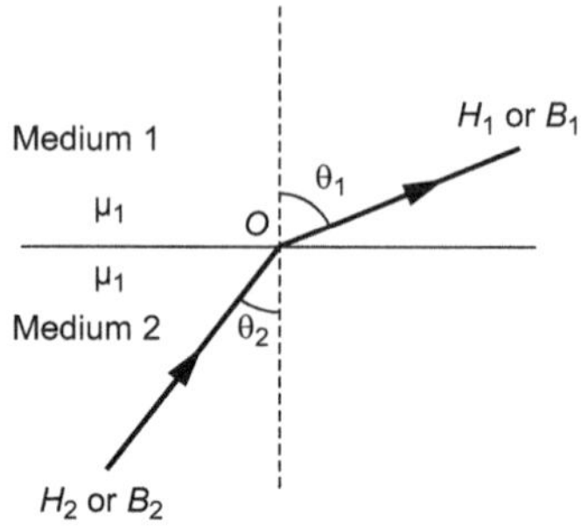

Fig. (28)

$$H_{t_1} = H_{t_2} \quad \text{or} \quad H_1 \sin\theta_1 = H_2 \sin\theta_2 \qquad \text{...(8)}$$

and
$$B_{n_1} = B_{n_2}$$

or
$$B_1 \cos\theta_1 = B_2 \cos\theta_2 \qquad \text{...(9)}$$

Dividing eq. (8) by eq. (9), we get

$$\frac{H_1}{B_1}\tan\theta_1 = \frac{H_2}{B_2}\tan\theta_2 \quad \text{or} \quad \frac{\tan\theta_1}{\tan\theta_2} = \frac{B_1 H_2}{B_2 H_1} = \left(\frac{B_1}{H_1}\right)\left(\frac{H_2}{B_2}\right)$$

or
$$\boxed{\frac{\tan\theta_1}{\tan\theta_2} = \frac{\mu_1}{\mu_2}}$$

This is the law of refraction for magnetic flux lines at the boundary with no surface current.

5.15 INDUCTORS AND INDUCTANCE

We have studied a capacitor which stores energy in an electric field. Similarly, ***inductor is a device which stores energy in magnetic field***. Solenoids, coils, loops, etc. are examples of inductors. Like resistance and capacitance, the inductance is also a parameter which is attached to the electric circuit.

In order to understand the definition of inductance, let us consider a solenoid coil in which a current I is passing. The solenoid is shown in Fig. (29). Now, magnetic lines are produced by

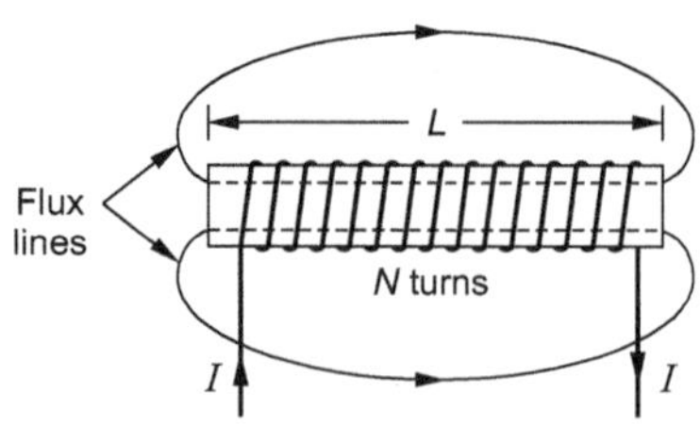

Fig. (29) Showing a solenoid

the current. Each line passes through the entire solenoid. Let N be the total number of turns in the solenoid. The total flux linked with the solenoid would be $N\,\Phi_m$ where Φ_m is the flux liked with a coil.

The inductance is defined as the ratio of total magnetic flux linked to the current producing the flux. Thus,

Inductance, $$L = \frac{N \Phi_m}{I}$$

The unit of inductance is henry

$$1 \text{ henry} = 1 \text{ wb/A}$$

This definition holds for materials having constant permeability.

5.16 INDUCTANCE OF A SOLENOID

Consider a long solenoid of length l metre and uniform cross-sectional area A metre^2 as shown in Fig. (29). Let N be the number of turns. Suppose a current I amp. flows through it. The magnetic field inside the solenoid is given by

$$B = \frac{\mu N I}{l} \quad ...(1)$$

where μ is the permeability of the material (usually air).

The total flux linked with the solenoid is given by

$$\text{Total flux} = N \Phi_m = N \int_S \mathbf{B} \cdot d\mathbf{S} = N B A \quad ...(2)$$

Substituting the value of B from eq. (1) in eq. (2), we get

$$\text{Total flux} = N \left(\frac{\mu N I}{l} \right) A = \frac{\mu N^2 A I}{l}$$

Now, Inductance, $$L = \frac{\text{Total flux}}{\text{Current}} = \frac{\mu N^2 A I}{I l}$$

or $$\boxed{L = \frac{\mu N^2 A}{l}} \quad ...(3)$$

5.17 INDUCTANCE OF A TOROID

A toroid may be regarded as a solenoid bent into a circular form so that its ends are joined together. Thus, it may be called as endless solenoid. Fig. (30a) shows a toroid having circular cross-section. Let R be the mean radius of the toroid and N, the number of turns in it. The magnetic flux lines form concentric circles as shown in Fig. (30b).

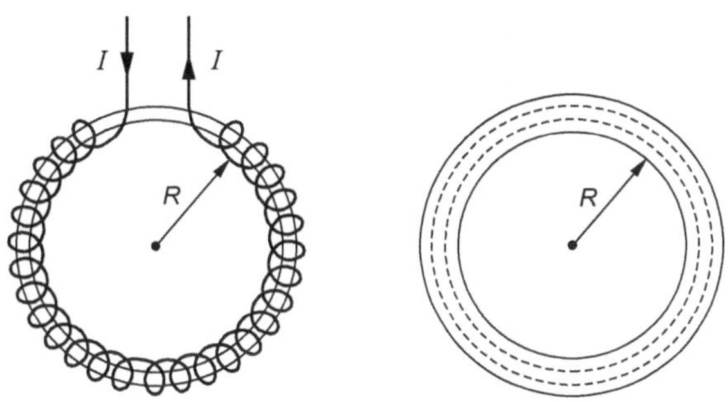

Fig. (30)

Let I amp. be the current flowing in the toroid and B, the magnetic field produced due to this current on the circular path. Then, according to Ampere's law

$$\oint \mathbf{B} \cdot d\mathbf{l} = \mu \times \text{current enclosed by the path} = \mu N I$$

As the magnetic field **B** is tangential around the circular path of radius R, we have

$$B \times 2 \pi R = \mu N I \quad \text{or} \quad B = \frac{\mu N I}{2 \pi R} \qquad ...(1)$$

Let A is the area of cross-section of each turn of the toroid, then magnetic flux limked with the toroid is

$$\text{Total flux} = B N A \qquad ...(2)$$

Substituting the value of B from eq. (1) in eq. (2), we get

$$N\Phi_m = \left(\frac{\mu N I}{2 \pi R}\right) N A \quad \text{or} \quad N\Phi_m = \frac{\mu_0 N^2 I A}{2 \pi R} \qquad ...(3)$$

Now,
$$L = \frac{N \Phi_m}{I} = \frac{\mu N^2 I A}{2 \pi R}$$

or
$$\boxed{L = \frac{\mu N^2 I A}{2 \pi R}} \text{ henry} \qquad ...(4)$$

or
$$L = \frac{\mu N^2 r^2}{2 \pi R}$$

or
$$\boxed{L = \frac{\mu N^2 r^2}{2 R}}$$

where r is radius of cross-section of the toroid.

5.18 INDUCTANCE PER UNIT LENGTH OF COAXIAL CONDUCTORS

Consider a coaxial cable having conducting cylinders of radii a and b as shown in Fig. (31). The internal conductor carries a current I in Z-direction while the outer conductor carries the same current I in $-Z$ direction. The aim of this article is to calculate the inductance of this system.

Consider a distance r between the conductors. The magnetic induction **B** between the conductors at a distance r is given by

$$\mathbf{B} = \frac{\mu_0 I}{2 \pi r} \hat{\mathbf{a}}_\phi$$

The currents in the two conductors are linked by the flux across surface Φ = constant. For a length l of the cable

$$\Phi_m = \int_S \mathbf{B} \cdot d\mathbf{S} \text{ where } d\mathbf{S} = dr\ dz\ \hat{\mathbf{a}}_\phi$$

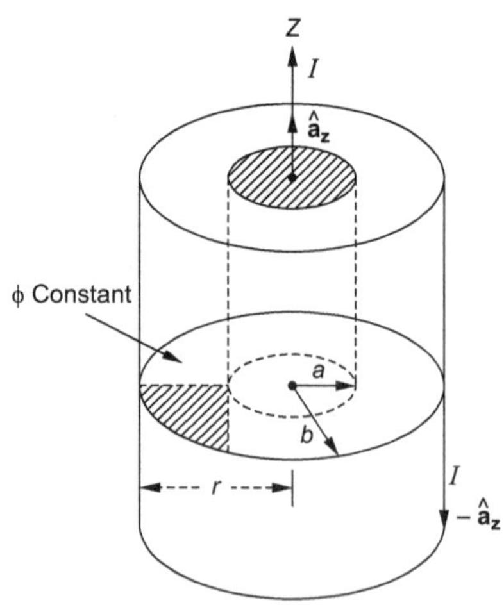

Fig. (31) Coaxial cable

$$\Phi_m = \int_0^1\int_a^b \left(\frac{\mu_0 I \hat{\mathbf{a}}_\phi}{2\pi r}\right)(dr\,dz\,\hat{\mathbf{a}}_\phi) = \int_0^1\int_a^b \frac{\mu_0 I}{2\pi r}\,dr\,dz$$

$$= \frac{\mu_0 I}{2\pi}\int_0^1 dz \int_a^b \frac{dr}{r} = \frac{\mu_0 I}{2\pi} \times l \times \log_e\left(\frac{b}{a}\right)$$

Now inductance, $L = \dfrac{\Phi_m}{I} = \dfrac{\mu_0 I}{2\pi}\log_e\left(\dfrac{b}{a}\right)$

Inductance per unit length of the cable

$$\boxed{L = \frac{\mu_0}{2\pi}\log_e\left(\frac{b}{a}\right)}\frac{\text{H}}{\text{m}} \qquad (\because l = 1)$$

5.19 ENERGY DENSITY IN MAGNETIC FIELD

Consider a very long solenoid of length l and cross-sectional area A. When a current flows in it, a magnetic field is established. This field is uniform inside and negligible outside. So, the volume associated with the magnetic field is Al.

We have seen that the amount of workdone in establishing a current I_0 in the solenoid is $\left(\frac{1}{2}\right)LI_0^2$ where L is the inductance of solenoid. The workdone is stored as energy in the magnetic field.

$$\therefore \qquad U = \text{energy stored} = \frac{1}{2}LI_0^2$$

The inductance of the solenoid is given by

$$L = \mu_0 n^2 A\, l.$$

where n is number of turns in solenoid per metre.

$$\therefore \qquad U = \frac{1}{2}(\mu_0 n_2 A\, I)LI_0^2 = \frac{1}{2}\frac{(\mu_0 n I_0)^2}{\mu_0}A\,l$$

The magnetic field inside the solenoid $= B = \mu_0 n I_0$

$$\therefore \qquad U = \frac{1}{2}\cdot\frac{B^2}{\mu_0}A\,l$$

This energy is uniformly distributed throughout the volume $A\,l$ of the magnetic field. So, the energy density (energy per unit volume) u in magnetic field is given by

$$u = \frac{U}{A\,l} = \frac{1}{2}\cdot\frac{B^2}{\mu_0}\cdot\frac{A\,l}{A\,l} = \frac{1}{2}\frac{B^2}{\mu_0}\ \text{joule/metre}^3$$

$$\boxed{u = \frac{B^2}{2\mu_0}}\ \frac{\text{joule}}{\text{metre}^3}$$

It should be noted that inductance in a circuit plays the same role as the mass or inertia does in mechanical motion.

Exercises and Problems

EXERCISES

1. State and explain Biot-Savart's law
2. Find the magnetic field due to a long straight conductor carrying current.
3. Apply Biot-Savart's law to calculate magnetic field of a circular current carrying loop.
4. Calculate the magnetic field due to current carrying solenoid.
5. State and explain Ampere's law of forces.
6. State and explain Ampere's law both in integral and differential forms as used in magnetic fields.
7. Using Ampere's circuital law, find the magnetic field intensity produced by an infinitely long filament carrying a current.
8. Explain the Ampere's circuital law in point form.
9. Determine the magnetic force in terms of moving charge.
10. Calculate the Lorentz force on a current carrying conductor.
11. Determine the force between two long and parallel current carrying conductors.
12. Define:
    (*i*) Magnetic dipole and magnetic dipole moment
    (*ii*) Intensity of magnetisation (**M**)
    (*iii*) Magnetic induction (**B**)
    (*iv*) Magnetic susceptibility (χ_m)
13. Derive the following relation between relative permeability (μ_r) and susceptibility (χ_m).
    $$\chi_m = (\mu_r - 1)$$
14. Derive a relation between magnetising current density ($\mathbf{J}_m$) and intensity of magnetisation (**M**).
15. Give a classification of magnetic materials.
16. Discuss about diamagnetic, paramagnetic and ferromagnetic substances.
17. Give the domain theory of ferromagnetic substances.
18. Derive general expressions for the boundary relations for static magnetic field for:
    (*i*) tangential components, (*ii*) normal components.
19. Show that normal component of B is continuous across the boundary between two media.
20. What are inductors? Define inductance.
21. Derive expression for inductance of a solenoid.
22. What do you mean by inductance? Obtain an expression for the inductance of a toroid.
23. Derive an expression for inductance of solenoid and toroid.
24. Using Biot-Savart's law, derive an expression for inductance per unit length of a long coaxial cable with radii of inner and outer conductors 'a' and 'b'.
25. Derive an expression for the energy stored in an inductance.

PRACTICE PROBLEMS

1. A current element $I\,\Delta l = 2\pi\,(0.6\,\hat{\mathbf{a}}_x - 0.8\,\hat{\mathbf{a}}_y)$ µA is situated at point (4, – 2, 3). Find the incremental field $\Delta\mathbf{H}$ at a point (1, 3, 2).

[**Ans.** $\Delta\mathbf{H} = 1.932\,\hat{\mathbf{a}}_x + 1.449\,\hat{\mathbf{a}}_y + 1.449\,\hat{\mathbf{a}}_z$ µA/m]

[**Hint.** $\Delta\mathbf{H} = \dfrac{I\,dl \times \hat{\mathbf{a}}_r}{4\pi r^2}$

Given
$$I\,\Delta\mathbf{l} = 2\pi\,(0.6\,\hat{\mathbf{a}}_x - 0.8\,\hat{\mathbf{a}}_y)$$
$$\mathbf{r} = (1-4)\,\hat{\mathbf{a}}_x + (3+2)\,\hat{\mathbf{a}}_y + (2-3)\,\hat{\mathbf{a}}_z = -3\,\hat{\mathbf{a}}_x + 5\,\hat{\mathbf{a}}_y - \hat{\mathbf{a}}_z$$
$$\hat{\mathbf{a}}_r = \frac{3\,\hat{\mathbf{a}}_x + 5\,\hat{\mathbf{a}}_y - \hat{\mathbf{a}}_z}{\sqrt{(3)^2 + (5)^2 + (1)^2}} = -\frac{3\,\hat{\mathbf{a}}_x + 5\,\hat{\mathbf{a}}_y - \hat{\mathbf{a}}_z}{\sqrt{35}}$$
$$r = \sqrt{35}\ ]$$

2. Find the vector magnetic field intensity in cartesian coordinates at a point $P(1.5, 2, 3)$ caused by a current filament of 24 amp. in $\hat{\mathbf{a}}_z$ direction on Z-axis extending from $z = 6$ to ∞

[**Ans.** $\mathbf{H}_z = -0.1417\,\hat{\mathbf{a}}_x + 0.1062\,\hat{\mathbf{a}}_y$ A/m]

[**Hint.** $r = (1.5-0)\,\hat{\mathbf{a}}_x + (2-0)\,\hat{\mathbf{a}}_y + (z-z)\,\hat{\mathbf{a}}_z$
$$d\mathbf{H}_z = \frac{I\,dl \times \hat{\mathbf{a}}_z}{4\pi r^2}\ ]$$

3. Consider the volume current density distribution in cylindrical coordinates as
$$\mathbf{J}(\rho, \phi, z) = 0 \qquad 0 < \rho < a$$
$$= J_0\left(\frac{\rho}{a}\right)\hat{\mathbf{a}}_z \qquad 0 < \rho < b$$
$$= 0 \qquad b < \rho < \infty$$
Find the magnetic field intensity $\mathbf{H}$ in region $a < \rho < b$.

[**Ans.** $\mathbf{H} = \dfrac{J_0\,(\rho^3 - a^3)}{3\,a\,\rho}\,\hat{\mathbf{a}}_\phi$ and at b, $\mathbf{H} = \dfrac{J_0\,(\rho^3 - a^3)}{3\,a\,\rho}\,\hat{\mathbf{a}}_\phi$]

[**Hint.** $\oint \mathbf{H}\cdot dl = \oint_S \mathbf{J}\cdot d\mathbf{S}$ where $d\mathbf{S} = \rho\,d\rho\,d\phi\,\hat{\mathbf{a}}_z$

or
$$\oint \mathbf{H}\cdot dl = \int_a^\rho \int_0^{2\pi} J_0\left(\frac{\rho}{a}\right)\hat{\mathbf{a}}_z \cdot \rho\,d\rho\,d\phi\,\hat{\mathbf{a}}_z$$

or
$$H\,2\pi\rho = \frac{J_0}{a}\left[\frac{\rho^3}{3}\right]_a^\rho [\phi]_0^{2\pi}$$
$$\mathbf{H} = \frac{J_0}{3\,a\,\rho}\,(\rho^3 - a^3)\,\hat{\mathbf{a}}_\phi$$

At $\rho = b$,
$$\mathbf{H} = \frac{J_0}{3\,a\,b}\,(b^3 - a^3)\,\hat{\mathbf{a}}_\phi$$

4. Given a current distribution in cylindrical coordinates

$$\mathbf{J}(\rho, \phi, z) = 45\, e^{-2\rho}\, \hat{\mathbf{a}}_z \qquad 0 < \rho < 0.5$$
$$= 0 \qquad 0.5 < \rho < \infty$$

Find the field intensity at different points.

[**Ans.** $\mathbf{H} = \dfrac{1.125}{\rho}[1 - e^{-2\rho} - 2\rho\, e^{-2\rho}]\, \hat{\mathbf{a}}_\phi$ for $0 < \rho < 0.5$

and $\mathbf{H} = \dfrac{0.297}{\rho}\, \hat{\mathbf{a}}_\phi$ for $\rho \geq 0.5$]

5. Calculate the magnetic flux density at the centre of a current carrying loop when the radius of the loop is 2 cm, loop current is 1 mA and loop is placed in air.

[**Ans.** 6.28 Tesla]

[**Hint.** $B = \dfrac{\mu_0 I}{2r} = \dfrac{(4\pi \times 10^{-7}(10^{-3}))}{2 \times 10^{-2}}$]

6. For a flux density

$$\mathbf{B} = \frac{0 \cdot 2}{\rho}(\sin^2 \phi)\, \hat{\mathbf{a}}_z$$

Calculate the magnetic flux across $z = 0$ plane in cylindrical coordinates for $r \leq 0.05$ m.

[**Ans.** 0.0314 Wb/m^2]

[**Hint.** $\Phi = \int_S \mathbf{B} \cdot d\mathbf{S}$

Here, $d\mathbf{S} = \rho\, d\rho\, d\phi\, \hat{\mathbf{a}}_z$

$$\therefore \quad \Phi = 0.2 \int_0^{0.05} \int_0^{2\pi} \frac{\sin^2 \phi}{\rho}\, \hat{\mathbf{a}}_z\, \rho\, d\rho\, d\rho\, \hat{\mathbf{a}}_z$$

$$= 0.2\, [\rho]_0^{0.05} \int_0^{2\pi} \left(\frac{1 - \cos 2\phi}{2}\right) d\phi$$

$$= 0.2 \times 0.05 \times \left[\frac{\phi}{2} - \frac{\sin 2\phi}{2}\right]_0^{2\pi}\Bigg]$$

7. The interface between two different regions is normal to one of the three cartesian axes. If

$$\mathbf{B}_1 = \mu_0\, (43.5\, \hat{\mathbf{a}}_x + 24\, \hat{\mathbf{a}}_y) = B_{t_1} + B_{n_1}$$

and $$\mathbf{B}_2 = \mu_0\, (22\, \hat{\mathbf{a}}_x + 24\, \hat{\mathbf{a}}_y) = B_{t_2} + B_{n_2}$$

Find $\dfrac{\tan \theta_1}{\tan \theta_2}$.

[**Ans.** 0.506]

[**Hint.** Here, $B_{n_1} = B_{n_2}$]

$$\frac{\tan \theta_1}{\tan \theta_2} = \frac{B_{t_2}}{B_{t_1}} = \frac{22}{43.5}\,]$$

8. Calculate the inductance of a solenoid of 2000 turns wound uniformly over a length of 0.5 m on a cylindrical paper tube of 0.04 in diameter. The medium is air.

[**Ans.** 12.6 mH]

9. Find the inductance of an ideal solenoid with 300 turns, $l = 0.50$ m and a circular cross-section of radius 0.02 m.

[**Ans.** 284 μH]

[**Hint.** $B = \dfrac{\mu_0 N I}{l} = \dfrac{\mu_0 \times 300 \times I}{0.50} = \mu_0 (600\, I)$

Inductance per unit length

$$\frac{L}{l} = \frac{N\Phi}{I} = \frac{N B A}{I} = \frac{N \mu_0 (600\, I)\, \pi (r^2)}{I}$$

$$= 300 \times \mu_0 \times 600 \times \pi(4 \times 10^{-4})$$

$$= 568\ \mu\text{H/m}$$

$$L = 568 \times 0.50\ \mu\text{H}]$$

10. Calculate the inductance of 10 cm long co-axial cable filled with a material for which $\mu_r = 80$, having inner and outer diameters as 1 mm and 4 mm respectively.

[**Ans.** 2.218×10^{-5} henry]

[**Hint.** $L = \dfrac{\mu_0 \mu_r}{2p} \log_e \left(\dfrac{b}{a}\right) l$

or $L = \dfrac{(4\,p \times 10^{-7})\, 80}{2\pi} \log_e \left(\dfrac{4}{1}\right) \times (10 \times 10^{-2})$]

6

Time Varying Fields and Maxwell's Equations

INTRODUCTION

In preceding chapters, we have studied the basic laws of electrostatics and magnetostatics. The static electric and magnetic fields are independent of time. The fields are not coupled with each other, *i.e.*, they can exist without each other. The aim of this chapter is to consider the electric and magnetic fields that change with time, *i.e.*, time varying fields. The fields are also known as dynamic fields. These fields cannot exist without the other. Due to mutual dependency on each other, we can also called them as electromagnetic fields. The electromagnetic fields can be produced by time varying currents or charges or static sources in motion.

In time varying fields, the electric field produced by changing magnetic field is a result of experimental researches of Michael Faraday. On the other hand, a magnetic field produced by changing electric field is a result of theoretical researches of James Clark Maxwell.

Before discussing the time varying fields, we list the equations of electrostatic fields and magnetostatic fields for ready reference.

Electrostatic Fields

Law	Differential Form	Integral Form
Gauss's law	$\vec{\nabla}\cdot\mathbf{D} = \rho v$	$\oint_S \mathbf{D}\cdot d\mathbf{S} = q_{enc}$
Faraday's law	$\vec{\nabla}\times\mathbf{E} = 0$	$\oint_S \mathbf{E}\cdot d\mathbf{l} = 0$

Magnetostatic Field

Law	Differential Form	Integral Form
Gauss's law	$\vec{\nabla}\cdot\mathbf{B} = 0$	$\oint_S \mathbf{B}\cdot d\mathbf{S} = 0$
Ampere's law	$\vec{\nabla}\times\mathbf{H} = \mathbf{J}$	$\oint \mathbf{H}\cdot d\mathbf{l} = I_{enc}.$

6.1 FARADAY'S LAW

When a constant or dc current passes through a conductor, a magnetostatic field (constant magnitude field) gets produce in vicinity of conductor. However, When alternating current (changing with time) passes through conductor, it produces magnetic filed in vicinity of conductor whose magnitude is changing with time. When a conducting circuit is placed in a time varying magnetic field, an induced emf is produced in it. Due to this induced emf a current flows through the circuit.

Faraday observed the following points:

(*i*) ***Whenever the magnetic flux linked with a circuit is changed, an emf is induced in the circuit.***

(*ii*) ***The magnitude of induced emf is directly proportional to the negative rate of magnetic flux linked with the circuit.***

If Φ be the magnetic flux linked with the circuit at any instant and V_e be the induced emf, then

$$V_e = -\frac{d\Phi}{dt} \qquad ...(1)$$

If there are N turns in a coil, then induced emf becomes

$$V_e = -N\frac{d\Phi}{dt} \qquad ...(2)$$

The negative sign indicates that the direction of induced emf (or current) in a closed circuit is such that it opposes the original cause that produces it. Thus, is known Lenz's law.

Consider a time varying magnetic field is produced and closed circuit C of any shape is placed in this field. The circuit encloses a surface S in the field as shown in Fig. (1). Let **B** be the magnetic flux density in the neighbourhood of the circuit.

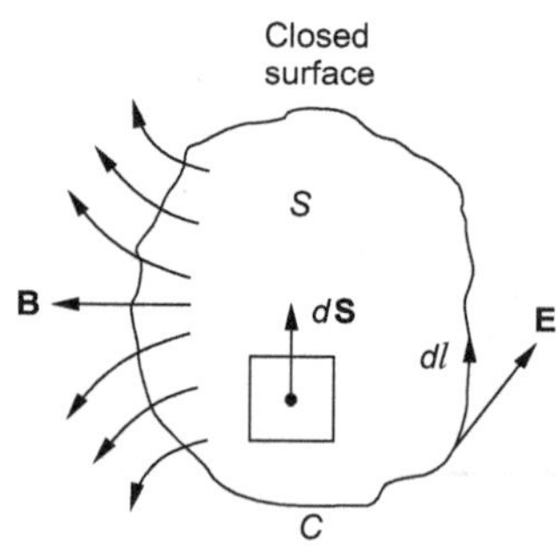

Fig. (1) Closed surface in time varying magnetic field

The magnetic flux through a small area dS will be $\mathbf{B} \cdot d\mathbf{S}$. Now, the flux, through the entire circuit is

$$\Phi = \int_S \mathbf{B} \cdot d\mathbf{S} \qquad ...(3)$$

When magnetic flux is changed, an electric field is induced around the circuit. The line integral of the electric field gives the induced emf in the closed circuit. Thus,

$$V_e = \oint \mathbf{E} \cdot dl \qquad \text{...(4)}$$

From eqs. (2) and (3),
$$V_e = -\frac{d\Phi}{dt} = -\frac{d}{dt}\int_S \mathbf{B} \cdot d\mathbf{S} \qquad \text{...(5)}$$

From eqs. (4) and (5), we get

$$\boxed{V_e = \oint \mathbf{E} \cdot dl = -\frac{d}{dt}\int_S \mathbf{B} \cdot d\mathbf{S}} \qquad \text{...(6)}$$

When the loop or closed circuit is stationary, then eq. (6) reduces to

$$V_e = \int_S \frac{\partial \mathbf{B}}{\partial t} d\mathbf{S} \qquad \text{...(7)}$$

The emf produced by changing field within a stationary circuit is celled as transformer emf. Eq. (7) is known as *transformer induction equation.*

Faraday's law in *integral form* can be expressed as

$$\boxed{V_e = \oint \mathbf{E} \cdot dl = -\int_S \frac{\partial \mathbf{B}}{\partial t} d\mathbf{S}} \qquad \text{...(8)}$$

By Stoke's theorem
$$\oint \mathbf{E} \cdot d\mathbf{l} = \int_S (\vec{\nabla} \times \mathbf{E}) \cdot d\mathbf{S} \qquad \text{...(9)}$$

From eqs. (8) and (9), we get

$$\int_S (\vec{\nabla} \times \mathbf{E}) \cdot d\mathbf{S} = -\int_S \frac{\partial \mathbf{B}}{\partial t} d\mathbf{S}$$

or
$$\boxed{\vec{\nabla} \times \mathbf{E} = -\frac{\partial \mathbf{B}}{\partial t}} \qquad \text{...(10)}$$

This is the *differential form* of Faraday's law. The eq. (10) gives a relation at a point.

6.2 GENERAL CASE OF INDUCTION

Here we shall consider a general case in which **B** is changing with time and the circuit is also in motion.

We have studied that an induced emf in a closed circuit due to time rate of change of **B** (tranformer induction) is given by

$$V_e = -\int_S \frac{\partial \mathbf{B}}{\partial t} \cdot d\mathbf{S} \qquad \text{...(1)}$$

When a wire forming a closed circuit is in motion, then induced emf due to motional induction is given by

$$V_e = \oint \mathbf{E} \cdot d\mathbf{l} = \oint (\mathbf{v} \times \mathbf{B}) \cdot d\mathbf{l} \qquad \text{...(2)}$$

Adding the two effects simultaneously, we get

$$\boxed{V_e = \oint (\mathbf{v} \times \mathbf{B}) \cdot d\mathbf{l} - \int_S \frac{\partial \mathbf{B}}{\partial t} \cdot d\mathbf{S}} \qquad \text{...(3)}$$

Eq. (3) gives the general case of emf produced.

Let us express the induction relations in **integral from.**

1. General case,	$V_e = \oint (\mathbf{v} \times \mathbf{B}) \cdot dl - \int_S \frac{\partial \mathbf{B}}{\partial t} \cdot d\mathbf{S}$
2. Motional enf,	$V_e = \oint (\mathbf{v} \times \mathbf{B}) \cdot dl$
3. Transformer emf,	$V_e = -\int_S \frac{\partial \mathbf{B}}{\partial t} \cdot d\mathbf{S}$

Now, we shall express the results in differential form:

From eq. (2), $\mathbf{E} = \mathbf{v} \times \mathbf{B}$

Taking the curl of both sides, we get

$$\vec{\nabla} \times \mathbf{E} = \vec{\nabla} \times (\mathbf{v} \times \mathbf{B})$$

The induction relations in **differential form** becomes

1. General case,	$\vec{\nabla} \times \mathbf{E} = \vec{\nabla} \times (\mathbf{v} \times \mathbf{B}) - \frac{\partial \mathbf{B}}{\partial t}$
2. Motional emf,	$\vec{\nabla} \times \mathbf{E} = \vec{\nabla} \times (\mathbf{v} \times \mathbf{B})$
3. Transformer emf,	$\nabla \times \mathbf{E} = -\frac{\partial \mathbf{B}}{\partial t}$

EXAMPLE 1. *A uniform metal rod placed parallel to X-axis is moving with a constant velocity* **v** *in a direction perpendicular to its length along Y-axis. A uniform magnetic field is applied along the Z-axis. Calculate the electric field produced due to the motion of the rod.*

Solution. Consider a uniform metal rod PQ placed parallel to X-axis is moving with a constant velocity **v** in a direction perpendicular to its length along Y-axis as show in Fig. (2). Let a uniform magnetic field is applied along the Z-axis. The rod PQ being a conductor, contains free charged particles (electrons). The charged particles are acted upon by the Lorentz force $q(\mathbf{v} \times \mathbf{B})$, where q is the charge on the particles. The direction of the force will be along X-axis in accordance with the Fleming's left hand rule. So, there will be a shifting of charged particles (electrons) of the conductor towards end P. In this way a negative charge is formed at P where a positive charge is formed at Q. Now, the two ends of the rod are oppositely charged and hence, a potential difference is developed between P and Q. This produces an electrostatic field E_e which is directed from P to Q. Finally a steady state is reached by moving the rod at constant velocity **v** when the Lorentz force $q(\mathbf{v} \times \mathbf{B})$, is equal to the force due to the electric field created inside the rod in opposite direction. Then the motion of charged particles relative to the rod ceases. Thus,

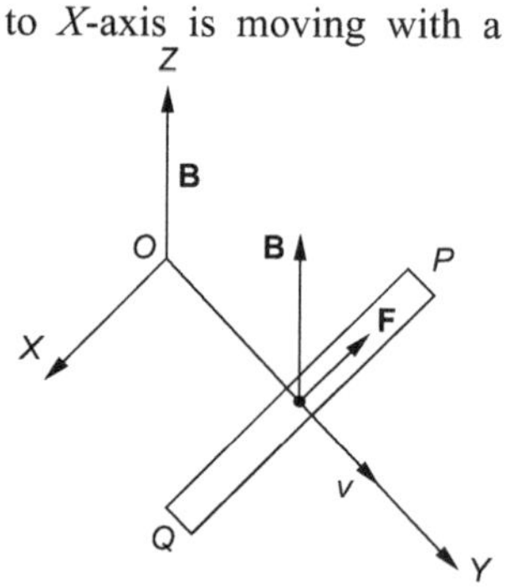

Fig. (2)

$$\mathbf{E}_e = -\mathbf{E}_L = -\frac{\mathbf{F}}{q} = -(\mathbf{v} \times \mathbf{B}) \qquad ...(1)$$

where $\mathbf{E}_L$ is the field due to Lorentz force.

Eq. (1) gives the expression for the elastic field produced due to the motion of conductor in uniform magnetic field.

EXAMPLE 2. *Calculate the motional emf developed in a rod when moved on a U-shaped stationary conductor placed in uniform magnetic field.*

Solution. Consider that a conducting rod *PQ* is sliding along a *U*-shaped stationary conductor placed in a uniform magnetic field of induction **B** as shown in Fig. (3). As *U*-shaped conductor is at rest, there will be no force acting on the charged particles. But when the conductor *PQ* is moved in the magnetic field, a potential difference is developed between *P* and *Q* as explained in example 1. Due to *U*-shaped conductor, the circuit is now completed as *PQRS*. Thus, a current passes in the circuit as shown in the figure. Due to current flow, there will be reduction in charge carriers at the ends but this a immediately compensated by the further charge carriers caused by Lorentz force. Hence, as long as the conductor is in motion, a current flows through the closed circuit *PQRS*. Thus, an emf is induced due to the motion of the conductor in a magnetic field. This emf is known as motional induced emf and the current produced by it is known as induced current. Let $d\mathbf{l}$ be the small length of the moving conducting rod, then motional emf is given by

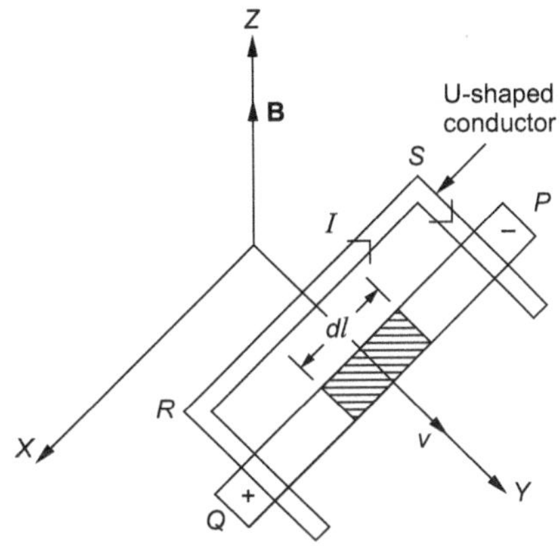

Fig. (3)

$$V_e = -\int \mathbf{E}_e \cdot dl = \int \mathbf{E}_L \cdot dl \qquad (\because \mathbf{E}_e = -\mathbf{E}_L)$$

$$= \int (\mathbf{v} \times \mathbf{B}) \cdot dl = \int \mathbf{B} \cdot d\mathbf{S}$$

where $d\mathbf{S}$ is the area swept in one second.

EXAMPLE 3. *A sliding conductor of length l is situated in X-Y plane as shown in Fig. (4). The conductor is moving with velocity $v = v_0\, \hat{\mathbf{a}}_x$ where v_0 is constant. Find the induced emf.*

Solution. The situation is shown in Fig. (4)

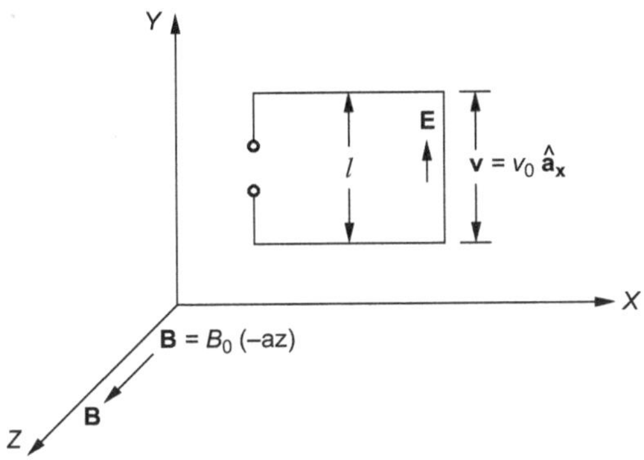

Fig. (4)

As shown in example 2, $V_e = \int (\mathbf{v} \times \mathbf{B}) \cdot dl$ ...(1)

Given $\mathbf{v} = v_0 \hat{\mathbf{a}}_x$ and $\mathbf{B} = B_0(-\hat{\mathbf{a}}_z)$

$$\therefore \quad \mathbf{v} \times \mathbf{B} = (v_0 \hat{\mathbf{a}}_x) \times (-B_0 \hat{\mathbf{a}}_z) = v_0 B_0 \hat{\mathbf{a}}_y$$

and $d\mathbf{l} = dy\, \hat{\mathbf{a}}_y$

$$\therefore \quad V_e = \int (v_0 B_0 \hat{\mathbf{a}}_y) \cdot dy\, \hat{\mathbf{a}}_y = v_0 B_0 \int_0^l dy = v_0 B_0 l$$

EXAMPLE 4. *A straight conductor of* 0.2 m *lies on the X-axis with one end at origin as shown in Fig.* (5). *The conductor is subjected to a magnetic flux density* $\mathbf{B} = 0.04\, \hat{\mathbf{a}}_y$ tesla *and velocity* $\mathbf{v} = 2.5 \sin 10^3 t\, \hat{\mathbf{a}}_z$ m/s. *Calculate the motional electric field intensity and emf induced in the conductor.*

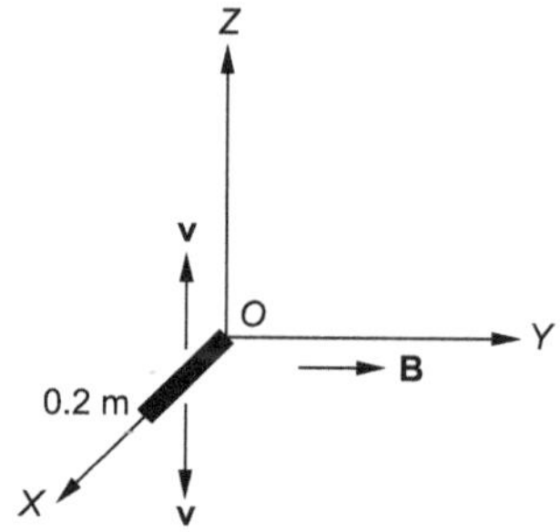

Fig. (5)

Solution. The motional emf, *i.e.*, induced voltage producing field intensity

$$\mathbf{E} = \mathbf{v} \times \mathbf{B} = (2.5 \sin 10^3 t\, \hat{\mathbf{a}}_z) \times (0.04\, \hat{\mathbf{a}}_y)$$

$$= 0.10 \sin 10^3 t(-\hat{\mathbf{a}}_x) \text{ V/m}$$

Therefore, emf induced in the conductor

$$V_e = \int (\mathbf{v} \times \mathbf{B}) \cdot d\mathbf{l}$$

$$= \int_0^{0.20} 0.10 \sin 10^3 t(-\hat{\mathbf{a}}_x) \cdot dx\, \hat{\mathbf{a}}_x$$

$$= -0.10 \sin 10^3 t[x]_0^{0.20} = -0.10 \sin 10^3 t(0.20)$$

$$= -0.020 \sin 10^3 t \text{ volt}$$

The conductor first moves in $\hat{\mathbf{a}}_z$ direction and afterward it moves in $-\hat{\mathbf{a}}_z$ direction.

EXAMPLE 5. *Consider* $\mathbf{B} = B_0 e^{kt} \hat{\mathbf{a}}_z$ *for cylindrical region* $\rho < b$. *Find electric field intensity* **E** *using Faraday's law.*

Solution. According to Faraday's law

$$\oint \mathbf{E} \cdot d\mathbf{l} = -\frac{d}{dt} \int_S \mathbf{B} \cdot d\mathbf{S}$$

$$\int_0^{2\pi} E_\phi\, \rho\, d\phi = -\frac{d}{dt} \int_0^{\rho} \int_0^{2\pi} B_0 e^{kt} \rho\, d\rho\, d\phi$$

$$E_\phi (2\,\pi\rho) = -\frac{d}{dt}\left[B_0 e^{kt}\left(\frac{\rho^2}{2}\right)\times 2\,\pi\right]$$

$$E_\phi (2\,\pi\rho) = -B_0\, k\, e^{kt}\, \pi\rho^2$$

or
$$E_\phi = -\frac{1}{2} B_0\, k\, e^{kt}\, \rho$$

or
$$\mathbf{E} = -\frac{1}{2} B_0\, k\, e^{kt}\, \rho\, \hat{\mathbf{a}}_\phi$$

6.3 CONDUCTION CURRENT AND DISPLACEMENT CURRENT

Physical Interpretation

Consider a parallel plate capacitor is connected to an alternating generator through a resistance in series circuit as shown in Fig. (6). We know that in a series circuit, the current has the same value at all cross-sections. The conduction current which is due to electrical charges is same at all cross-sections of the circuit except at all cross-sections (such as *AB*) of the dielectric inside capacitor. The dielectric is air with permittivity ε_0. It is important to mention here that conduction electrons (free electrons) are flowing into one plate of the capacitor and forcing other free electrons out of the second plate. Therefore, *there is no flow of free electrons through the dielectric.* In other words, we can say that there is a discontinuity of the current in the space between the plates of a capacitor. Now, the question is that how this discontinuity of the current can be explained. In order to remove this discontinuity J.C. Maxwell revised and extended the definition of current by introducing the idea of *displacement current.*

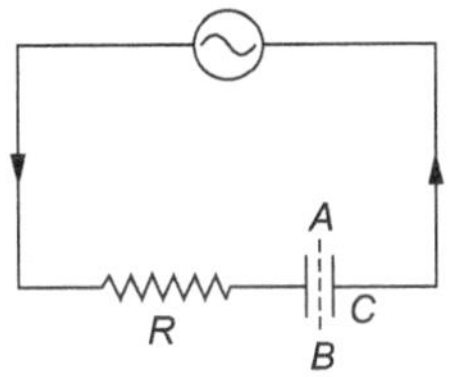

Fig. (6)

Let at any particular instant, q be the charge on capacitor plate. The conduction current (i_C) is defined as the time rate of flow of charge, *i.e.*,

$$i_C = \frac{dq}{dt} \qquad ...(1)$$

We have studied the electrical displacement D in dielectrics. The electrical displacement D is given by

$$D = \sigma = \frac{q}{A} \qquad ...(2)$$

where σ is the surface charge density and A is area of each plate.

From eq. (2), $q = DA$ ...(3)

Substituting the value of q from eq. (3) in eq. (1), we get

$$i = \frac{d}{dt}(D\,A) = A\frac{dD}{dt} \qquad ...(4)$$

Maxwell suggested that the term $A\left(\frac{dD}{dt}\right)$ should be considered as the current inside the dielectric. This current is called as *displacement current* and is denoted by i_d.

Therefore,
$$\boxed{i_d = A\left(\frac{dD}{dt}\right)} \qquad ...(5)$$

Consequently, the displacement current density $\mathbf{J}_d$ is given by

$$\boxed{\mathbf{J}_d = \left(\frac{d\mathbf{D}}{dt}\right)} \quad ...(6)$$

The vector **D** may vary with space, hence,

$$\boxed{\mathbf{J}_d = \left(\frac{\partial \mathbf{D}}{\partial t}\right)} \quad ...(7)$$

Thus, inside the dielectric, there will be a displacement current i_d which is equal to conduction current i in the line.

Important Points

1. The concepts of conduction current and displacement current are shown in Fig. (7)

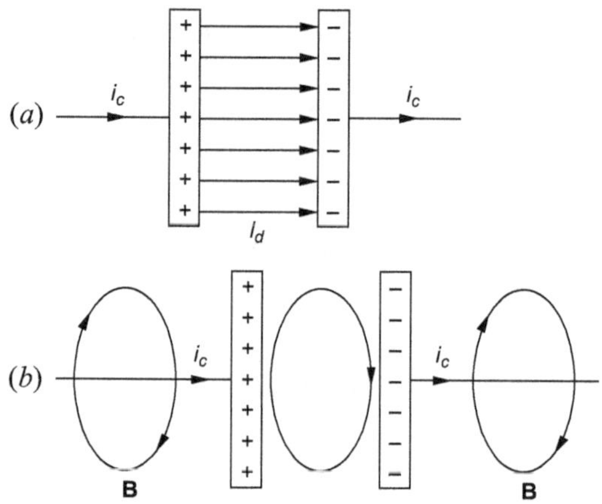

Fig. (7) Concepts of conduction and displacement current

2. The conduction current i_C flows from one plate to another plate of the capacitor through conducting wires. Maxwell suggested that during time varying electric field between the plates of the capacitor, an electric current, called the displacement current i_d, also flows across the space between the plates of the capacitor.
3. The displacement current density $\mathbf{J}_d$ does not represent a current which directly passes through the capacitor. It is only the apparent current representing the rate at which flow of charge takes place from one plate to another plate. Hence, the term displacement is justified.
4. The conduction current and displacement current are equal, *i.e.*,

$$i_C = i_d$$

5. Like conduction current, the displacement current is also a source of magnetic field.
6. In good conductors, the displacement current is negligible small as compared to conduction current at frequencies less than optical frequencies. At ultra high frequencies, the displacement current in conductors become quite important.
7. The displacement current exists only so long as electric field **E** is changing with time, *i.e.*, displacement **D** is changing with time. When the capacitor is fully charged upto

the value of applied emf, the current in the line drops to zero. Now, the electric field **E** between the plates of the capacitor attains at steady value. As a result, $\left(\frac{d\mathbf{E}}{dt}\right)$ and $\left(\frac{d\mathbf{D}}{dt}\right)$ becomes zero.

8. The displacement current is called a current in the sense that it has same dimensions as that of current.

6.4 MODIFIED AMPERE'S LAW AND DISPLACEMENT CURRENT

We have studied that a current in conductor produces a magnetic field. Maxwell proved that a changing electric field in vacuum or in a dielectric also produces a magnetic field. So, a changing electric field is equivalent to a current which flows as long as the electric field is changing. This produces the same magnetic effect as the ordinary conduction current. This is known as ***displacement current.*** Now, we shall show that how the Ampere's law is modified by using the idea of displacement current.

Ampere's law in vector form can be expressed as

$$\vec{\nabla} \times \mathbf{B} = \mu_0 \mathbf{J} \quad \text{...(1)}$$

where **J** is current density.

Taking divergence of this equation

$$\vec{\nabla} \cdot (\vec{\nabla} \times \mathbf{B}) = \text{div. curl } \mathbf{B} = \vec{\nabla} \cdot (\mu_0 \mathbf{J})$$

or $$\text{div. curl } \mathbf{B} = \mu_0 \text{ div. } \mathbf{J} \quad \text{...(2)}$$

We know that div. of curl of a vector is always zero and hence,

$$0 = \mu_0 \text{ div.}\mathbf{J}$$

$$\therefore \quad \text{div. } \mathbf{J} = 0 \qquad (\because \mu_0 \neq 0) \quad \text{...(3)}$$

This is in contradiction with the continuity equation, which states that

$$\text{div. } \mathbf{J} + \frac{\partial \rho}{\partial t} = 0 \quad \text{...(4)}$$

where ρ represents the charge density.

A close inspection of eqs. (3) and (4) shows that the Ampere's law in the form of div. $\mathbf{J} = 0$ is valid only when the charge density is static and not varying with time.

Maxwell concluded that eq. (1) is incomplete. He suggested that something must be added in **J** of eq. (1) such that the divergence of both side is same. Thus,

$$\vec{\nabla} \times \mathbf{B} = \mu_0 \mathbf{J} + \text{Something} \quad \text{...(5)}$$

In order to know this something, Maxwell postulated that similar to electric field due to changing magnetic field (Faraday's law of induction), there must be a magnetic field due to changing electric field. Thus, a changing electric field is equivalent to a current which flows as long as the electric field is changing. This produces the same magnetic effect as ordinary conduction current. Let us add an unknown term $\mathbf{J}_d$ for something in eq. (5). Therefore,

$$\vec{\nabla} \times \mathbf{B} = \mu_0 (\mathbf{J} + \mathbf{J}_d)$$

Taking div. of both sides,

$$\vec{\nabla} \cdot (\vec{\nabla} \times \mathbf{B}) = \mu_0 [\vec{\nabla} \cdot \mathbf{J} + \vec{\nabla} \cdot \mathbf{J}_d]$$

$$0 = \mu_0 [\vec{\nabla} \cdot \mathbf{J} + \vec{\nabla} \cdot \mathbf{J}_d]$$

$$\therefore \quad \vec{\nabla}\cdot\mathbf{J}+\vec{\nabla}\cdot\mathbf{J}_d = 0 \qquad (\because \mu_0 \neq 0)$$

or
$$\vec{\nabla}\cdot\mathbf{J}_d = -\vec{\nabla}\cdot\mathbf{J} \qquad ...(6)$$

Using eq. (4), we get $\vec{\nabla}\cdot\mathbf{J}_d = \dfrac{\partial\rho}{\partial t}$ ...(7)

According to Gauss's law in electrostatic

$$\vec{\nabla}\cdot\mathbf{D} = \rho \qquad ...(8)$$

Substituting the value of ρ from eq. (8) in eq. (7), we get

$$\vec{\nabla}\cdot\mathbf{J}_d = \frac{\partial}{\partial t}(\vec{\nabla}\cdot\mathbf{D}) = \vec{\nabla}\cdot\frac{\partial D}{\partial t}$$

$$\therefore \quad \boxed{\mathbf{J}_d = \frac{\partial \mathbf{D}}{\partial t}} \qquad ...(9)$$

where $\mathbf{J}_d$ is displacement current density.

Now, the modified form of Ampere's law is

$$\boxed{\vec{\nabla}\times\mathbf{B} = \mu_0\left(\mathbf{J}+\frac{\partial \mathbf{D}}{\partial t}\right)} \qquad ...(10)$$

or
$$\boxed{\vec{\nabla}\times\mathbf{H} = \mathbf{J}+\frac{\partial \mathbf{D}}{\partial t}} \qquad ...(11)$$

Displacement Current

The displacement current i_d is given by

$$i_d = \int_S \mathbf{J}_d\cdot d\mathbf{S} = \int_S \frac{\partial \mathbf{D}}{\partial t}\cdot d\mathbf{S} = \frac{\partial}{\partial t}\int_S \mathbf{D}\cdot d\mathbf{S}$$

$$= \varepsilon_0\frac{\partial}{\partial t}\int_S \mathbf{E}\cdot d\mathbf{S} = \varepsilon_0\frac{d\Phi_E}{dt}$$

$$\therefore \quad \boxed{i_d = \varepsilon_0\frac{d\Phi_E}{dt}} \qquad ...(12)$$

where Φ_E is the electric flux.

Modified Ampere's Law in Integral Form

Integrating eq. (11) over the surface, we have

$$\int_S(\vec{\nabla}\times\mathbf{H})\cdot d\mathbf{S} = \int_S \mathbf{J}\cdot d\mathbf{S}+\int_S\frac{\partial \mathbf{D}}{\partial t}\cdot d\mathbf{S} \qquad ...(13)$$

Applying Stoke's theorem, we get

$$\boxed{\oint \mathbf{H}\cdot d\mathbf{l} = i+i_d = \int_S \mathbf{J}\cdot d\mathbf{S}+\int_S\frac{\partial \mathbf{D}}{\partial t}\cdot d\mathbf{S}} \qquad ...(14)$$

In this way, Maxwell succeeded in bringing the complete symmetry between electric and magnetic fields by introducing the concept of displacement current.

EXAMPLE 1. *A dielectric medium is kept in variable field. Show that the value of displacement current is equal to the conduction current.*

Solution. The displacement current density $\mathbf{J}_d$ is given by

$$\mathbf{J}_d = \varepsilon_0 \frac{\partial E}{\partial t}$$

The displacement current i_d is give by

$$i_d = \int_S \mathbf{J}_d \cdot d\mathbf{S} = \varepsilon_0 \frac{\partial}{\partial t} \int_S \mathbf{E} \cdot d\mathbf{S}$$

According to Gauss's theorem

$$\int_S \mathbf{E} \cdot d\mathbf{S} = \frac{q}{\varepsilon_0}$$

$$\therefore \quad i_d = \varepsilon_0 \frac{\partial}{\partial t}\left(\frac{q}{\varepsilon_0}\right) = \frac{\partial q}{\partial t} = i$$

Thus, displacement current is equal to conduction current.

EXAMPLE 2. *A parallel plate capacitor with plate area of* 5 cm^2 *and plate separation of* 3 mm *has a voltage of* 50 sin 10^3 *t applied to its plates. Calculate the displacement current assuming* $\varepsilon = 2\,\varepsilon_0$.

Solution. We know that $\mathbf{D} = \varepsilon\,\mathbf{E}$

Further, $\mathbf{E} = \dfrac{V}{d}$

$$\therefore \quad D = \frac{\varepsilon V}{d}$$

Now
$$j_d = \frac{\partial D}{\partial t} = \frac{E}{d}\frac{dV}{dt}$$

$$\therefore \quad i_d = j_d \times A = \frac{\varepsilon\, A}{d}\frac{dV}{dt} = C\frac{dV}{dt}$$

$$= \frac{2\varepsilon_0\, A}{d}\frac{d}{dt}\,[50 \sin 10^3 t]$$

$$= \frac{2\varepsilon_0\, A}{d}\,[(10)^3\, 50 \cos 10^3 t]$$

$$= \frac{2\times(8.83\times10^{-12})\,(5\times10^{-4})}{3\times10^{-3}}\,[5\times10^4 \cos 10^3\, t]$$

$$= \mathbf{147 \cos 10^3\, t\ nA}$$

6.5 A REVIEW OF BASIC LAWS

The basic equations which we have studied are summarised in the following four equations:

(1) $\oint \mathbf{E} \cdot d\mathbf{S} = \left(\dfrac{q}{\varepsilon_0}\right)$ or $\oint \mathbf{D} \cdot d\mathbf{S} = q$

This is ***Gauss's law of electrostatic.*** This law states that the electric flux through a closed surface is equal to the net charge enclosed by the surface divided by the permittivity constant ε_0.

(2) $\oint \mathbf{B} \cdot d\mathbf{S} = 0$

This is ***Gauss's law for magnetism.*** This law states that the magnetic flux through a closed surface is zero.

(3) $\oint \mathbf{E} \cdot dl = -\dfrac{d\Phi_B}{dt}$

This is ***Faraday's law*** of electromagnetic induction. This law states than an electric field is produced by changing magnetic field.

(4) $\oint \mathbf{B} \cdot dl = \mu_0 I$ or $\oint \mathbf{H} \cdot dl = \mu_0$

This is ***Ampere's law*** for magnetic field due to steady current. This law states that the amount of workdone in carrying a unit magnetic pole one around a closed arbitrary path linked with the current is μ_0 times the current I.

The different symbols used are:

E = electric field intensity vector:

D = electric flux density or electric displacement vector

B = magnetic flux density vector

H = magnetic field intensity vector

For a linear and isotropic media, **E** and **D** are related as

$$\mathbf{D} = \varepsilon \mathbf{E}$$

For linear isotropic media $\mathbf{B} = \mu \mathbf{H}$

In non time varying fields (static case) **E** and **D** and **B** and **H** and independent pairs while in time varying fields vectors **E** and **D** are properly related to **B** and **H.**

6.6 MAXWELL'S EQUATIONS

We have studied that when the electric and magnetic fields are changing very rapidly in space with time, then the varying electric fields give magnetic field and vice-versa. Maxwell in in 1862 formulated the basic laws of electromagnetic in the form of four fundamental equations. These equation are known as Maxwell's electromagnetic equations. These equations are based upon the well known laws such as Gauss's law of electrostatic, Gauss's law of magnetostatic, Faraday's law of electromagnetic induction and Ampere's circuital law.

The integral forms of these equations are given below:

$$\oint \mathbf{E} \cdot d\mathbf{S} = \left(\frac{q}{\varepsilon_0}\right) \text{ or } \oint \mathbf{D} \cdot d\mathbf{S} = q \quad \text{...(1)}$$

$$= \oint \mathbf{B} \cdot d\mathbf{S} = 0 \quad \text{...(2)}$$

$$\oint \mathbf{E} \cdot dl = -\frac{d\Phi_B}{dt} = -\int_S \frac{\partial \mathbf{B}}{\partial t} \cdot d\mathbf{S} \quad \text{...(3)}$$

$$\left.\begin{array}{ll} & \oint \mathbf{B}\cdot dl = \mu_0\left(\mathbf{J}+\varepsilon_0\dfrac{\partial \mathbf{E}}{\partial t}\right) \\ \text{or} & \oint \mathbf{H}\cdot dl = I + \int_S \dfrac{\partial \mathbf{D}}{\partial t}\cdot d\mathbf{S} \end{array}\right] \quad ...(4)$$

The differential forms of these equations are given below:

$$\text{div. } \mathbf{E} = \frac{\rho}{\varepsilon_0} \quad \text{or} \quad \vec{\nabla}\cdot\mathbf{D} = \rho \quad ...(1)$$

$$\text{div. } \mathbf{B} = 0 \quad \text{or} \quad \vec{\nabla}\cdot\mathbf{B} = 0 \quad ...(2)$$

$$\text{curl } \mathbf{E} = -\frac{\partial \mathbf{B}}{\partial t} \quad \text{or} \quad \vec{\nabla}\times\mathbf{E} = -\frac{\partial \mathbf{B}}{\partial t} \quad ...(3)$$

$$\text{curl } \mathbf{B} = \mu_0\left(\mathbf{J}+\varepsilon_0\frac{\partial \mathbf{E}}{\partial t}\right) \quad \text{or} \quad \vec{\nabla}\times\mathbf{H} = \mathbf{J}+\frac{\partial \mathbf{D}}{\partial t} \quad ...(4)$$

6.7 DEVIATION OF MAXWELL'S EQUATIONS IN DIFFERENTIAL FORM

1. Maxwell's first equation:

$$\vec{\nabla}\cdot\mathbf{D} = \rho$$

According to Gauss's law $\oint \mathbf{E}\cdot d\mathbf{S} = \dfrac{\rho}{\varepsilon_0}$

If ρ be the charge density and dV, the small volume considered, then

$$q = \int_V \rho dV$$

$$\therefore \quad \oint \mathbf{E}\cdot d\mathbf{S} = \frac{1}{\varepsilon_0}\int_V \rho dV \quad \text{or} \quad \varepsilon_0\oint \mathbf{E}\cdot d\mathbf{S} = \int_V \rho dV$$

$$\text{or} \quad \oint \mathbf{D}\cdot d\mathbf{S} = \int_V \rho\, dV \quad (\because \varepsilon_0\,\mathbf{E} = \mathbf{D}) \quad ...(1)$$

According to divergence theorem,

$$\oint \mathbf{D}\cdot d\mathbf{S} = \int_V (\vec{\nabla}\cdot\mathbf{D})dV \quad ...(2)$$

From eqs. (1) and (2), we get

$$\int_V (\vec{\nabla}\cdot\mathbf{D})dV = \int_V \rho dV$$

$$\boxed{\vec{\nabla}\cdot\mathbf{D} = \rho} \quad ...(a)$$

From eq. (a), $\vec{\nabla}\cdot\mathbf{E} = \dfrac{\rho}{\varepsilon_0}$ or div. $\mathbf{E} = \dfrac{\rho}{\varepsilon_0}$ ...(3)

2. Maxwell's second equation:

$$\vec{\nabla}\cdot\mathbf{B} = 0$$

According to Gauss's law for magnetism

$$\oint \mathbf{B}\cdot d\mathbf{S} = 0$$

Transforming the surface integral into volume integral by Gauss's divergence theorem, we have

$$\oint_S \mathbf{B}\cdot d\mathbf{S} = \int_V (\vec{\nabla}\cdot\mathbf{B})\, dV$$

$$\int_V (\vec{\nabla}\cdot\mathbf{B})\, dV = 0$$

As the volume is arbitrary, the integral must be zero. Hence,

$$\boxed{\vec{\nabla}\cdot\mathbf{B} = 0} \qquad \text{...}(b)$$

or $\quad \text{div. } B = 0$

3. Maxwell's third equation:

$$\vec{\nabla}\times\mathbf{E} = -\left(\frac{\partial \mathbf{B}}{\partial t}\right)$$

We know that $\quad \oint \mathbf{E}\cdot dl = -\dfrac{d\Phi_B}{dt} = -\dfrac{\partial}{\partial t}\displaystyle\int_S \mathbf{B}\cdot d\mathbf{S}$

or $$\oint \mathbf{E}\cdot dl = -\int_S \frac{\partial \mathbf{B}}{\partial t}\, d\mathbf{S} \qquad \text{...}(4)$$

Applying Stoke's theorem

$$\oint \mathbf{E}\cdot dl = \int_S (\vec{\nabla}\times\mathbf{E})\cdot d\mathbf{S} \qquad \text{...}(5)$$

From eqs. (4) and (5), we get

$$\int_S (\vec{\nabla}\times\mathbf{E})\cdot d\mathbf{S} = -\int_S \frac{\partial \mathbf{B}}{\partial t}\, d\mathbf{S} \qquad \text{...}(6)$$

Eq. (6) is true for all surfaces, therefore,

$$\boxed{\vec{\nabla}\times\mathbf{E} = -\frac{\partial \mathbf{B}}{\partial t}} \qquad \text{...}(c)$$

or $$\text{curl } \mathbf{E} = -\frac{\partial \mathbf{B}}{\partial t}$$

4. Maxwell' fourth equation:

$$\nabla\times\mathbf{H} = \mathbf{J} + \frac{\partial \mathbf{D}}{\partial t}$$

This equation is already proved (when displacement current) is discussed. For ready reference

$$\oint \mathbf{B}\cdot dl = \int_S (\vec{\nabla}\times\mathbf{B})\cdot d\mathbf{S}$$

But $$\int_S (\vec{\nabla}\times\mathbf{B})\cdot d\mathbf{S} = \mu_0 \int_S (\mathbf{J}\cdot d\mathbf{S})$$

or $$\vec{\nabla}\times\mathbf{B} = \mu_0 \mathbf{J}$$

Now, we replace **J** by $\left(\mathbf{J} + \varepsilon_0 \dfrac{\partial \mathbf{E}}{\partial t}\right)$

$$\vec{\nabla} \times \mathbf{B} = \mu_0 \left(\mathbf{J} + \varepsilon_0 \frac{\partial \mathbf{E}}{\partial t} \right)$$

or $$\boxed{\nabla \times \mathbf{H} = \mathbf{J} + \frac{\partial \mathbf{D}}{\partial t}}$$

EXAMPLE 1. *Use Maxwell's equation to establish the charge-current equation of continuity.*

Solution. The Maxwell's fourth equation is

$$\text{curl } \mathbf{B} = \mu_0 \mathbf{J} + \mu_0 \varepsilon_0 \frac{\partial \mathbf{E}}{\partial t}$$

Taking divergence of both sides, we get

$$\text{div. curl } \mathbf{B} = \text{div. } \mu_0 \mathbf{J} + \mu_0 \varepsilon_0 \frac{\partial}{\partial t} (\vec{\nabla} \cdot \mathbf{E})$$

Now, $$\text{div. curl } \mathbf{B} = \vec{\nabla} \cdot (\vec{\nabla} \cdot \mathbf{B}) = 0$$

$\therefore$ $$0 = \text{div. } \mu_0 \mathbf{J} + \mu_0 \varepsilon_0 \frac{\partial}{\partial t} (\vec{\nabla} \cdot \mathbf{E})$$

For Maxwell's first equation

$$\vec{\nabla} \cdot \mathbf{E} = \frac{\rho}{\varepsilon_0}$$

$\therefore$ $$0 = \mu_0 \vec{\nabla} \cdot \mathbf{J} + \mu_0 \varepsilon_0 \frac{\partial}{\partial t} \left(\frac{\rho}{\varepsilon_0} \right)$$

or $$\vec{\nabla} \cdot \mathbf{J} = -\frac{\partial \rho}{\partial t}$$

EXAMPLE 2. *Using the Maxwell's relation*

$$\textit{curl } \mathbf{B} = \mu_0 \left(\mathbf{J} + \frac{\partial \mathbf{D}}{\partial t} \right)$$

Prove that $$\textit{div. } \mathbf{D} = \rho$$

Solution. Taking the divergence of the given Maxwell's equation, we have

$$\frac{1}{\mu_0} [\text{div. curl } \mathbf{B}] = \text{div. } \mathbf{J} + \frac{\partial}{\partial t} (\text{div. } \mathbf{D})$$

But $$\text{div. curl B} = \vec{\nabla} \cdot \vec{\nabla} \times \mathbf{B} = 0$$

$\therefore$ $$\text{div. } \mathbf{J} = -\frac{\partial}{\partial t} (\text{div. } \mathbf{D})$$

or $$-\frac{\partial \rho}{\partial t} = -\frac{\partial}{\partial t} (\text{div . } \mathbf{D})$$ (Using continuity equation)

or $$\text{div. } \mathbf{D} = \rho$$

EXAMPLE 3. *Use Maxwell's equation*

$$\textit{Curl}\ \mathbf{E} = -\frac{\partial \mathbf{B}}{\partial t}$$

Prove that $\quad \textit{div.}\ \mathbf{B} = 0.$

Solution. Taking the divergence of the give equation, we get

$$\text{div. curl}\ \mathbf{E} = -\frac{\partial}{\partial t}(\text{div}\,.\,\mathbf{B})$$

or
$$0 = -\frac{\partial}{\partial t}(\text{div}\,.\,\mathbf{B})$$

or
$$\text{div.}\ \mathbf{B} = 0$$

Exercises

1. Give field based interpretation of Faraday's law.
2. Discuss the general case of induction.
3. Given the physical interpretation of conduction current and displacement current.
4. Explain the significance of displacement current
5. What do you mean by displacement current?
6. What do you mean by displacement current and show how it led to the modification of Ampere's law?
7. Obtain the generalized form of Ampere's law.
8. Write and explain differential and integral forms of Maxwell's equations.
9. Give the physical interpretation of Maxwell's equation.
10. Are all Maxwell's equation independent? Explain.

UNIT - IV

1. Uniform Plane Waves
2. Reflection, Refraction and Dispersion of Plane Waves

7

Uniform Plane Waves

INTRODUCTION

In previous chapter, we have studied the Maxwell's equations. According to Maxwell, an oscillating moving charge continuously emits waves which are known as *electromagnetic waves*. The reason is that the oscillating charge has an accelerating motion and a magnetic field is produced around the accelerating charge. As the velocity of charge changes, the magnetic field also changes with time. This produces an electric field. The electric field also changes with time. The two varying fields are mutually perpendicular to each other. The frequency of these fields is same as the frequency of oscillations of charged particle. The wave associated with these oscillations is called as electromagnetic wave. Therefore, ***an electromagnetic wave is a wave of oscillations of electric and magnetic fields in mutually perpendicular planes and these oscillations are perpendicular to the direction of propagation of wave***.

In this chapter, we shall apply Maxwell's equation to introduce the fundamental theory of wave motion and basic principles of wave propagation. The wave equation, for a particular case when the electric field is independent of y and z and is a function of x and t is called a ***uniform plane wave***. The uniform plane wave, based on Maxwell's equation, illustrates the principles behind the propagation of energy. After this, Poynting theorem will be discussed to find the power carried by a wave.

7.1 WAVE EQUATIONS IN FREE-SPACE

Here we shall derive the wave equations using Maxwell's equations for homogeneous medium (Free-space). In homogeneous medium, we have no charges and no conduction current, *i.e.*, $\rho_V = 0$, $\sigma = 0$ and $\mathbf{J} = 0$.

The Maxwell's equations in free-space are:

$$\vec{\nabla} \times \mathbf{H} = \frac{\partial \mathbf{D}}{\partial t} = \varepsilon_0 \frac{\partial \mathbf{E}}{\partial t} \quad ...(1)$$

$$\vec{\nabla} \times \mathbf{E} = \frac{\partial \mathbf{B}}{\partial t} = -\mu_0 \frac{\partial \mathbf{H}}{\partial t} \quad ...(2)$$

$$\vec{\nabla} \cdot \mathbf{D} = 0, \qquad \vec{\nabla} \cdot \mathbf{E} = 0 \quad ...(3)$$

$$\vec{\nabla} \cdot \mathbf{B} = 0, \qquad \vec{\nabla} \cdot \mathbf{H} = 0 \quad ...(4)$$

(1) Taking the curl of both sides of eq. (2), we get

$$\vec{\nabla}\times\vec{\nabla}\times\mathbf{E} = -\mu_0\,\vec{\nabla}\times\frac{\partial\mathbf{H}}{\partial t}$$

or
$$\vec{\nabla}\times\vec{\nabla}\times\mathbf{E} = -\mu_0\,\frac{\partial}{\partial t}(\vec{\nabla}\times\mathbf{H}) \qquad ...(5)$$

We know the vector identity

$$\vec{\nabla}\times\vec{\nabla}\times\mathbf{E} = \vec{\nabla}\times(\vec{\nabla}\cdot\mathbf{E}) - \nabla^2\mathbf{E} = \vec{\nabla}\times(0) - \nabla^2\mathbf{E} \qquad \text{[Using eq. (3)]}$$

or
$$\vec{\nabla}\times\vec{\nabla}\times\mathbf{E} = -\nabla^2\mathbf{E} \qquad ...(6)$$

From eqs. (5) and (6), we get

$$-\nabla^2\mathbf{E} = -\mu_0\frac{\partial}{\partial t}(\vec{\nabla}\times\mathbf{H}) = -\mu_0\frac{\partial}{\partial t}\left(\varepsilon_0\frac{\partial\mathbf{E}}{\partial t}\right) \qquad \text{[(Using eq. (1)]}$$

$$= -\mu_0\,\varepsilon_0\,\frac{\partial^2\mathbf{E}}{\partial t^2}$$

∴
$$\boxed{\nabla^2\mathbf{E} = \mu_0\varepsilon_0\,\frac{\partial^2\mathbf{E}}{\partial t^2}} \qquad ...(a)$$

This expression is the wave equation for propagation of electric field **E** in free space. In cartesian coordinates, eq. (*a*) can be expressed as

$$\left.\begin{aligned}\frac{\partial^2 E_x}{\partial x^2}+\frac{\partial^2 E_x}{\partial y^2}+\frac{\partial^2 E_x}{\partial z^2} &= \mu_0\varepsilon_0\,\frac{\partial^2 E_x}{\partial t^2}\\ \frac{\partial^2 E_y}{\partial x^2}+\frac{\partial^2 E_y}{\partial y^2}+\frac{\partial^2 E_y}{\partial z^2} &= \mu_0\varepsilon_0\,\frac{\partial^2 E_y}{\partial t^2}\\ \frac{\partial^2 E_z}{\partial x^2}+\frac{\partial^2 E_z}{\partial y^2}+\frac{\partial^2 E_z}{\partial z^2} &= \mu_0\varepsilon_0\,\frac{\partial^2 E_z}{\partial t^2}\end{aligned}\right\} \qquad ...(a)$$

(**2**) Similarly, taking curl of both sides of eq. (1), we get

$$\vec{\nabla}\times\vec{\nabla}\times\mathbf{H} = \varepsilon_0\,\frac{\partial}{\partial t}(\vec{\nabla}\times\mathbf{E})$$

or
$$-\nabla^2\mathbf{H} = \varepsilon_0\frac{\partial}{\partial t}\left(-\mu\,\frac{\partial\mathbf{H}}{\partial t}\right) \qquad \text{[Using eq. (2)]}$$

or
$$-\nabla^2\mathbf{H} = \mu\varepsilon\,\frac{\partial^2\mathbf{H}}{\partial t^2}$$

$$\boxed{\nabla^2\mathbf{H} = \mu_0\varepsilon_0\,\frac{\partial\mathbf{H}}{\partial t}} \qquad ...(b)$$

This expression is the wave equation for the propagation of **H** in free-space. Replacing μ_0 by μ and ε_0 by ε in eqs. (*a*) and (*b*), we get wave equations in perfect dielectric (or non-conducting) medium.

It is obvious from eqs. (*a*) and (*b*) that electromagnetic wave propagation involves electric and magnetic fields with more than one component. Each component depends on the three coordinates, in addition to time. However, if the electromagnetic waves consist of electric and magnetic fields which are perpendicular to each other and to the direction of propagation and are uniform in planes perpendicular to the direction of propagation, the waves are called as ***uniform plane waves***.

7.2 PROPAGATION OF A PLANE WAVE

(**1**) The velocity of propagation of **E** and **H** is the same as the velocity of light.

We know that general wave equation is represented by

$$\nabla^2 y = \frac{1}{v_0^2}\frac{\partial^2 y}{\partial t^2}$$

Comparing it with eq. (a), we get

$$\frac{1}{v_0^2} = \mu\,\varepsilon \quad \text{or} \quad v_0^2 = \frac{1}{\mu\varepsilon} \quad \text{or} \quad v_0 = \frac{1}{\sqrt{(\mu\,\varepsilon)}}$$

For free space, $\mu = \mu_0 = 4\,\pi \times 10^{-7}$ and $\varepsilon = \varepsilon_0$

and $$\varepsilon = \varepsilon_0 = \frac{1}{4\,\pi \times (9\times 10^9)} \qquad \left(\because \frac{1}{4\,\pi\,\varepsilon_0} = 9\times 10^9\right)$$

$\therefore$ $$v_0 = \frac{1}{\sqrt{(4\,\pi\times 10^{-7})\times\left[4\,\pi\times 9\times 10^9\right]}}$$

$$= \sqrt{[(9\times 10^{16})]} = 3\times 10^8 \text{ m/s}$$

(**2**) Let a plane wave is propagating in X-direction. In this case, **E** will have components E_y and E_z but not E_x. Without loss of generality, we focus our attention say on E_y. Knowing the result of E_y, we can obtain E_z. The equation to be solved has the value.

$$\boxed{\frac{\partial^2 E_y}{\partial x^2} = \mu_0\varepsilon_0\frac{\partial^2 E_y}{\partial t^2}} \qquad ...(1)$$

The solution of eq. (1) is given by

$$E_y = f_1\,(x - v_0\,t) + f_2\,(x + v_0\,t)$$

The functions $f_1\,(x - v_0\,t)$ and $f_2\,(x + v_0\,t)$ describe such a wave mathematically. The variation of wave is confined to one dimension in space as shown in Fig. (1).

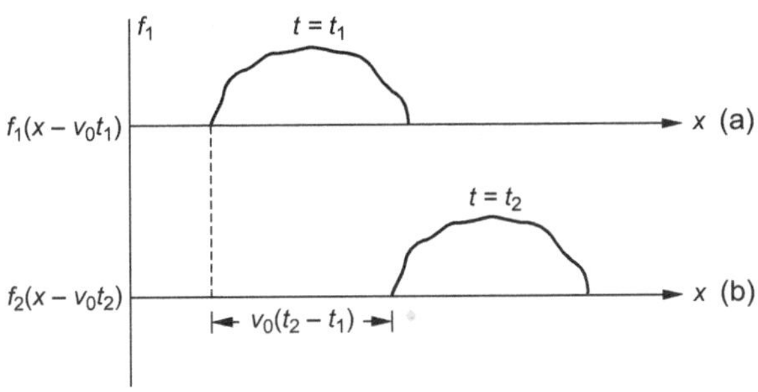

Fig. (1) A wave travelling in positive X-direction

At any particular instant t_1, the function $f_1\ (x - v_0\ t_1)$ becomes a function of x because $v_0\ t_1$ becomes constant. The function is represented by Fig. (1a). Let us consider the position of the function at another time t_2. The time interval will be $(t_2 - t_1)$. The function has displaced through a distance $v_0\ (t_2 - t_1)$, of course, the shape remains the same. The function is shown in Fig. (1*b*).

The function $f_2\ (x + v_0\ t)$ represents a wave travelling in opposite direction, *i.e.*, negative direction. If this function exists, then the solution of eq. (1) consists of two waves. One wave will be travelling to the right, *i.e.*, away from the source while the second wave will be travelling to the left, *i.e.*, back towards source. The wave is known as reflected wave. In absence of any reflecting wave, the solution of eq. (1) is given by

$$\boxed{E_y = f_1\ (x - v_0 t)} \qquad ...(2)$$

For example, we can consider a sinusoidal wave of type

$$\boxed{E_y = \sin\ (x - v_0 t)} \qquad ...(3)$$

with velocity $v_0 = 3 \times 10^8$ m/s

(3) At any particular instant of time, **E** and **H** will be mutually perpendicular to each other and also perpendicular to the direction of propagation as shown in Fig. (2).

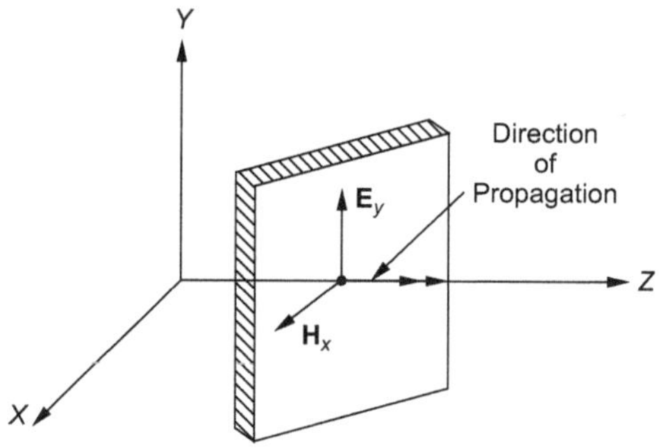

Fig. (2)

As the time passed, the magnitude of **E** and **H** will vary with time and time and the vectors will be also propagating in space with time as shown in Fig. (3).

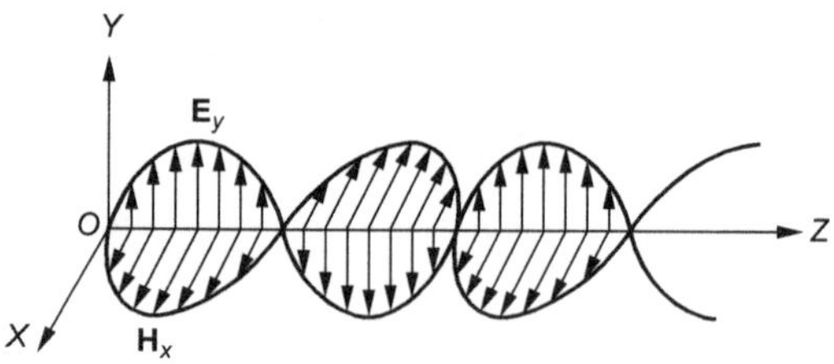

Fig. (3)

7.3 WAVE EQUATION FOR CONDUCTING MEDIA

In previous section, we have solved Maxwell's equations for perfect dielectric such as free space in which there are no charges, *i.e.*, no conduction currents. The aim of this article is to consider the wave equation for a conducting medium, *i.e.*, conduction current exists. Let us consider a uniform linear medium having dielectric constant ε, permeability μ and conductivity σ. Now,

$$\mathbf{D} = \varepsilon\,\mathbf{E} \qquad \text{and} \qquad \mathbf{J} = \sigma\,\mathbf{E}$$

Consider the following Maxwell equations

$$\vec{\nabla} \times \mathbf{E} = -\mu \frac{\partial \mathbf{E}}{\partial t} \qquad ...(1)$$

and

$$\vec{\nabla} \times \mathbf{H} = \varepsilon \frac{\partial \mathbf{E}}{\partial t} + \mathbf{J} = \varepsilon \frac{\partial \mathbf{E}}{\partial t} + \sigma\,\mathbf{E} \qquad ...(2)$$

Taking curl of both sides of eq. (1), we get

$$\vec{\nabla} \times \vec{\nabla} \times \mathbf{E} = -\mu\,\vec{\nabla} \times \frac{\partial \mathbf{H}}{\partial t}$$

or

$$\vec{\nabla} \times \vec{\nabla} \times \mathbf{E} = -\mu \times \frac{\partial}{\partial t}[\vec{\nabla} \times \mathbf{H}] \qquad ...(3)$$

Substituting the value of $(\vec{\nabla} \times \mathbf{H})$ in eq. (3) from eq. (2), we get

$$\vec{\nabla} \times \vec{\nabla} \times \mathbf{E} = -\mu \frac{\partial}{\partial t}\left[\varepsilon \frac{\partial \mathbf{E}}{\partial} + \sigma\,\mathbf{E}\right] \qquad ...(4)$$

or

$$\vec{\nabla} \times (\vec{\nabla} \cdot \mathbf{E}) - \nabla^2 \mathbf{E} = -\mu\varepsilon \frac{\partial^2 \mathbf{E}}{\partial t^2} - \mu\sigma \frac{\partial \mathbf{E}}{\partial t}$$

As

$$\mathbf{D} = \varepsilon\,\mathbf{E},\ \mathbf{E} = \frac{1}{\varepsilon}\mathbf{D}$$

or

$$\vec{\nabla} \cdot \mathbf{E} = \frac{1}{\varepsilon}\vec{\nabla} \cdot \mathbf{D} = \frac{1}{\varepsilon}(\rho) \qquad (\because\ \vec{\nabla} \cdot \mathbf{D} = \rho)$$

We know that there is no net charge within a conductor because the charge resides on the surface of the conductor, $\rho = 0$

$$\therefore \qquad \vec{\nabla} \cdot \mathbf{E} = \frac{1}{\varepsilon}(0) = 0 \qquad ...(5)$$

Substituting the value of $(\vec{\nabla} \cdot \mathbf{E})$ from eq. (5) in eq. (4), we get

$$\vec{\nabla}(0) - \nabla^2 \mathbf{E} = -\mu\varepsilon \frac{\partial^2 \mathbf{E}}{\partial t^2} + \mu\sigma \frac{\partial \mathbf{E}}{\partial t}$$

or

$$\nabla^2 \mathbf{E} = \mu\varepsilon \frac{\partial^2 \mathbf{E}}{\partial t^2} + \mu\sigma \frac{\partial \mathbf{E}}{\partial t}$$

$$\therefore \qquad \boxed{\nabla^2 \mathbf{E} - \mu\varepsilon \frac{\partial^2 \mathbf{E}}{\partial t^2} - + \mu\sigma \frac{\partial \mathbf{E}}{\partial t} = 0} \qquad ...(a)$$

This is the wave equation for **E**.

Similarly, taking curl of both sides of eq. (2), we get

$$\vec{\nabla} \times \vec{\nabla} \times \mathbf{H} = \varepsilon \frac{\partial}{\partial t}(\vec{\nabla} \times \mathbf{E}) + \sigma(\vec{\nabla} \times E) \qquad ...(6)$$

$$= \varepsilon \frac{\partial}{\partial t}\left(-\mu \frac{\partial \mathbf{H}}{\partial t}\right) + \sigma\left(-\mu \frac{\partial \mathbf{H}}{\partial t}\right) \qquad \text{[Using eq. (1)]}$$

or $$\vec{\nabla}(\vec{\nabla}\cdot \mathbf{H}) - \nabla^2 \mathbf{H} = \mu\varepsilon \frac{\partial^2 \mathbf{H}}{\partial t^2} - \sigma\mu \frac{\partial \mathbf{H}}{\partial t}$$

As $\nabla \cdot \mathbf{B} = 0$, therefore, $\vec{\nabla}\cdot \mathbf{H} = 0$

$$\therefore \quad -\nabla^2 \mathbf{H} = -\mu\varepsilon \frac{\partial^2 \mathbf{H}}{\partial t^2} - \sigma\mu \frac{\partial \mathbf{H}}{\partial t}$$

$$\therefore \quad \boxed{\nabla^2 \mathbf{H} - \mu\varepsilon \frac{\partial^2 \mathbf{H}}{\partial t^2} - \sigma\mu \frac{\partial \mathbf{H}}{\partial t} = 0}$$

This is wave equation for **H**.

EXAMPLE 1. *Using eqs. (a) and (b) obtain wave equations of* **E** *and* **H** *for free space.*

Solution. For free space, $\rho = 0$, $\sigma = 0$, $\varepsilon = \varepsilon_0$ and $\mu = \mu_0$.

Substituting these values in equations (*a*) and (*b*), we get wave equations of **E** and **H** respectively

$$\nabla^2 \mathbf{E} = \mu_0 \varepsilon_0 \frac{\partial^2 \mathbf{E}}{\partial t^2} \quad \text{and} \quad \nabla^2 \mathbf{H} = \mu_0 \varepsilon_0 \frac{\partial^2 \mathbf{H}}{\partial t^2}$$

EXAMPLE 2. *Using eqs. (a) and (b) obtain wave equations of* **E** *and* **H** *for isotropic dielectric or non-conducting medium.*

Solution. For a non-conducting medium

$$\rho = 0 \quad \text{and} \quad \sigma = 0$$

Therefore, for non-conducting medium, the wave equations are

$$\nabla^2 \mathbf{E} = \mu\varepsilon \frac{\partial^2 \mathbf{E}}{\partial t^2} \quad \text{and} \quad \nabla^2 \mathbf{H} = \mu\varepsilon \frac{\partial^2 \mathbf{H}}{\partial t^2}$$

7.4 PHASOR NOTATION (SINUSOIDAL TIME VARYING FIELD REPRESENTATION)

Let us consider that electric and magnetic fields vary sinusoidally with time. This can be expressed as

$$E = E_0 \cos(\omega t + \phi) \quad \text{or} \quad E = E_0 \sin(\omega t + \phi) \qquad ...(1)$$

where $\omega = 2\pi f$ is angular frequency of vibration and ϕ is phase angle.

It is clear that a sinusoidal time factor has to be attached to every term in the equation. However, it is not necessary because this may be suppressed by the use of *phasor natation*.

By Euler's identity $e^{j\omega t} = \cos(\omega t) + j\sin(\omega t)$ ...(2)

Using Euler's identity, a time varying field $E(r, t)$ can be expressed as

$$E_x(r, t) = \text{Re}\,[E_x(r)e^{j(\omega t + t)}] = \text{Re}\,[E_x(r)e^{j\omega t}\, e^{j\phi}]$$

Here, Re expresses real part of the quantity inside bracket. Simplifying by dropping Re and suppressing $e^{j\omega t}$, the field quantity $E_x(r)$ becomes a phasor. This can be identified by putting a symbol (~) over the time varying quantity. Therefore,

$$\tilde{E}_x(r) = E_x(r)\, e^{j\phi}$$

7.5 MAXWELL'S EQUATIONS USING PHASOR NOTATION

We have already derived the general time varying Maxwell's equations. Now, we shall derive the corresponding set of phasor equations which are applicable for sinusoidal time variations. Let us consider the following Maxwell's equations in phasor form

$$\vec{\nabla} \times \tilde{\mathbf{H}} = \frac{\partial \tilde{\mathbf{D}}}{\partial t} + \tilde{\mathbf{J}} \qquad ...(1)$$

For sinusoidal steady state, we may substitute the phasor relations as follows

$$\vec{\nabla} \times \mathrm{Re}[H\, e^{j\omega t}] = \frac{\partial}{\partial t} \mathrm{Re}[D\, e^{j\omega t}] + \mathrm{Re}[J\, e^{j\omega t}]$$

$$\vec{\nabla} \times \mathrm{Re}[H\, e^{j\omega t}] = \mathrm{Re}[j\omega D\, e^{j\omega t}] + \mathrm{Re}[J\, e^{j\omega t}]$$

or $$\mathrm{Re}[(\vec{\nabla} \times \mathbf{H} - j\omega\mathbf{D} - \mathbf{J})\, e^{j\omega t}] = 0$$

If this result is valid for all t, then

$$\vec{\nabla} \times \mathbf{H} = j\,\omega\,\mathbf{D} + \mathbf{J} \qquad ...(2)$$

This is maxwell's equation in phasor form

Similarly, $$\vec{\nabla} \times \mathbf{E} = j\,\omega\,\mathbf{B}$$

The Maxwell's equation in differential form using phasor notation can be expressed as

$$\boxed{\begin{aligned} &\vec{\nabla} \times \mathbf{H} = j\,\omega\,\mathbf{D} + \mathbf{j} \\ &\vec{\nabla} \times \mathbf{E} = -\,j\,\omega\,\mathbf{B} \\ &\vec{\nabla} \times \mathbf{D} = \rho \\ &\vec{\nabla} \times \mathbf{B} = 0 \end{aligned}} \qquad ...(3)$$

The Maxwell's equation in integral form using phasor notation can be expressed as

$$\boxed{\begin{aligned} &\oint \mathbf{H} \cdot dl = \int_S J \cdot d\mathbf{S} + j\omega \int_S \mathbf{D} \cdot d\mathbf{S} \\ &\oint \mathbf{E} \cdot dl = -\,j\,\omega \int_S \mathbf{B} \cdot d\mathbf{S} \\ &\oint \mathbf{B} \cdot d\mathbf{S} = 0 \\ &\oint \mathbf{D} \cdot d\mathbf{S} = \int \rho_V \, dV \end{aligned}} \qquad ...(4)$$

7.6 WAVE PROPAGATION IN A LOSSLESS MEDIUM

For a uniform plane wave in which there is no variation with respect to y and z in phasor form can be expressed as

$$\frac{\partial^2 \tilde{\mathbf{E}}}{\partial x^2} = -\,\omega^2 \mu\varepsilon\, \tilde{\mathbf{E}} = -\,\beta^2\, \tilde{\mathbf{E}} \qquad ...(1)$$

where $$\beta^2 = \omega^2\, \mu\, \varepsilon \quad \text{or} \quad \beta = \omega\sqrt{(\mu\, \varepsilon)}$$

Consider the E_y component of eq. (1), it becomes

$$\frac{\partial^2 E_y}{\partial x^2} + \beta^2 E_y = 0 \qquad ...(2)$$

The solution of eq. (2) can be expressed as

$$E_y(x) = C_1 e^{-j\beta x} + C_2 e^{+j\beta x} \qquad ...(3)$$

where C_1 and C_2 constants.

The corresponding time varying field is

$$\tilde{\mathbf{E}}_y (x, t) = \text{Re} [E_y (x) e^{j\omega t}] = \text{Re} [(C_1 e^{-j\beta x} + C_2 e^{+j\beta x}) e^{j\omega t}]$$
$$= \text{Re} [C_1 e^{j(\omega t - \beta x)} + C_2 e^{j(\omega t + \beta x)}$$

Taking the real part, we have

$$\boxed{\tilde{\mathbf{E}}_y (x, t) = C_1 \cos (\omega t - \beta x) + C_2 \cos (\omega t + \beta x)} \qquad ...(4)$$

Eq. (4) gives the combination of two travelling waves in which one is travelling in forward direction while the other is moving in backward direction, *i.e.*, reflected wave.

When $C_1 = C_2$, the two waves combine to produce a standing wave (which does not progress).

Expressing eq. (4) as a function of $(x \pm v\, t)$ we observe that

$$v = \frac{\omega}{\beta}$$

Clearly, v is ***wave velocity***.

For a wave travelling in X-direction, the point is given by

$$(\omega t - \beta x) = \text{constant}$$

Now,
$$\frac{dx}{dt} = \frac{\omega}{\beta} = v_p \qquad i.e., \boxed{v_p = \frac{\omega}{\beta}} \qquad ...(5)$$

This gives the velocity at some point in sinusoidal waveform. It is called as ***phasor velocity***. Here, β is called as ***phasor shift constant***. Thus, phasor velocity is equal to wave velocity.

The wavelength λ is defined as the distance covered in full cycle of radian, *i.e.*,

$$\beta\lambda = 2\pi \qquad \text{or} \qquad \beta = \frac{2\pi}{\lambda}$$

Now,
$$v = \frac{\omega}{\beta} = \frac{2\pi f}{\left(\frac{2\pi}{\lambda}\right)} = f\lambda$$

$\therefore$
$$\boxed{v = f\lambda} \quad \text{or} \quad \text{velocity} = \text{frequency} \times \text{wavelength} \qquad ...(6)$$

Further, wave velocity,

$$v = \frac{\omega}{\beta} = \frac{1}{\sqrt{\mu\,\varepsilon}}$$

$\therefore$
$$v = \frac{1}{\sqrt{\mu_0\,\varepsilon_0}\sqrt{\mu_r\,\varepsilon_r}} = \frac{c}{\sqrt{\varepsilon_r}} \qquad \text{where } \mu_r = 1 \qquad ...(7)$$

7.7 WAVE PROPAGATION IN LOSSY DIELECTRIC MEDIUM (CONDUCTING MEDIUM)

The wave equation for a conducting medium is given by

$$\Delta^2\mathbf{E} - \mu\,\varepsilon\,\frac{\partial^2\mathbf{E}}{\partial t^2} - \mu\,\sigma\,\frac{\partial\mathbf{E}}{\partial t} = 0 \qquad ...(1)$$

(see eq. (*a*) of article 7.3)

Let us consider a plane electromagnetic wave propagating in *Z*-direction with vector **E** in *X*-direction and vector **H** in *Y*-direction. Therefore,

$$\frac{\partial}{\partial x} = \frac{\partial}{\partial y} = 0$$

and $\mathbf{E} = E_x\,\hat{\mathbf{a}}_x$ and $\mathbf{H} = H\,\hat{\mathbf{a}}_y$

Wave eq. (1), can be written as

$$\frac{\partial^2 E_x}{\partial z^2} = \mu\,\varepsilon\,\frac{\partial^2 E_x}{\partial t^2} + \mu\,\sigma\,\frac{\partial E_x}{\partial t} \qquad ...(2)$$

Similarly, we can write

$$\frac{\partial^2 H_y}{\partial z^2} = \mu\,\varepsilon\,\frac{\partial^2 H_y}{\partial t^2} + \mu\,\sigma\,\frac{\partial H_y}{\partial t} \qquad ...(3)$$

If electric field E_x varies harmonically with time, then

$$E_x = E_0\,e^{j\omega t} \qquad ...(4)$$

Now, $$\frac{\partial E_x}{\partial t} = E_0\,(j\omega)\,e^{j\omega t} = j\,\omega\,Ex$$

and $$\frac{\partial^2 E_x}{\partial t^2} = E_0\,(j\omega)\,(j\omega)\,e^{j\omega t} = -\,\omega^2\,Ex$$

Substituting these values in eq. (2), we get

$$\frac{\partial^2 E_x}{\partial z^2} = -\,\mu\,\varepsilon\,\omega^2\,Ex + j\,\omega\,\mu\,\sigma\,E_x$$

or $$\frac{\partial^2 E_x}{\partial z^2} - j\,\omega\,\mu\,\sigma\,E_x + \omega^2\,\mu\,\varepsilon\,E_x = 0$$

or $$\frac{\partial^2 E_x}{\partial z^2} - (j\,\omega\,\mu\,\sigma - \omega^2\,\mu\,\varepsilon)E_x = 0 \qquad ...(5)$$

Let us put γ (propagation constant) as

$$\gamma^2 = (j\,\omega\,\mu\,\sigma - \omega^2\,\mu\varepsilon) \qquad ...(6)$$

Substituting in eq. (5), we get

$$\boxed{\frac{\partial^2 E_x}{\partial z^2} - \gamma^2\,E_x = 0} \qquad ...(7)$$

The propagation constant γ is a complex quantity and can be expressed as

$$\gamma = \alpha + j\,\beta \qquad ...(8)$$

where α is attenuation constant and β is phasor constant.

Eq. (7) shows the propagation of uniform plane electromagnetic wave through a conducting medium of permittivity ε and permeability μ. The solution of this equation is of the form

$$\boxed{E_x = E_m\, e^{-\gamma z}} \quad ...(9)$$

where $\gamma^2 = [j\,\omega\mu\sigma - \omega^2\,\mu\varepsilon]$

Attenuation Constant (α) and Phase Constant (β)

Now, $(\alpha + j\beta)^2 = j\,\omega\,\mu\,\sigma - \omega^2\,\mu\varepsilon$

or $a^2 - \beta^2 + 2j\,\alpha\,\beta = j\,\omega\,\mu\,\sigma - \omega^2\,\mu\varepsilon$ ...(10)

Comparing the real and imaginary parts on both sides of eq. (10), we get

$$\alpha^2 - \beta^2 = -\,\omega^2\,\mu\varepsilon \quad ...(11)$$

and $\alpha^2 - \beta = \omega\mu\sigma$ ...(12)

Solving eqs. (11) and (12) for α and β, we get

$$\alpha = \omega\sqrt{\left(\frac{\mu\,\varepsilon}{2}\right)\left[\sqrt{1+\left(\frac{\sigma}{\omega\,\varepsilon}\right)^2} - 1\right]} \quad ...(13)$$

and

$$\beta = \omega\sqrt{\left(\frac{\mu\,\varepsilon}{2}\right)\left[\sqrt{1+\left(\frac{\sigma}{\omega\,\varepsilon}\right)^2} + 1\right]} \quad ...(14)$$

Intrinsic Impedance

Let us consider E_x component of **E** varying with respect to z. We have

$$\vec{\nabla}\times\mathbf{E} = \frac{\partial E_x}{\partial z}\hat{\mathbf{a}}_y = \frac{\partial}{\partial z}(E_m\, e^{-\gamma z})\,\hat{\mathbf{a}}_y = -\,\gamma\, E_m\, e^{-\gamma z}\,\hat{\mathbf{a}}_y$$

Substituting $\vec{\nabla}\times\mathbf{E} = -j\,\omega\,\mu\,\mathbf{H}$

We get $-j\,\omega\,\mu\,\mathbf{H} = -\,\gamma\,E_m\,e^{-\gamma z}\,\hat{\mathbf{a}}_y$

H has component H_y corresponding to E_x, hence,

$$H_y = \frac{\gamma}{j\,\omega\,\mu}E_x$$

The intrinsic impedance η is defined as

$$\eta = \frac{E_x}{H_y} = \frac{j\,\omega\,\mu}{\gamma}$$

Substituting the value of γ, we get

$$\eta = \frac{j\,\omega\,\mu}{j\,\omega\sqrt{\mu\,\varepsilon}\sqrt{1+\dfrac{\sigma}{j\,\omega\,\varepsilon}}} = \frac{\sqrt{\mu}}{\sqrt{\varepsilon\left(1+\dfrac{\sigma}{j\,\omega\,\varepsilon}\right)}}$$

$$\therefore \quad \boxed{\eta = \sqrt{\left(\frac{j\,\omega\,\mu}{\sigma + j\,\omega\,\varepsilon}\right)}} \quad ...(15)$$

7.8 WAVE PROPAGATION IN GOOD DIELECTRICS

In previous article, we have derived the wave equation for a conducting medium. This is expressed as

$$\frac{\partial^2 E_x}{\partial z^2} - \gamma^2 E_x = 0 \qquad \text{[eq. (7)]} \quad ...(1)$$

where γ is propagation constant. Further,

$$\gamma = \alpha + j\beta \qquad ...(2)$$

where α is attenuation constant and β is phase constant. The values of α and β are given by eq. (13) and eq. (14) respectively.

The aim of this article is to calculate the values of (1) attenuation constant (α), (2) phase constant (β), (3) wave velocity (v) and (4) intrinsic impedance in case of good dielectric.

For good dielectric, the ratio of conduction current density to displacement current density is always less than one in radio frequency range, *i.e.*,

$$\frac{\sigma}{\omega\varepsilon} << 1 \qquad ...(3)$$

$\sigma/\omega\varepsilon$ is known as loss tangent.

(1) Attenuation Constant (α)

According to eq. (13) of previous article, we have

$$\alpha = \omega\sqrt{\left(\frac{\mu\varepsilon}{2}\right)\left[\sqrt{1+\left(\frac{\sigma}{\varepsilon\omega}\right)^2} - 1\right]} \qquad ...(4)$$

In case of good dielectrics, $\left(\frac{\sigma}{\varepsilon\omega}\right) << 1$. Therefore, using Binomial theorem, we can write

$$\left(1+\frac{\sigma^2}{\varepsilon^2\omega^2}\right)^{1/2} = \left(1+\frac{\sigma^2}{2\varepsilon^2\omega^2}\right)$$

Substituting this value in eq. (4), we get

$$\alpha \approx \omega\sqrt{\left(\frac{\mu\varepsilon}{2}\right)\left[\left(1+\frac{\sigma^2}{2\varepsilon^2\omega^2}\right)-1\right]} \approx \omega\sqrt{\left(\frac{\mu\varepsilon}{2}\right)\times\left(\frac{\sigma^2}{2\varepsilon^2\omega^2}\right)} \approx \frac{\sigma}{2}\sqrt{\left(\frac{\mu}{\varepsilon}\right)}$$

$$\therefore \qquad \boxed{\alpha \approx \frac{\sigma}{2}\sqrt{\left(\frac{\mu}{\varepsilon}\right)}} \qquad ...(5)$$

(2) Phase constant (β)

According to eq. (14) of previous article, we have

$$\beta = \omega\sqrt{\left(\frac{\mu\varepsilon}{2}\right)\left[\sqrt{1+\left(\frac{\sigma}{\omega\varepsilon}\right)^2} + 1\right]} \qquad ...(6)$$

$$\approx \omega \sqrt{\left(\frac{\mu \varepsilon}{2}\right)\left[1+\frac{\sigma^2}{2\,\omega^2\,\varepsilon^2}+1\right]} \approx \omega \sqrt{(\mu \varepsilon)}\left[1+\frac{\sigma^2}{4\,\omega^2\,\varepsilon^2}\right]^{1/2}$$

$$\approx \omega \sqrt{\mu \varepsilon}\left[1+\frac{\sigma^2}{8\,\omega^2\,\varepsilon^2}\right]$$

$$\therefore \qquad \beta \approx \omega \sqrt{\mu \varepsilon}\left[1+\frac{\sigma^2}{8\,\omega^2\,\varepsilon^2}\right] \qquad ...(7)$$

(3) Wave Velocity

The wave velocity is given by

$$v = \frac{\omega}{\beta} = \frac{1}{\sqrt{\mu \varepsilon}\left[1+\frac{\sigma^2}{8\,\omega^2\,\varepsilon^2}\right]} \qquad \text{[Using eq. (6)]}$$

$$= \frac{1}{\sqrt{\mu \varepsilon}} \times \left[1+\frac{\sigma^2}{8\,\omega^2\,\varepsilon^2}\right] = \frac{1}{\sqrt{\mu \varepsilon}}\left[1-\frac{\sigma^2}{8\,\omega^2\,\varepsilon^2}\right]$$

$$\therefore \qquad \boxed{v \approx \frac{1}{\sqrt{\mu \varepsilon}}\left[1-\frac{\sigma^2}{8\,\omega^2\,\varepsilon^2}\right]} \qquad ...(8)$$

In case of a perfect dielectric, the conductivity is zero. The wave velocity $v_0 = \frac{1}{\sqrt{(\mu \varepsilon)}}$. The effect of small loss is to reduce the velocity of propagation slightly.

(4) Intrinsic Impedance

The intrinsic impedance η is given by

$$\eta = \sqrt{\left[\frac{j\,\omega\,\mu}{(\sigma + j\,\omega\,\varepsilon)}\right]} \qquad \text{(see eq. (15) of last article)}$$

$$= \sqrt{\left[\frac{(j\,\omega\,\mu)}{j\,\omega\,\varepsilon\left(1+\frac{\sigma}{j\,\omega\,\varepsilon}\right)}\right]} = \sqrt{\left(\frac{\mu}{\varepsilon}\right)} \times \frac{1}{\left(1+\frac{\sigma}{j\,\omega\,\varepsilon}\right)^{1/2}}$$

$$= \sqrt{\left(\frac{\mu}{\varepsilon}\right)} \times \left[1+\frac{\sigma}{j\,\omega\,\varepsilon}\right]^{-1/2} = \sqrt{\left(\frac{\mu}{\varepsilon}\right)} \times \left[1-\frac{\sigma}{2j\,\omega\,\varepsilon}\right] = \sqrt{\left(\frac{\mu}{\varepsilon}\right)} \times \left[1+\frac{j\,\sigma}{2\,\omega\,\varepsilon}\right]$$

$$\therefore \qquad \boxed{\eta \approx \sqrt{\left(\frac{\mu}{\varepsilon}\right)} \times \left[1+\frac{j\,\sigma}{2\,\omega\,\varepsilon}\right]}$$

For a dielectric with $\sigma = 0$, the intrinsic impedance

$$\eta = \sqrt{\left(\frac{\mu}{\varepsilon}\right)} = \eta_0 \qquad \text{(intrinsic impedance of medium)}$$

Now, $$\eta \approx \eta_0\left[1+\frac{j\sigma}{2\,\omega\,\varepsilon}\right] = \left(\eta_0 + \frac{j\,\sigma\,\eta_0}{2\,\omega\,\varepsilon}\right)$$

So, for dielectric for $\sigma \neq 0$, a reactive component $\left(\frac{\sigma\,\eta_0}{2\,\omega\,\varepsilon}\right)$ is added to intrinsic impedance of medium (η_0). This is the main effect of loss.

7.9 WAVE PROPAGATION IN GOOD CONDUCTORS

The wave equation for a conducting medium is given by

$$\frac{\partial^2 E_x}{\partial z^2} - \gamma^2 E_x = 0 \qquad ...(1)$$

where γ is propagation constant $\gamma = \alpha + j\,\beta$ where α is attenuation constant and β is phase constant. Further,

$$\alpha = \omega\sqrt{\left(\frac{\omega\,\varepsilon}{2}\right)\sqrt{1+\left(\frac{\sigma}{\omega\,\varepsilon}\right)^2} - 2}$$

and $$\beta = \omega\sqrt{\left(\frac{\omega\,\varepsilon}{2}\right)\sqrt{1+\left(\frac{\sigma}{\omega\,\varepsilon}\right)^2} + 1}$$

The aim of this article is to calculate the values of eq. (1) attenuation constant (α), eq. (2) phase constant β, eq. (3) wave velocity (v) and (4) intrinsic impedance (η) for good conductor. For good conductors σ and ε are nearly independent of frequency and $\frac{\sigma}{\omega\,\varepsilon} >> 1$. $\frac{\sigma}{\varepsilon}$ is known as loss tangent.

We know that γ^2 $(j\,\omega\,\mu\,\sigma - \omega^2\,\mu\,\varepsilon)$ (see article 7.6)

$$\therefore \qquad \gamma = \left[j\,\omega\,\mu\,\sigma\left(1+\frac{j\,\omega\,\varepsilon}{\sigma}\right)\right]^{1/2} \qquad ...(2)$$

$$= \sqrt{(j\,\omega\mu\sigma)} \qquad \left(\because \text{ for good conductor } \frac{\sigma}{\omega\varepsilon} >> 1\right)$$

$$= \sqrt{(\omega\mu\sigma)}\,\sqrt{j} = \sqrt{\omega\mu\sigma}\ \angle 45^\circ$$

$$= \sqrt{(\omega\mu\sigma)}\,(\cos 45^\circ + j\sin 45^\circ)$$

$$= \sqrt{(\omega\mu\sigma)}\left[\frac{1}{\sqrt{2}} + j\frac{1}{\sqrt{2}}\right] = \sqrt{\left(\frac{\omega\mu\sigma}{2}\right)}\,[1+j]$$

$$= \sqrt{\left(\frac{\omega\mu\sigma}{2}\right)} + j\sqrt{\left(\frac{\omega\mu\sigma}{2}\right)}$$

(1) Attenuation Constant (α)

We know that $\gamma = \alpha + j\,\beta$

Comparing eqs. (1) and (2), we get

$$\boxed{\alpha = \sqrt{\left(\frac{\omega\mu\sigma}{2}\right)}} \qquad ...(3)$$

In case of conductors σ is very large. Thus, α is very large. This means that the wave is attenuated greatly as it progresses through good conductor.

(2) Phase Constant (β)

Comparing the imaginary parts of eqs. (1) and (2), we get

$$\boxed{\beta = \sqrt{\left(\frac{\omega\mu\sigma}{2}\right)}} \quad \text{...(4)}$$

Thus, α and β are equal. β is also large, *i.e.*, phase shift is very large.

(3) Wave velocity (v)

$$v = \frac{\omega}{\beta} = \frac{\omega}{\sqrt{\left(\frac{\omega\mu\sigma}{2}\right)}} = \sqrt{\left(\frac{2\,\omega}{\mu\,\sigma}\right)}$$

$$= \sqrt{\left(\frac{4\,\pi\,f}{\mu\,\sigma}\right)} \qquad \text{(where } \omega = 2\,\pi f, f = \text{frequency)}$$

$$\therefore \qquad \boxed{v = \sqrt{\left(\frac{2\,\omega}{\mu\,\sigma}\right)}}$$

As $v \propto \frac{1}{\beta}$, the wave velocity in the good conductors is very small ...(5)

(4) Intrinsic Impedance

$$\eta = \sqrt{\left(\frac{j\,\omega\,\mu}{\sigma + j\,\omega\,E}\right)}$$

$$\approx \sqrt{\left(\frac{j\,\omega\,\mu}{\sigma}\right)} \qquad (\because\ \sigma >> \omega\,\varepsilon)$$

$$\therefore \qquad \boxed{\eta \approx \sqrt{\left(\frac{j\,\omega\,\mu}{\sigma}\right)} \angle 45^\circ} \quad \text{...(6)}$$

Eq. (6) shows that intrinsic impedance or characteristic impedance is very small. The characteristic impedance has a reactive component with impedance angle of 45°. The refractive index of conducting medium

$$n = \frac{c}{v} = \frac{c}{\sqrt{\left(\frac{2\,\omega}{\mu\,\sigma}\right)}} = \sqrt{\left(\frac{\mu\,\sigma}{2\,\omega}\right)} \quad \text{...(7)}$$

7.10 DEPTH OF PENETRATION (SKIN DEPTH)

We have seen that in a medium which has conductivity, the wave suffers attenuation as it progresses. Therefore, we find the penetration of the wave into the conducting medium, called the depth of penetration or skin depth. ***The depth of penetration* δ *is defined as that depth in***

***which the wave attenuates* (1/*e*) *times of its strength or* 37% *of the original value as shown in* Fig. (4).**

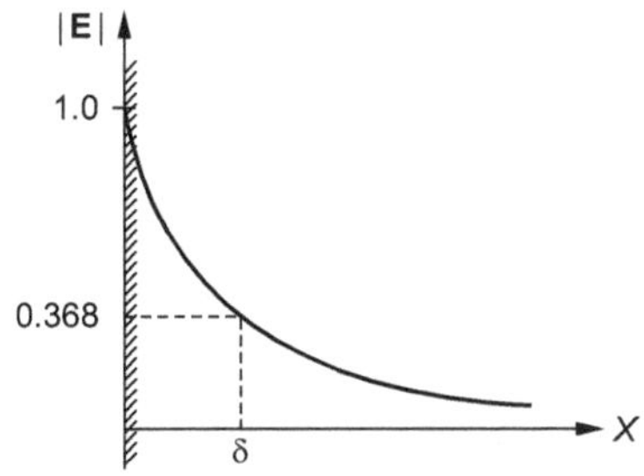

Fig. (4) Showing skin depth

We know that, $$E_x = E_m e^{-\alpha x}$$

If $\alpha x = 1$, then $$\left(\frac{E_x}{E_m}\right) = \frac{1}{e}$$

Thus, on travelling a distance $x = (1/\alpha)$, the amplitude falls to $(1/e)$ times its value. By definition, such a distance is equal to skin depth. Therefore,

$$\delta = \frac{1}{\alpha} = \sqrt{\left(\frac{2}{\omega\mu\sigma}\right)} = \sqrt{\left(\frac{2}{\pi f \mu\sigma}\right)}$$

So, higher is the frequency, smaller will be δ, *i.e.*, less penetration. For a good conductor, at a frequency of 10^4 cycles, the skin distance is 0.667×10^{-3} m. So, at high frequencies, the current flows on the surface of conductor.

In case of copper, with $\sigma = 5.8 \times 10^7$ S/m, $\mu = \mu_0$ and at a frequency of 1 MHz, we have

$$\delta = \frac{1}{\sqrt{(3.14 \times 10^6 \times (4\pi \times 10^{-7}) \times 5.8 \times 10^7)}}$$
$$= 0.0661 \text{ mm}$$

EXAMPLE 1. *For a non-magnetic material having* $\varepsilon_r = 2.25$, $\sigma = 10^{-4}$ seimen/m, *find* (*i*) *loss tangent*, (*ii*) *attenuation constant*, (*iii*) *phase constant and* (*iv*) *intrinsic impedance for a wave having frequency of* 2.5 MHz. *Assume the material to be a good dielectric.*

Solution.

(*i*) Loss tangent $= \frac{\sigma}{\omega\varepsilon} = \frac{\sigma}{2\pi f \varepsilon_r \varepsilon_0} = \frac{10^{-4}}{2 \times 3.14 \times (2.5 \times 10^6) \times 2.25 \times 8.854 \times 10^{-12}}$

$= \mathbf{0.320}$

(*ii*) Attenuation constant $\alpha = \frac{\sigma}{2}\sqrt{\left(\frac{\mu}{\varepsilon}\right)} = \frac{10^{-4}}{2}\sqrt{\left[\frac{4\pi \times 10^{-7}}{2.25 \times 8.854 \times 10^{-12}}\right]}$

$= \mathbf{0.01256}$ **Np/m**

(*iii*) Phase constant $\beta = \omega\sqrt{(\mu\varepsilon)}\left[1+\dfrac{\sigma^2}{8\,\omega^2\,\varepsilon^2}\right]$

$$= \{2 \times 3.14 \times (2.5 \times 10^6)\}\{(4 \times 3.14 \times 10^{-7}) \times (2.25 \times 8.854 \times 10^{-12})\}^{1/2}\left[1+\frac{1}{8}(0.320)^2\right]$$

$$= \mathbf{0.0796\ rad/m}$$

(*iv*) Intrinsic impedance $\eta = \sqrt{\dfrac{\mu}{\varepsilon}}\left[1+j\dfrac{\sigma}{2\omega\varepsilon}\right]$

$$= \left[\frac{4\pi\times10^{-7}}{2.25\times8.854\times10^{-12}}\right]^{1/2}\left[1+j\frac{0.320}{2}\right]$$

$$= \mathbf{254.34 \angle 9.09^\circ\ \Omega}$$

EXAMPLE 2. *A uniform plane wave in a medium having* $\sigma = 10^{-3}$ S/m, $\varepsilon = 80\,\varepsilon_0$ *and* $\mu = \mu_0$ *is having a frequency of* 10 kHz. *Calculate the different parameters of the wave.*

Solution.

(*i*) Loss tangent $= \dfrac{\sigma}{\omega\varepsilon} = \dfrac{\sigma}{2\pi f\varepsilon} = \dfrac{10^{-3}}{2\pi\times10^4\times80\times8.854\times10^{-3}} = \mathbf{22.47 >> 1}$

So, the medium is a good conductor.

(*ii*) Attenuation constant $\alpha = \sqrt{\pi f \mu \sigma}$

$$= \sqrt{[\mu\times10^4\times(4\pi\times10^{-7})\times10^{-3}]}$$

$$= \mathbf{2\,\pi \times 10^{-3}\ Np/m}$$

(*iii*) $\beta = \alpha = \mathbf{2\,\pi \times 10^{-3}\ Np/m}$

(*iv*) Intrinsic impedance $\eta = \sqrt{\left(\dfrac{\omega\mu}{\sigma}\right)}\angle 45^\circ = \sqrt{\left(\dfrac{2\,p\,f\,\mu}{\sigma}\right)}45^\circ$

$$= \sqrt{\left[\frac{2\,\pi\times10^4\,(4\,\pi\times10^{-7})}{10^{-3}}\right]}\angle45^\circ = 2\,\pi\,(1+\mathbf{j})\,\Omega$$

(*v*) Wavelength $\gamma = \dfrac{2\,\pi}{\beta} = \dfrac{2\,\pi}{2\,\pi\times10^{-3}} = \mathbf{1000\ m}$

(*vi*) Velocity of wave $v = \dfrac{\omega}{\beta} = \dfrac{2\,\pi\,f}{\beta} = \dfrac{2\,\pi\times10^4}{2\,\pi\times10^{-3}} = \mathbf{10^7\ m/s}$

EXAMPLE 3. *The electric field amplitude of a uniform plane wave propagation in* $\hat{\mathbf{a}}_z$ *direction if* 250 V/m. *If* $\mathbf{E} = E_x\,\hat{\mathbf{a}}_x$ *and* ω = 1 M rad/s, *find* (*a*) *frequency,* (*b*) *wavelength,* (*c*) *period and* (*d*) *amplitude of* **H**.

Solution.

$$(a)\ f = \frac{\omega}{2\pi} = \frac{1\times10^6}{2\times3.14} = \mathbf{159.23\ kHz}$$

$$(b)\ \gamma = \frac{c}{f} = \frac{3\times10^8}{159.23\times10^3} = 1.88\times10^3\ \text{m} = \mathbf{1.88\ km}$$

$$(c)\ T = \frac{1}{f} = \frac{1}{159.23\times10^3} = \mathbf{6.28\ \mu s}$$

$$(d)\ \frac{E}{H} = \sqrt{\left(\frac{\mu_0}{\varepsilon_0}\right)} = 337 \qquad \therefore \quad H = \frac{250}{377} = \mathbf{0.663\ A/m}$$

EXAMPLE 4. *The constitution parameter of aluminium are given by*

$\mu_r = 1, \varepsilon_r = 1$ and $\sigma = 3.54 \times 107$ mho/m

Find the frequency for which the skin depth of aluminium is 0.01 mm.

Solution. The skin depth of a good conductor is given by

$$\delta = \sqrt{\left(\frac{1}{\pi f \mu \sigma}\right)}$$

or

$$f = \frac{1}{\pi \delta^2 \mu \sigma}$$

Substituting the given values, we get

$$f = \frac{1}{3.14\times(0.01\times10^{-3})^2\times(4\pi\times10^{-7})\times(3.54\times10^{-7})}$$

$$= 71.56\times10^6\ \text{Hz} = \mathbf{71.56\ MHz}$$

EXAMPLE 5. *For sea water* $\mu = \mu_0 = 4\pi \times 10–7$ N/ A^2, $\varepsilon = 70\ \varepsilon_0$ *and conductivity* $\sigma = 5$ S/m. .*Find the skin depth and attenuation constant for sea water.*

Solution. As sea water is a poor conductor and hence, the skin depth is given by

$$\delta = \frac{1}{\alpha} = \frac{2}{\sigma}\sqrt{\left(\frac{\varepsilon}{\mu}\right)}$$

Substituting the given values, we get

$$\delta = \frac{2}{5}\sqrt{\left[\frac{70\times8.854\times10^{-12}}{4\pi\times10^{-7}}\right]} = \mathbf{0.0089\ m}$$

Attenuation constant $\alpha = \frac{1}{\delta} = \frac{1}{0.0089} = \mathbf{112.36\ Np/m}$

EXAMPLE 6. *Calculate the refractive index of copper at* 10 MHz *assuming that conductivity of copper is* 5.8×10^7 mho/m *and its relative permeability and permitivity is unity.*

Solution. We know that

$$v_P = \sqrt{\left(\frac{2\,\omega}{\sigma\,\mu}\right)} = \sqrt{\left[\frac{2\times 2\,\pi\,(10\times 10^6)}{(5.8\times 10^7)\times(4\,\pi\times 10^{-7})}\right]} \quad (\because\ \mu = \mu_0\mu_r = \mu_0\times 1 = \mu_0)$$

$$= \mathbf{0.131 \times 10^4\ m/s}$$

Now, refractive index

$$n = \frac{c}{v_p} = \frac{3\times 10^8}{0.131\times 10^4} = \mathbf{0.0229 \times 10^7}$$

EXAMPLE 7. *Calculate the skin depth, propagation constant and wave velocity at frequency of* 1.6 MHz *in aluminium where* σ = 38.2 MS/m *and* μ_r = 1.

Solution. The skin depth is given by

$$\delta = \frac{1}{\sqrt{\pi\,f\,\mu\,\sigma}} = \frac{1}{\sqrt{\pi\,f(\mu_0\,\mu_r)\,\sigma}}$$

$$= \frac{1}{\sqrt{\pi\,f\times(4\pi\times 10^{-7}\times 1)\sigma}} = \frac{1}{2\pi\sqrt{f\times 10^{-7}\times\sigma}}$$

$$= \frac{1}{2\times 3.14\sqrt{(1.6\times 10^6)\times 10^{-7}\times(38.2\times 10^6)}}$$

$$= \frac{10^{-2}}{6.28\times\sqrt{611.2}} = \frac{10^{-2}}{6.28\times 24.7224}$$

$$= 64.409\times 10^{-6}\ \text{m} = \mathbf{64.409\ \mu m}$$

$$\alpha = \beta = \frac{1}{\delta} = \frac{1}{64.409\times 10^{-6}} = \mathbf{1.5525 \times 10^4\ Np/m}$$

$$\gamma = \alpha + j\,\beta = 1.5525\times 10^4 + j(1.5525\times 10^4)$$

$$= 10^4\times\sqrt{(1.5525)^2 + (1.5525)^2}\ \angle 45°$$

$$= 10^4\sqrt{4.8204}\ \angle 45° = \mathbf{21.96 \times 10^3\ \angle\ 45\ m^{-1}}$$

$$v = \frac{\omega}{\beta} = \omega\,\delta = 2\,\pi\,f\,\delta$$

$$= 2\times 3.14\times(1.6\times 10^6)\times(64.409\times 10^{-6}) = \mathbf{647.2\ m/s.}$$

7.11 THE POYNTING VECTOR AND POWER FLOW

One important characteristic of electromagnetic waves is that they transfer energy through space from source to receiving point. The rate of transfer of this energy can be expressed in terms of electric and magnetic field strengths of the wave. The amount of energy passing through unit area of the surface perpendicular to the direction of propagation of energy is called as Poynting vector. This denoted by **P.** For example, in a plane electromagnetic wave, $\vec{E}$, $\vec{H}$ and $\vec{P}$ are perpendicular to each other. In addition, $\vec{P}$ shows the direction of wave propagation. Therefore,

$\mathbf{P} = \mathbf{E} \times \mathbf{H}$. ***Poynting theorem relates the rate of energy transfer and the amplitudes of electric and magnetic field strengths of electromagnetic wave***. We shall derive the desired result by using Maxwell's equation.

By Maxwell's equation $\vec{\nabla} \times \mathbf{H} = \mathbf{J} + \varepsilon \dfrac{\partial \mathbf{E}}{\partial t}$ ...(1)

Taking dot product with **E** on both sides, we get

$$\mathbf{E} \cdot (\vec{\nabla} \times \mathbf{H}) = \mathbf{E} \cdot \mathbf{J} + \varepsilon \, \mathbf{E} \cdot \frac{\partial \mathbf{E}}{\partial t}$$

or
$$\vec{\nabla} \cdot (\mathbf{E} \times \mathbf{H}) = \mathbf{E} \cdot \mathbf{J} + \varepsilon \, \mathbf{E} \cdot \frac{\partial \mathbf{E}}{\partial t} \qquad ...(2)$$

We know the vector identity

$$\vec{\nabla} \cdot (\boldsymbol{E} \times \boldsymbol{H}) = \vec{\boldsymbol{H}} \cdot (\vec{\nabla} \times \boldsymbol{E}) - \boldsymbol{E} \cdot (\vec{\nabla} \times \boldsymbol{H})$$

Substituting this value in eq. (2), we get

$$\mathbf{H} \cdot (\vec{\nabla} \times \mathbf{E}) - \vec{\nabla} \cdot (\mathbf{E} \times \mathbf{H}) = \mathbf{E} \cdot \mathbf{J} + \varepsilon \, \mathbf{E} \cdot \frac{\partial \mathbf{E}}{\partial t} \qquad ...(3)$$

Consider the following Maxwell's equation

$$\vec{\nabla} \times \mathbf{E} = -\frac{\partial \mathbf{B}}{\partial t} = -\mu \frac{\partial \mathbf{H}}{\partial t}$$

Substituting in eq. (3), we get

$$-\mu \left(\mathbf{H} \cdot \frac{\partial \mathbf{H}}{\partial t} \right) - \vec{\nabla} \cdot (\mathbf{E} \times \mathbf{H}) = \mathbf{E} \cdot \mathbf{J} + \varepsilon \left(\mathbf{E} \cdot \frac{\partial \mathbf{E}}{\partial t} \right)$$

$$-\mu \left(\frac{1}{2} \frac{\partial}{\partial t} H^2 \right) - \vec{\nabla} \cdot (\mathbf{E} \times \mathbf{H}) = \mathbf{E} \cdot \mathbf{J} + \varepsilon \left[\frac{1}{2} \frac{\partial}{\partial t} E^2 \right]$$

or
$$-\vec{\nabla} \cdot (\mathbf{E} \times \mathbf{H}) = \mathbf{E} \cdot \mathbf{J} + \frac{\partial}{\partial t} \left[\frac{\varepsilon E^2}{2} + \frac{\mu H^2}{2} \right]$$

or
$$\vec{\nabla} \cdot (\mathbf{E} \times \mathbf{H}) = -\mathbf{E} \cdot \mathbf{J} - \frac{\partial}{\partial t} \left[\frac{\varepsilon E^2}{2} + \frac{\mu H^2}{2} \right] \qquad ...(4)$$

Integrating over volume V, we have

$$\int_V \vec{\nabla} \cdot (\mathbf{E} \times \mathbf{H}) dV = -\int_V \mathbf{E} \cdot \mathbf{J} \, dV - \frac{\partial}{\partial t} \int_V \left(\frac{\varepsilon E^2}{2} + \frac{\mu H^2}{2} \right) dV \qquad ...(5)$$

Using divergence theorem

$$\int_V \vec{\nabla} \cdot (\mathbf{E} \times \mathbf{H}) \, dV = \oint_S (\mathbf{E} \times \mathbf{H}) \, dS$$

Now, eq. (5) becomes

$$\boxed{\oint_S (\mathbf{E} \times \mathbf{H}) \, dS = -\int_V \mathbf{E} \cdot \mathbf{J} \, dV - \frac{\partial}{\partial t} \int_V \left(\frac{\varepsilon E^2}{2} + \frac{\mu H^2}{2} \right) dV} \qquad ...(6)$$

This is an important equation which tells us about the flow of power in a wave. Let us discuss the terms in this equation one by one:

(*i*) $\oint_S (\mathbf{E} \times \mathbf{H})\, dS$ gives the rate of flow energy outward through the surface. The product $(\mathbf{E} \times \mathbf{H}) = \mathbf{P}$ is a measure of rate of energy flow per unit area. **P** is known as ***Poynting Vector***. The unit of **P** is Watt/m^2 and is in the direction of **E** × **H.** Therefore, Poynting vector

$$\mathbf{P} = \mathbf{E} \times \mathbf{H} \text{ watt/m}^2 \qquad \text{...(7)}$$

Thus, the integral of **P** or **E** × **H** over a closed surface represents the rate at which electromagnetic energy crosses the closed surface. In this relation of **E** and **H**. represents the instantaneous values.

(*ii*) The term $\int_V \mathbf{E} \cdot \mathbf{J}$ represents instantaneous power dissipated in volume *V.* In an electric field

$$E = \frac{\text{Voltage}}{\text{Distance}} = \frac{E}{d} \text{ and } J \text{ (Current density)} = \frac{I}{A}$$

$$\therefore \qquad \mathbf{E} \cdot \mathbf{J} = \frac{V}{A}\frac{I}{d} = \text{power loss in unit volume} = I^2R$$

(*iii*) $\left(-\frac{\partial}{\partial t}\right)$ represents the rate at which thc stored energy is decreasing

$\left(\frac{\mu H^2}{2}\right)$ represents the energy density due to magnetic field.

$\left(\frac{\varepsilon E^2}{2}\right)$ represent the energy density due to electric field.

Therefore, $\left(\frac{\mu H^2}{2} + \frac{\varepsilon E^2}{2}\right)$ represents the total stored energy in volume *V* due to both electric and magnetic fields. Thus,

$$-\frac{\partial}{\partial t}\int_V \left(\frac{\varepsilon E^2}{2} + \frac{\mu H^2}{2}\right) dV$$

represents the rate of decrease of stored energy.

Average Power for a Plane Wave

Consider a plane wave is travelling *Z*-direction. Then

$$\mathbf{P} = \mathbf{E} \times \mathbf{H}$$

or

$$P_z \hat{\mathbf{a}}_z = E_x \hat{\mathbf{a}}_x \times H_y \hat{\mathbf{a}}_y \qquad \text{...(8)}$$

For a perfect dielectric

$$E_x = E_m \cos(\omega t - \beta z)$$

and

$$H_y = \frac{E_m}{\eta} \cos(\omega t - \beta z)$$

$$\therefore \qquad P_z = E_x H_y = \frac{E_m^2}{\eta} \cos^2(\omega t - \beta z) \qquad \text{...(9)}$$

The average power density is given by

$$(P_z)_{\text{av.}} = \frac{1}{T}\int_0^T P_z\, dt = \frac{1}{T}\int_0^T \frac{E_m^2}{\eta}\cos^2(\omega t - \beta z)\, dt$$

$$= \frac{E_m^2}{2T\eta}\int_0^T [1 + \cos(2\omega t - 2\beta z)]\, dt$$

$$= \frac{E_m^2}{2T\eta}\left[t + \frac{1}{2\omega}\sin(2\omega t - 2\beta z)\right]_0^T$$

$$= \frac{1}{2}\frac{E_m^2}{\eta} \qquad \left[\because \frac{1}{2\omega}\sin(2\omega t - 2\beta z)\right]_0^T = 0$$

$\therefore$

$$\boxed{(P_z)_{\text{ave.}} = \frac{1}{2}\frac{E_m^2}{\eta}\ \text{W/m}^2} \qquad ...(10)$$

If we use root-mean square values instead of peak amplitudes, then the factor (1/2) would be absent from the expression of $(P_z)_{\text{av}}$.

For a surface of area A, normal to Z-axis, the average power flowing through it is given by

$$\boxed{(P_z)_{\text{ave.}} = \frac{1}{2}\frac{E_m^2}{\eta}A}\ \text{watt} \qquad ...(11)$$

Average Power in a Lossy Dielectric

Let us calculate the average power in a lossy dielectric. In this case, we have

$$E_x = E_m\, e^{-\alpha z}\cos(\omega t - \beta z) \qquad ...(12)$$

and

$$H_y = \frac{E_m}{|\eta|}\, e^{-\alpha z}\cos(\omega t - \beta z - \theta_n) \qquad ...(13)$$

where

$$\eta = |\eta| \angle \theta_n$$

Now,

$$P_z = E_x H_y \qquad ...(14)$$

$$= \frac{E_m^2}{|\eta|}\, e^{-\alpha z}\cos(\omega t - \beta z)\cos(\omega t - \beta z - \theta_n) \qquad ...(15)$$

Using the identity

$$\cos A \cos B = \frac{1}{2}\cos(A+B) + \frac{1}{2}\cos(A-B),$$

We have

$$\cos(\omega t - \beta z)\cos(\omega t - \beta z - \theta_\eta) = \frac{1}{2}\cos(2\omega t - 2\beta z - \theta_\eta) + \frac{1}{2}\cos\theta_\eta \qquad ...(16)$$

Substituting this value in eq. (15), we get

$$P_z = \frac{1}{2}\frac{E_m^2}{|\eta|}\, e^{-2\alpha z}[\cos(2\omega t - 2\beta z - \theta_\eta) + \cos\theta_\eta] \qquad ...(17)$$

Eq. (17) shows that power density has a second harmonic component [first bracketted term] and a d.c. component [second bracketted term]. The average value of first term is zero. Therefore, time average value of Poynting vector is

$$\boxed{(P_z)_{\text{av.}} = \frac{1}{2}\frac{E_m^2}{|\eta|}\, e^{-2\alpha z} \cos\theta_\eta} \qquad \text{...(18)}$$

The above expression shows that due to the presence of the term $e^{-2\alpha z}$, the power density attenuates.

EXAMPLE 1. *A plane wave propagating in free space with a peak electric field intensity* 0.75 V/m. *Find average power through the square area* 120 cm *on a side perpendicular to the direction of propagation.* $\eta = 377$.

Solution. The average power in *Z*-direction is given by

$$(P_z)_{\text{av.}} = \frac{1}{2}\frac{E_m^2}{\eta} A \text{ watt}$$

Here, $E_m = 0.75$ V/m, $\eta = 377$ and $A = (120 \times 10^{-2}) = 1.44\ \text{m}^2$

$$\therefore \quad (P_z)_{\text{av}} = \frac{1}{2} \times \frac{(0.75)^2}{377} \times (1.44) = 10.7 \times 10^{-3} \text{ watt}$$

$$= \mathbf{10.7\ mW}$$

EXAMPLE 2. *A sinusoidal plane wave is transmitting through a medium whose breakdown voltage is* 20 kV (rms) *with* $\eta = 60\,\pi$. *Find average power density and the peak value of magnetic field intensity.*

Solution. We know that $\quad P_{\text{av}} = \dfrac{E_m^2}{2\eta} \quad$ or $\quad (P_{\text{av.}})_{\text{rms}} = \dfrac{E_{\text{rms}}^2}{\eta}$

Given $\quad \eta = 60\,\pi \quad$ and $\quad E_{\text{rms}} = (20 \times 10^3)$ V

$$\therefore \quad (P_{\text{av.}})_{\text{rms}} = \frac{(20\times 10^3)^2}{60\,\pi} = \mathbf{2.123 \times 106\ W/m^2}$$

Now,

$$H_{\text{rms}} = \frac{E_{\text{rms}}}{\eta} = \frac{20\times 10^3}{60\,\pi} = 0.1061 \times 10^3 \text{ A/m}$$

$$= \mathbf{106.1\ A/m}$$

EXAMPLE 3. *In free space* $\mathbf{E}(x, t) = 50\cos(\omega t - \beta x)\,\hat{\mathbf{a}}_y$ V/m. *Find the average power crossing a circular area of radius* 5 m *in the plane* x = *constant.*

Solution. In complex form

$$\mathbf{E}(x, t) = 50\, e^{j(\omega t - \beta x)}\hat{\mathbf{a}}_y \text{ V/m}$$

$$\eta = 120\,\pi\,\Omega$$

$$\mathbf{H}(z, t) = \frac{50}{120\,\pi}\, e^{j(\omega t - \beta x)}\hat{\mathbf{a}}_z \text{ A/m}$$

The average Poynting vector

$$P_{av.} = \frac{1}{2} \text{Re}\,(\mathbf{E} \times \mathbf{H})\,\hat{\mathbf{a}}_x$$

where Re represents the real part

$$= \frac{1}{2}(50)\left(\frac{50}{120\,\pi}\right)\hat{\mathbf{a}}_x$$

Average power crossing the circular area

$$P_{av.} = \frac{1}{2}(50)\left(\frac{50}{120\,\pi}\right)\pi\,(5)^2 = \mathbf{260.4\ W}$$

EXAMPLE 4. *A plane wave 200 MHz is travelling in a medium for which* $\sigma = 0$, $\mu_r = 2$, $\varepsilon_r = 4$. *If the average Poynting vector is* 5 W/m^2, *find* E_{rms}, H_{rms}, *phase velocity and impedance of the medium.*

Solution. The phase velocity is given by

$$v_p = \frac{c}{\sqrt{(\varepsilon_r\,\mu_r)}} = \frac{3 \times 10^8}{\sqrt{4 \times 2}} = \mathbf{1.06 \times 10^8\ m/s}$$

$$\eta = \sqrt{\left(\frac{\mu}{\varepsilon}\right)} = \sqrt{\left(\frac{\mu_0\,\varepsilon_r}{\varepsilon_0\,\varepsilon_r}\right)} = \sqrt{\left(\frac{\mu_0}{\varepsilon_0}\right)}\sqrt{\left(\frac{\mu_r}{\varepsilon_r}\right)}$$

$$= 377\sqrt{\left(\frac{2}{4}\right)} = \mathbf{266.6\ \Omega}$$

$$P_{av.} = \frac{1}{2} \times \frac{E_0^2}{\eta} \quad \text{or} \quad E_0^2 = 2\,\eta\,P_{av.}$$

$$\therefore \quad E_0^2 = 2 \times 266.5 \times 5 = 2666 \quad \text{or} \quad E_0 = \mathbf{51.6}$$

Now,

$$E_{rms} = \frac{E_0}{\sqrt{2}} = \frac{51.6}{\sqrt{2}} = \mathbf{36.51\ V/m}$$

and

$$H_{rms} = \frac{E_{rms}}{\eta} = \frac{36.51}{266.6} = \mathbf{136.9\ mA/m}$$

EXAMPLE 5. *An elliptically polarized electromagnetic wave travelling in positive Z-direction in free space has x and y components as*

$$E_x = 2\sin(\omega\,t - \beta\,z)\ \text{V/m}$$

and

$$E_y = 4\sin(\omega\,t - \beta\,z + 75°)\ \text{V/m}$$

Calculate the average power per unit conveyed by the electromagnetic wave.

Solution. We know that

$$P_{av} = \frac{1}{2}\frac{E_0^2}{\eta}\ \text{W/m}^2$$

Now,

$$E_0^2 = E_1^2 + E_2^2 = (2)^2 + (4)^2 = 20$$

$$\eta = \sqrt{\left(\frac{\mu_0}{\varepsilon_0}\right)} = \sqrt{\left(\frac{4\pi\times10^{-7}}{8.854\times10^{-12}}\right)} = 377\ \Omega$$

$$\therefore \qquad P_{av} = \frac{1}{2}\times\frac{20}{377} = \mathbf{26.52 \times 10^{-3}\ W/m^2}$$

7.12 WAVE POLARIZATION

We have consider the case of a plane electromagnetic wave say propagating in Z-direction. Then the electric vector **E** is in X-direction while magnetic vector **H** in Y-direction. There is an orthogonal (perpendicular) relationship between **E, H** and Poynting vector **P**. To describe an electromagnetic wave completely, we should not only include the parameters such as its wavelength, phase velocity and power but also *the instantaneous orientation of its field vectors*. It is important to mention here that if the magnitude and direction of **E** are known then the magnitude and direction of **H** are automatically specied. Thus, it is sufficient to specify the quantities related to electric vector **E**.

The wave polarization is defined as the orientation of electric field vector* E *at a given instant of time in space.

There are three types of polarization:

(1) Linear Polarization

A plane electromagnetic wave in which electric vector vibrates simple harmonically along a fixed straight line perpendicular to the direction of propagation without changing its orientation is called *plane polarized* electromagnetic wave as shown in Fig. (5a) The wave is travelling out of page in Z-direction.

(2) Elliptical Polarization

Let us consider the case of two plane polarized electromagnetic waves which superimpose over each other Now, the resultant vector rotates under certain conditions. If the magnitude and

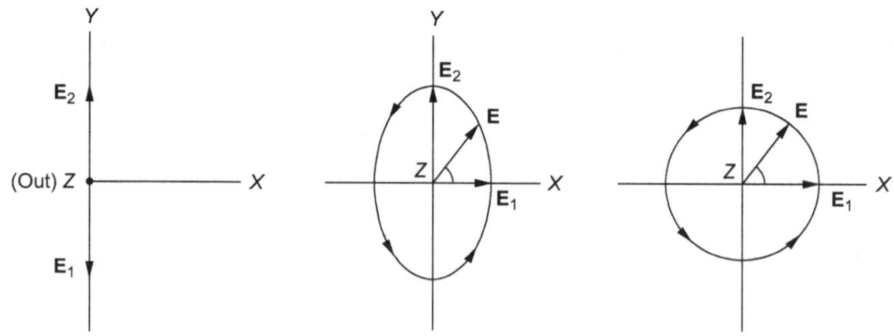

(*a*) Linear polarization

(*b*) Elliptical polarization

(c) Circular polarization

Fig. (5)

orientation of the electric field vector **E** both varies continuously, then the tip vector traces an ellipse as shown in Fig. (5*b*). The wave is said to be *elliptically polarized.*

(3) Circular Polarization

In the above case, if the magnitude of resultant vector remains consider but its orientation varies continuously then the tip of vector **E** traces a circle as shown in Fig. (5c). The wave is said to be *circularly polarized.*

Mathematical Consideration

A plane polarized wave propagating in *Z*-direction can be considered to be the sum of two polarized components in perpendicular directions but in phase as shown in Fig. (6). Here **E** has been resolved into two components $\mathbf{E}_1$ and $\mathbf{E}_2$. The two components are given by

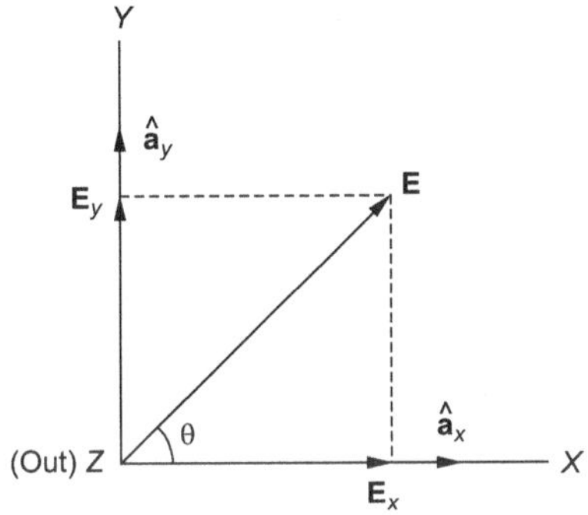

Fig. (6)

$$\mathbf{E}_x = \mathbf{E}_x^0 \, e^{j(\omega t - kx)} \quad \text{or} \quad \mathbf{E}_x = \hat{\mathbf{a}}_x E_x^0 \, e^{j(\omega t - kx)} \qquad ...(1)$$

Similarly,

$$\mathbf{E}_y = \hat{\mathbf{a}}_y E_y^0 \, e^{j(\omega t - kx)} \qquad ...(2)$$

Here $\hat{\mathbf{a}}_x$ and $\hat{\mathbf{a}}_y$ are unit vectors along *X* and *Y*-axes respectively.

(1) Plane Polarized Wave

The general solution of plane polarized wave is given by

$$\mathbf{E}(r, t) = (\hat{\mathbf{a}}_x E_x^0 + \hat{\mathbf{a}}_y E_y^0) \, e^{j(\omega t - kz)} \qquad ...(3)$$

Here, E_x^0 and E_y^0 must have the same phase if **E** (**r**, *t*) is to be plane polarized. The magnitude E^0 is given by

$$E^0 = \sqrt{[(E_x^0)^2 + (E_y^0)^2]} \qquad ...(4)$$

The polarization vector E makes an angle

$$\theta = \tan^{-1}\left(\frac{E_y^0}{E_x^0}\right) \text{ with } \hat{\mathbf{a}}_x \qquad ...(5)$$

(2) Elliptically Polarized Wave

In elliptically polarized wave, the two linearly polarized components (one in *X*-direction and other in *Y*-direction) are not equal in magnitude. As the magnitude is real, we can express them as

$$E_x(\mathbf{r}, t) = E_1 \cos(\omega t - kz) \qquad ...(6)$$

and $$E_y(\mathbf{r}, t) = E_2 \sin(\omega t - kz) \quad ...(7)$$

where E_1 amplitude of linearly polarized light in X-direction and E_2 is the amplitude of wave linearly polarized in Y-direction.

The resultant of eqs. (6) and (7) is given by

$$\frac{E_x^2}{E_1^2} + \frac{E_y^2}{E_2^2} = 1 \quad ...(8)$$

This is equation of ellipse. *Therefore, elliptically polarized wave is the resultant of two plane perpendicular polarized waves of unequal amplitudes.*

(3) Circularly Polarized Wave

The components of actual electric field obtained by taking real part of eq. (3) are given as

$$E_x(\mathbf{r}, t) = E_0 \cos(\omega t - kz) \quad ...(9)$$

and $$E_y(\mathbf{r}, t) = E_0 \sin(\omega t - kz) \quad ...(10)$$

In case of circularly polarized wave, the amplitude of two components is the same, *i.e.*,

$$E_1 = E_2 = E_0$$

The resultant of eqs. (9) and (10) is given by

$$\frac{E_x^2}{E_0^2} + \frac{E_y^2}{E_0^2} = 1$$

or $$E_x^2 + E_y^2 = E_0^2 \quad ...(11)$$

Eq. (11) represents a circle. Thus, electric vector **E**, at a fixed point in space, sweeps around a circle at a frequency of ω. Therefore, *circularly polarized wave is the resultant of two perpendicular plane polarized waves with equal amplitude.*

Exercises and Problems

EXERCISE

1. Define uniform plane wave. Derive the wave equation form Maxwell's equation for free space, charge free region.
2. Discuss the propagation of a plane wave.
3. Starting with Maxwell's equation, derive the wave equation for **E** and **H** in conducting media.
4. Solve the wave equation for a uniform plane wave in an isotropic homogeneous lossy dielectric medium with no sources.
5. Define attenuation constant, phase constant and propagation constant. Obtain their values in lossy dielectric medium.
6. Explain the propagation of a uniform wave in a perfect dielectric. Point out the values of attenuation constant, phase constant, wave velocity and intrinsic impedance.
7. Explain the propagation of uniform plane wave in a good conductor. List out the propagation parameters.

8. What do you understand by skin effect? Show that in case of a semi-infinite solid conductor, the skin depth δ is given by

$$\delta = \sqrt{\left(\frac{2}{\omega \mu \sigma}\right)}$$

where the symbols have their usual meanings.

9. What us Poynting vector? Derive the formula and explain the physical meaning of each term

10. Define the following:

(*i*) Skin depth (*ii*) Intrinsic impedance,

(*iii*) Phase velocity. (*iv*) Poynting vector.

11. Prove the formula

$$(P_z)_{\text{av.}} = \frac{1}{2}\frac{E_m^2}{\eta} \text{ watt/m}^2$$

where the symbols have their usual meaning.

12. Discuss the wave propagating in perfect dielectric.

13. Discuss the wave propagation in conducting media.

14. Discuss the wave propagating in good conductor.

15. Discuss the wave propagating in good dielectric.

16. What do you mean by polarization of a wave? Show that linear polarization is related to circular polarization of wave.

PRACTICE PROBLEMS

1. Determine the propagation constant γ for a material having $\mu_r = 1$ $\varepsilon_r = 8$ and $\sigma = 0.25$ ps/m, if the wave frequency is 1.6 MHz.

[**Ans.** $j(9.48 \times 10^{-2})\ \text{m}^{-1}$]

[**Hint.** $$\frac{\sigma}{\omega \varepsilon} = \frac{0.25 \times 10^{-12}}{2\pi(1.6 \times 10^6)\, 8\left(\frac{10^{-9}}{3\pi}\right)} \approx 10^{-9} \approx 0$$

So, $$\frac{\sigma}{\omega \varepsilon} << 1, \quad \text{(perfect dielectric)}$$

$$\alpha = 0 \text{ and } \gamma = \alpha + j\beta = j\beta$$

where $$\beta = \omega\sqrt{\mu\varepsilon} = 2\pi f\sqrt{\left(\frac{\mu_r \varepsilon_r}{c}\right)}$$]

2. Determine the following for a damp soil at 1 MHz, $\varepsilon_r = 12$, $\mu_r = 1$ and $\sigma = 2 \times 10^{-12}$ mho/m. Calculate α, δ, v_p and η.

[**Ans.** $\alpha = \beta = 2.808 \times 10^{-6}$/m, $\delta = 3.65 \times 10^5$ m, $v_p = 2.236 \times 10^{12}$ m/s

and $\eta = 140.425 \times 10^3 \angle 45°$]

[**Hint.** $\dfrac{\sigma}{\omega\,\varepsilon} >> 1$. So, material is a good conductor

$$\alpha = \beta = \sqrt{\left(\frac{\omega\,\mu\,\sigma}{2}\right)} = \sqrt{\left[\frac{2\,\pi\times 10^{6}\times 4\,\pi\times 10^{-7}\times 2\times 10^{-12}}{2}\right]}$$

$$\delta = \frac{1}{\alpha},\ v_p = \frac{\omega}{\beta} = \frac{2\,\pi\,f}{\beta}$$

and $$\eta = \sqrt{\left(\frac{j\,\omega\,\mu}{\sigma + j\,\omega\,\varepsilon}\right)} = \sqrt{\left(\frac{j\,\omega\,\mu}{2\,\sigma}\right)}\angle 45°\Bigg]$$

3. Find the intrinsic impedance η, propagation constant γ and the wave velocity for a conducting medium in which $\sigma = 58$ seimen/m, $\mu_r = 1$ at a frequency $f = 100 \times 10^6$ Hz.
[**Ans.** $3.6877 \times 10^{-3} \angle 45°$, $2.1389 \times 10^5 \angle 45°\ \text{m}^{-1}$ and 4152.336 m/sec]

4. For a non-magnetic material for which relative permittivity ε_r and conductivity σ are 2.5 and 10^{-4} seimen/m respectively. Calculate (*a*) loss tangent, (*b*) attenuation constant, (*c*) phase constant, (*d*) velocity of wave propagation and (*e*) intrinsic impedance for an electromagnetic wave having a frequency of 3 MHz. Assume the material to be a good dielectric.
[**Ans.** (*a*) 0.2398, (*b*) 0.01191 neper/m, (*c*) 0.1 rad/m, (*d*) 188.4×10^6 m/s
and (*e*) $239.92 \angle 6.84°\Omega$]
[**Hint.** Given that, $s = 10^{-4}$ seimen/m, $\varepsilon_0 = 8.854 \times 10^{-12}\ \text{C}^2/\text{N-m}^2$
$\varepsilon_r = 2.5$, $\varepsilon = \varepsilon_r\,\varepsilon_0$, $\mu_0 = \mu = 4\pi \times 10^{-7}\ \text{N-sec}^2/\text{C}^2$
and $f = 3 \times 10^6$ Hz

$$\text{Loss tangent} = \frac{\sigma}{\omega\varepsilon},\ \alpha = \frac{\sigma}{2}\sqrt{\left(\frac{\mu}{\varepsilon}\right)}$$

$$\beta = \omega\sqrt{\mu\varepsilon}\left[1 + \frac{\sigma^2}{8\omega^2\,\varepsilon^2}\right],\ v = \frac{\omega}{\beta}$$

and $$\eta = \sqrt{\left(\frac{\mu}{\varepsilon}\right)}\left[1 + j\frac{\sigma}{2\omega\varepsilon}\right]\Bigg]$$

5. Earth is a good conductor when $\dfrac{\sigma}{\omega\varepsilon} >> 1$. Determine the highest frequency for which earth can be considered good conductor when $\dfrac{\sigma}{\omega\varepsilon}$ is greater than 10. Assume the following constants $\sigma = 5 \times 10^{-3}$ S/m and $\varepsilon = 10\,\varepsilon_0$.
[**Ans.** $f_{max} = 900$ kHz]

8

Reflection, Refraction and Dispersion of Plane Waves

INTRODUCTION

In previous chapter, we have discussed about the plane waves and their propagation in different media. We know that when a wave strikes the boundary between two media, it is partly reflected and partly transmitted, depending upon the nature of boundary. In case of pure conductor, the wave is reflected and constitute a standing wave. When the refractive index of the medium varies with frequency, the medium is known as dispersive medium. The variation of refractive index of the medium with wavelength or frequency constitutes the phenomenon of dispersion. The aim of this chapter is to deal with the fundamentals of reflection, refraction and dispersion of plane waves.

8.1 REFLECTION OF UNIFORM PLANE WAVES BY PERFECT DIELECTRIC-NORMAL INCIDENCE

Consider a uniform plane electromagnetic wave is incident normally at the interface of a perfect dielectric media (such as light passes from air to glass) as shown in Fig. (1)

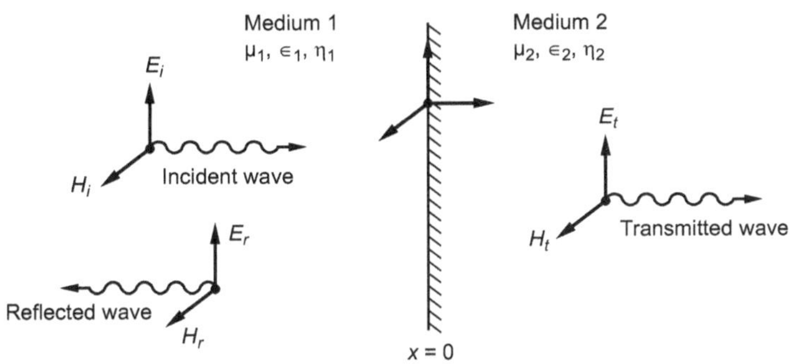

Fig. (1) Reflection and refraction of electromagnetic wave (Normal incidence) (*i*, signifies incident, *r*, reflected and *t* transmitted)

Let the wave is travelling in X-direction and the boundary of two dielectrics is at $x = 0$. Due to partial reflection and partial refraction, a reflected wave is obtained in medium 1 and refracted ray in medium 2. Let ε_1, μ_1 be the dielectric permittivity and magnetic permeability of medium 1 respectively. Similarly, ε_2, μ_2 are the corresponding quantities respectively in medium 2. Suppose η_1 and η_2 be the impedances of medium 1 and 2 respectively.

Symbols Used and Their Relationship

Incident wave $\begin{cases} E_i = \text{electric field of incident wave} \\ H_i = \text{magnetic field of incident wave} \end{cases}$

Reflected wave $\begin{cases} E_r = \text{electric field of reflected wave} \\ H_r = \text{magnetic field of reflected wave} \end{cases}$

Transmitted wave $\begin{cases} E_t = \text{electric field of transmitted wave} \\ H_t = \text{magnetic field of transmitted wave} \end{cases}$

The impedance of first medium

$$\eta_1 = \sqrt{\left(\frac{\mu_1}{\varepsilon_1}\right)}$$

The impedance of second medium

$$\eta_2 = \sqrt{\left(\frac{\mu_2}{\varepsilon_2}\right)}$$

We know that $$\eta = \frac{E}{H}$$

Therefore, the electric and magnetic fields associated with incident, reflected and transmitted waves are related as

$$E_i = \eta_1 H_i \quad ...(1)$$

$$E_r = -\eta_1 H_r \quad ...(2)$$

(negative sign is used to show that reflected energy direction is changed)

$$E_t = \eta_2 H_t \quad ...(3)$$

Boundary Conditions

The continuity of tangential components of E and H require that

$$E_i + E_r = E_t$$

and $$H_i + H_r = H_t$$

From eq. (3), $$H_t = \frac{E_t}{\eta_2} = \frac{1}{\eta_2}(E_i + E_r) \quad ...(4)$$

From eqs. (1) and (2), $$H_i = \frac{E_i}{\eta_1}$$

and $$H_r = -\frac{E_r}{\eta_1}$$

$\therefore$ $$H_i + H_r = \frac{1}{\eta_1}(E_i - E_r) \quad ...(5)$$

Reflection Coefficient for *E*

The reflection coefficient is definded as the ratio of reflected electric field intensity to the electric intensity of incident wave. Therefore,

$$\Gamma_E = \frac{E_r}{E_i} \qquad \text{...(6)}$$

From eqs. (4) and (5), we have

$$\frac{1}{\eta_2}(E_i + E_r) = \frac{1}{\eta_1}(E_i + E_r)$$

or $$\eta_1 (E_i + E_r) = \eta_2 (E_i - E_r)$$

or $$\eta_1 E_i + \eta_1 E_r = \eta_2 E_i - \eta_2 E_r$$

or $$\eta_1 E_r + \eta_2 E_r = \eta_2 E_i - \eta_1 E_i$$

or $$E_r(\eta_1 + \eta_2) = E_i (\eta_2 - \eta_1)$$

or $$\frac{E_r}{E_i} = \left(\frac{\eta_2 - \eta_1}{\eta_1 + \eta_2}\right) \qquad \text{...(7)}$$

From eqs. (6) and (7), we get

$$\boxed{\Gamma_E = \left(\frac{\eta_2 - \eta_1}{\eta_1 + \eta_2}\right)} \qquad \text{...(8)}$$

Transmission Coefficient For *E*

The transmission coefficient is defined as the ratio of transmitted electric intensity to the incident electric intensity. Therefore,

$$\tau_E = \frac{E_t}{E_i} \qquad \text{...(9)}$$

Now, $$\tau_E = \frac{E_i + E_r}{E_i} = 1 + \frac{E_r}{E_i} \qquad \text{...(10)}$$

Substituting the value of $\left(\frac{E_r}{E_i}\right)$ from eq. (7), we get

$$\tau_E = 1 + \frac{\eta_2 - \eta_2}{\eta_1 + \eta_2} = \frac{2\eta_2}{\eta_1 + \eta_2}$$

$\therefore$ $$\boxed{\tau_E = \frac{2\eta_2}{\eta_1 + \eta_2}} \qquad \text{...(11)}$$

Reflection and Transmission Coefficients for *H*

Similarly, we can get reflection and transmission coefficients for *H*.

From eqs. (1) and (2),

$$\frac{H_r}{H_i} = -\frac{E_r}{E_i} = -\left(\frac{\eta_2 - \eta_1}{\eta_1 + \eta_2}\right) = \left(\frac{\eta_2 - \eta_2}{\eta_1 + \eta_2}\right) \qquad \text{...(12)}$$

Therefore, reflection coefficient for H is given by

$$\boxed{\tau_H = \frac{2\eta_1}{\eta_1 + \eta_2}} \qquad \text{...(13)}$$

Reflection and Transmission Coefficients in Terms of Dielectric Permittivity (ε)

For perfect dielectric $\qquad \mu_1 = \mu_2 = \mu_0$

$$\therefore \qquad \eta_1 = \sqrt{\left(\frac{\mu_0}{\varepsilon_1}\right)} \text{ and } \eta_2 = \sqrt{\left(\frac{\mu_0}{\varepsilon_2}\right)}$$

Substituting these values in eqs. (8), (11), (12) and (13), we get

$$\Gamma_E = \left(\frac{\eta_2 - \eta_1}{\eta_1 + \eta_2}\right) = \frac{\sqrt{\left(\frac{\mu_0}{\varepsilon_2}\right)} - \sqrt{\left(\frac{\mu_0}{\varepsilon_1}\right)}}{\sqrt{\left(\frac{\mu_0}{\varepsilon_1}\right)} + \sqrt{\left(\frac{\mu_0}{\varepsilon_2}\right)}} = \frac{\sqrt{\varepsilon_1} - \sqrt{\varepsilon_2}}{\sqrt{\varepsilon_1} + \sqrt{\varepsilon_2}} \qquad \text{...(14)}$$

$$\tau_E = \frac{2\eta_2}{\eta_1 + \eta_2} = \frac{2\sqrt{\left(\frac{\mu_0}{\varepsilon_0}\right)}}{\sqrt{\left(\frac{\mu_0}{\varepsilon_1}\right)} + \sqrt{\left(\frac{\mu_0}{\varepsilon_2}\right)}} = \frac{2\sqrt{\varepsilon_1}}{\sqrt{\varepsilon_1} + \sqrt{\varepsilon_2}} \qquad \text{...(15)}$$

$$\Gamma_H = \frac{\eta_1 - \eta_2}{\eta_1 + \eta_2} = \frac{\sqrt{\left(\frac{\mu_0}{\varepsilon_1}\right)} - \sqrt{\left(\frac{\mu_0}{\varepsilon_2}\right)}}{\sqrt{\left(\frac{\mu_0}{\varepsilon_1}\right)} + \sqrt{\left(\frac{\mu_0}{\varepsilon_2}\right)}} = \frac{\sqrt{\varepsilon_2} - \sqrt{\varepsilon_1}}{\sqrt{\varepsilon_1} + \sqrt{\varepsilon_2}} \qquad \text{...(16)}$$

$$\tau_H = \frac{2\eta_1}{\eta_1 + \eta_2} = \frac{2\sqrt{\left(\frac{\mu_0}{\varepsilon_1}\right)}}{\sqrt{\left(\frac{\mu_0}{\varepsilon_1}\right)} + \sqrt{\left(\frac{\mu_0}{\varepsilon_2}\right)}} = \frac{2\sqrt{\varepsilon_2}}{\sqrt{\varepsilon_1} + \sqrt{\varepsilon_2}} \qquad \text{...(17)}$$

8.2 PROPERTIES OF REFLECTION AND TRANSMISSION COEFFICIENTS

We observe the following properties of reflection and transmission coefficients:

(1) For $\sigma_1 = \sigma_2 = 0$ (perfect dielectrics), then η_1 and η_2 are real. Therefore, Γ and τ are real.

(2) $\eta_1 = \eta_2$, then $\Gamma = 0$ and $\tau = 1$, *i.e.*, incident wave is entirely transmitted. The situation corresponds to a matched condition.

(3) If medium 2 is perfect conductor ($\eta_2 = 0$), then $\Gamma = -1$ and $\tau = 0$. In this case, the incident wave will be totally reflected and a standing wave will be produced in the medium.

(4) Consider the case when $\eta_2 > \eta_1$. For this case $\Gamma > 0$. So, there will be a standing wave in medium 1 and transmitted wave in medium 2.

(5) Consider the case when $\eta_1 > \eta_2$. For this case $\Gamma > 0$. There will be a standing wave in medium 1.

(6) The ratio of maximum value of electric field intensity to the minimum value of electric field intensity is called as standing ratio (*SWR*) *S*. Therefore,

$$S = \frac{(E_1)_{\max}}{(E_1)_{\min}} = \frac{(H_1)_{\max}}{(H_1)_{\min}} = \frac{1+\Gamma}{1-\Gamma}$$

$$\therefore \quad \boxed{S = \frac{1+\Gamma}{1-\Gamma}} \quad ...(1)$$

The inverse relation is $\boxed{\Gamma = \frac{S-1}{S+1}}$ ...(2)

EXAMPLE 1. *Determine the amplitudes of the reflected and transmitted* **E** *and* **H** *at the interface shown in Fig.* (2).

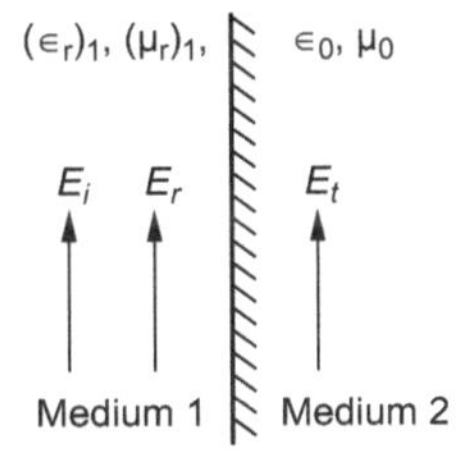

Fig. (2)

Given that $E_i = 1.5 \times 10^{-3}$ V/m *in medium* 1 *in which* $(\varepsilon_r)_1 = 8.5$, $(\mu_r)_1 = 1$ *and* $\sigma = 0$. *Medium* 2 *is free space. Assume normal incidence.*

Solution. We know that

$$\eta_1 = \sqrt{\left[\frac{\mu_0 (\mu_r)_1}{\varepsilon_0 (\varepsilon_r)_1}\right]} = \frac{\mu_0 \sqrt{(\mu_r)_1}}{\sqrt{(\varepsilon_r)_1}} = \frac{377 \times 1}{\sqrt{8.5}} = 129\ \Omega$$

$$\eta_2 = 377\ \Omega$$

Now,

$$E_r = \frac{\eta_2 - \eta_1}{\eta_1 + \eta_2} \times E_i = \frac{377-129}{129+377} \times (1.5 \times 10^{-3})$$

$$= \frac{248}{506} \times (1.5 \times 10^{-3}) = \mathbf{7.34 \times 10^{-4}\ V/m}$$

$$E_t = \frac{2\,\eta_2}{\eta_1 + \eta_1} \times E_i$$

$$= \frac{2 \times 377}{129+377} \times (1.5 \times 10^{-3}) = 1.49 \times (1.5 \times 10^{-3})$$

$$= \mathbf{2.33 \times 10^{-3}\ V/m}$$

$$H_i = \frac{E_i}{\eta_1} = \frac{1.5\times10^{-3}}{129} = \mathbf{1.159 \times 10^{-5}\ A/m}$$

$$H_r = \frac{\eta_1-\eta_2}{\eta_1+\eta_2}\times H_i = \frac{129-377}{129+377}\times(1.159\times10^{-5})$$

$$= \mathbf{-\ 5.69 \times 10–6\ A/m}$$

$$H_t = \frac{2\eta_1}{\eta_1+\eta_2}\times H_i = \frac{2\times129}{129+377}\times(1.159\times10^{-5})$$

$$= \mathbf{5.91 \times 10^{-6}\ A/m}$$

EXAMPLE 2. *Determine the amplitudes of reflected and transmitted* **E** *and* **H** *at the interface between two regions. The characteristics of region* 1 *are* $(\varepsilon_r)_1 = 8$, $(\mu_r)_1 = 1$ *and* $\sigma_1 = 0$. *Region* (2) *is free space. The incident wave in region* 1 *has* $E_i = 1.5$ V/m. *Assume normal incidence. Also, find the average powers in two regions.*

Solution. The situation is shown in Fig. (3).

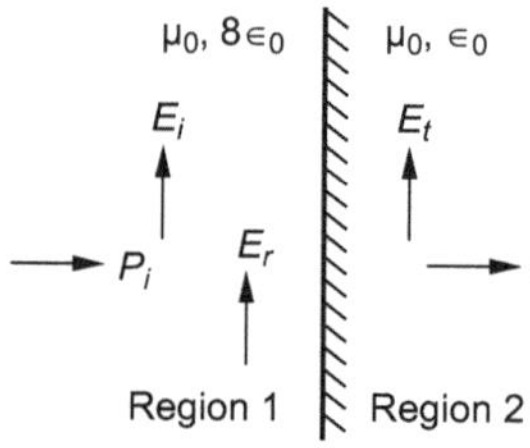

Fig. (3)

Here,
$$\eta_1 = \sqrt{\left(\frac{\mu_1}{\varepsilon_1}\right)} = \sqrt{\left(\frac{\mu_0\,\mu_r}{\varepsilon_0\,\varepsilon_r}\right)} = \sqrt{\frac{(4\pi\times10^{-7})(1)}{8\times(8.854\times10^{-12})}} = 133.3\ \Omega$$

$$\eta_2 = \sqrt{\left(\frac{\mu_0}{\varepsilon_0}\right)} = 120\,\pi = 377\ \Omega$$

Now,
$$E_r = \frac{\eta_1-\eta_1}{\eta_1+\eta_2}\times E_i = \frac{377-133.3}{133.3+377}\times1.5 = \frac{243.7}{510.3}\times1.5$$

$$= \mathbf{0.716\ V/m}$$

$$E_t = \frac{2\eta_2}{\eta_1+\eta_2}\times E_i = \frac{2\times377}{133.3+377}\times1.5 = \frac{754}{510.3}\times1.5$$

$$= \mathbf{2.216\ V/m}$$

$$H_i = \frac{E_i}{\eta_1} = \frac{1.5}{133.3} = \mathbf{1.13 \times 10^{-2}\ A/m}$$

$$H_r = \left(\frac{n_1-n_2}{n_1+n_2}\right)H_i = \frac{133.3-377}{133.3+377}\times(1.13\times10^{-2})$$

$$= -\frac{243.7}{510.3}\times(1.13\times10^{-2}) = \mathbf{5.4 \times 10^{-3}\ A/m}$$

$$H_t = \frac{2\eta_1}{\eta_1+\eta_2} \times H_i = \frac{2\times 133.3}{133.3+377} \times (1.13\times 10^{-2})$$

$$= \frac{266.6}{510.3} \times (1.13\times 10^{-2}) = \mathbf{5.90 \times 10^{-3}\ A/m}$$

The incident average power in the two regions are

$$P_i = \frac{1}{2} E_i H_i = \frac{1}{2}(1.5)(1.13\times 10^{-2}) = \mathbf{8.46 \times 10^{-3}\ W/m^2}$$

$$P_r = \frac{1}{2} E_r H_r = \frac{1}{2}(0.716)(5.4\times 10^{-3}) = \mathbf{1.93 \times 10^{-3}\ W/m^2}$$

$$P_t = \frac{1}{2} E_t H_t = \frac{1}{2}(2.216)(5.90\times 10^{-3}) = \mathbf{6.53 \times 10^{-3}\ W/m^2}$$

EXAMPLE 3. *A uniform plane wave in air is partially reflected from the surface of a material whose properties are unknown. Measurements of the electric field in the region in front of interface yield a* 1.5 m *spacing between maxima, with first maximum occurring* 0.75 m *form the interface. A standing wave ratio of* 5 *is measured. Determine the intrinsic impedance* η_2 *of unknown material.*

Solution. We know that the spacing between two maxima is $\lambda/2$.

$\therefore$ Wavelength $\lambda = 2 \times 1.5 = 3$ m

This corresponds to a frequency $f = 100$ MHz

Given that the first maximum occurs at 0.75 m, *i.e.*, $\lambda/4$ distance from the interface. This shows that the field minimum occurs at the boundary. Thus, Γ will be real and negative. Hence,

$$|\Gamma| = \frac{S-1}{S+1} = \frac{5-1}{5+1} = \frac{4}{6} = \frac{2}{3}$$

$$\Gamma = -\frac{2}{3}$$

We know that

$$\Gamma = \frac{\eta_2 - \eta_1}{\eta_1 + \eta_2}$$

or

$$-\frac{2}{3} = \frac{\eta_2 - \eta_1}{\eta_1 + \eta_2}$$

or

$$-\frac{2}{3}(\eta_1 + \eta_2) = \eta_2 - \eta_1$$

or

$$\eta_1\left(1 - \frac{2}{3}\right) = \left(1 + \frac{2}{3}\right)\eta_2$$

or

$$\frac{\eta_1}{3} = \frac{5}{3}\eta_2 \quad \text{or} \quad \eta_2 = \frac{\eta_1}{5}$$

or

$$\eta_2 = \frac{377}{5} = \mathbf{75.4\ \Omega}$$

8.3 REFLECTION OF A PLANE WAVE BY PERFECT DIELECTRIC OR INSULATOR-OBLIQUE INCIDENCE

Consider a boundary *MN* separating two media as shown in Fig. (4). Let a plane wave is incident obliquely on this boundary. A part of the wave is reflected in medium 1 and a part is refracted in medium 2. Let θ_1, θ_2 and θ_3 be angles of incidence, refraction and reflection respectively. Let v_1 and v_2 be the velocities of rays in medium 1 and 2 respectively.

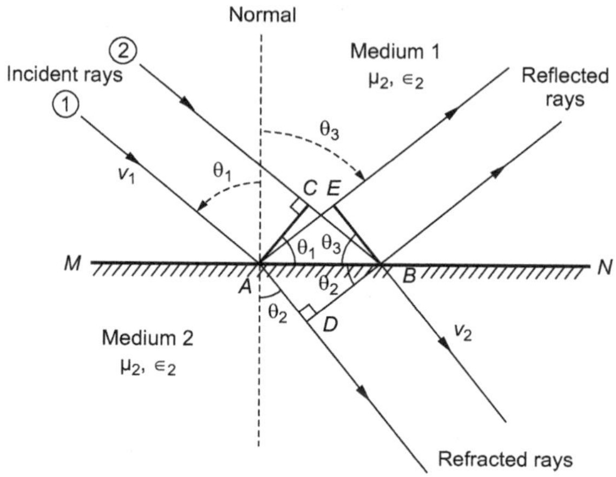

Fig. (4) Reflection and Refraction of a wave in medium 1 and 2 respectively.

In Fig. (4), we have taken two rays, *i.e.*, ray 1 and ray 2. In order to consider the additional distances covered by these rays in medium 1 and medium 2, we have drawn two perpendiculars *AC* and *BD*. Therefore, incident ray 2 travels a distance *CB* in medium 1 while the refracted ray 1 travels a distance *AD* in medium, 2. The time taken (say *t*) is the same. Hence,

$$CB = v_1 t \text{ and } AD = v_2 t \qquad (\because \text{ velocity} = \text{distance/time})$$

$$\therefore \qquad \frac{CB}{AD} = \frac{v_1}{v_2} \qquad ...(1)$$

From Δ *ACB*,
$$\frac{CB}{AB} = \sin\theta_1 \quad \text{or} \quad CB = AB\sin\theta_1 \qquad ...(2)$$

From Δ *ADB*,
$$\frac{AD}{AB} = \sin\theta_2 \quad \text{or} \quad AD = AB\sin\theta_2 \qquad ...(3)$$

Substituting the values of *CB* and *AD* from eqs. (2) and (3) in eq. (1), we get

$$\frac{AB\sin\theta_1}{AB\sin\theta_2} = \frac{v_1}{v_2}.$$

or
$$\frac{\sin\theta_1}{\sin\theta_2} = \frac{v_1}{v_2} \qquad ...(4)$$

In terms of constants of media, v_1 and v_2 are expressed as

$$v_1 = \frac{1}{\sqrt{(\mu_1\,\varepsilon_1)}} = \frac{1}{\sqrt{(\mu_0\,\varepsilon_1)}}$$

and
$$v_2 = \frac{1}{\sqrt{(\mu_2\,\varepsilon_2)}} = \frac{1}{\sqrt{(\mu_0\,\varepsilon_2)}}$$

Substituting these values in eq. (4), we get

$$\frac{\sin\theta_1}{\sin\theta_2} = \frac{\dfrac{1}{\sqrt{(\mu_0\,\varepsilon_1)}}}{\dfrac{1}{\sqrt{(\mu_0\,\varepsilon_2)}}} = \sqrt{\left(\frac{\varepsilon_2}{\varepsilon_1}\right)}$$

Therefore,
$$\boxed{\frac{\sin\theta_1}{\sin\theta_2} = \sqrt{\frac{\varepsilon_2}{\varepsilon_1}}} \qquad \text{...(5)}$$

Considering Δ's ACB and AEB, it can be proved that

$$AE = CB$$

$$\therefore \qquad \frac{AE}{AB} = \frac{CB}{AB}$$

or
$$\sin\theta_1 = \sin\theta_3$$

or
$$\boxed{\theta_1 = \theta_3} \qquad \text{...(6)}$$

This shows that angle of incidence is equal to the angle of reflection. This is well known law of reflection.

Let n_1 and n_2 be refractive indices of medium 1 and medium 2 respectively, then

$$n_1 = \frac{c}{v_1} \text{ and } n_2 = \frac{c}{v_2} \qquad (c = \text{velocity of light})$$

$$\therefore \qquad v = \frac{c}{n_1} \text{ and } v_2 = \frac{c}{n_2}$$

Substituting these values in eq. (4), we get

$$\frac{\sin\theta_1}{\sin\theta_2} = \frac{c/n_1}{c/n_2} = \frac{n_2}{n_1}$$

or
$$\boxed{n_1\sin\theta_1 = n_2\sin\theta_2} \qquad \text{...(7)}$$

This is Snell's law of refraction.

8.3.1 ENERGY CONSIDERATION

According to the law of conservation of energy, the incident energy is equal to the sum of reflected energy and refracted energy, *i.e.*,

Incident energy = reflected energy + refracted energy

We know that power transmitted per square metre is equal to product of **E** and **H**, *i.e.*, **E** × **H**. Here, **E** and **H** are perpendicular to each other and hence, the power transmitted per square metre ^{2} is equal to E^2/η. Now,

Power in incident wave striking $AB \propto \dfrac{1}{\eta_1} E_i^{\,2} \cos\theta_1$

Power in reflected wave $\propto \dfrac{1}{\eta_1} E_r^{\,2} \cos\theta_2$

Power in refracted wave $\propto \dfrac{1}{\eta_2} E_t^{\,2} \cos\theta_2$

Applying the law of conservation of energy, we get

$$\frac{1}{\eta_1} E_i^2 \cos\theta_1 = \frac{1}{\eta_1} E_r^2 \cos\theta_1 + \frac{1}{\eta_2} E_t^2 \cos\theta_2$$

$$1 = \frac{E_r^2}{E_i^2} + \frac{\eta_1}{\eta_2} \frac{E_t^2}{E_i^2} \cdot \frac{\cos\theta_2}{\cos\theta_1}$$

or

$$\frac{E_r^2}{E_i^2} = 1 - \frac{\eta_1}{\eta_2} \cdot \frac{E_t^2}{E_i^2} \frac{\cos\theta_2}{\cos\theta_1} \quad ...(8)$$

We know that $\eta_1 = \sqrt{\frac{\mu_1}{\varepsilon_1}} = \sqrt{\frac{1}{\varepsilon_1}}$ and $\eta_2 = \sqrt{\frac{\mu_2}{\varepsilon_2}} = \sqrt{\frac{1}{\varepsilon_2}}$

Substitution these values in eq. (8), we get

$$\boxed{\frac{E_r^2}{E_i^2} = 1 - \frac{\sqrt{\varepsilon_2}}{\sqrt{\varepsilon_1}} \cdot \frac{E_t^2}{E_i^2} \cdot \frac{\cos\theta_2}{\cos\theta_1}} \quad ...(9)$$

Whenever a wave is incident obliquely on the interface between two media, it is necessary to consider the following two cases separately.

(1) Electric vector **E** is perpendicular to the plane of incidence*. This is known as horizontal polarization.

(2) Electric vector **E** is parallel to the plane of incidence. This is known as vertical polarization.

We shall discuss these cases one by one.

(1) Electric vector E is perpendicular to the plane of incidence (Perpendicular polarization). In this case, the electric vector **E** is perpendicular to the plane of incidence and magnetic vector **H** is parallel to the plane of incidence as shown in Fig. (5).

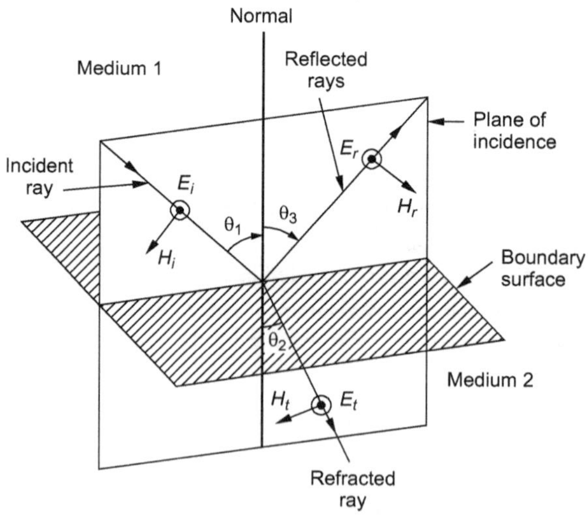

Fig. (5) Showing horizontal polarization

* A plane of incidence is defined as a plane which contains both the incident ray and line normal to the surface.

Let E_i be in positive direction of X-axis (outward in the figure). E_r and E_t will also be in positive X-direction.

According to boundary condition, the tangential component of **E** should be continuous across the boundary. Applying this condition, we get

$$E_i + E_r = E_t$$

or
$$1 + \frac{E_r}{E_i} = \frac{E_t}{E_i} \qquad \text{...(10)}$$

Substituting the value of $\left(\frac{E_t}{E_i}\right)$ from eq. (10) in eq. (9), we get

$$\frac{E_r^2}{E_i^2} = 1 - \frac{\sqrt{\varepsilon_2}}{\sqrt{\varepsilon_1}} \cdot \left(1 + \frac{E_r}{E_i}\right)^2 \cdot \frac{\cos\theta_2}{\cos\theta_1}$$

or
$$\left(1 - \frac{E_r^2}{E_i^2}\right) = \left(1 + \frac{E_r}{E_i}\right)^2 \cdot \frac{\sqrt{\varepsilon_2}\cos\theta_2}{\sqrt{\varepsilon_1}\cos\theta_1}$$

or
$$\left(1 + \frac{E_t}{E_i}\right)\left(1 - \frac{E_r}{E_i}\right) = \left(1 + \frac{E_r}{E_i}\right)\left(1 + \frac{E_r}{E_i}\right)\frac{\sqrt{\varepsilon_2}\cos\theta_2}{\sqrt{\varepsilon_1}\cos\theta_1}$$

or
$$\left(1 - \frac{E_r}{E_i}\right) = \left(1 + \frac{E_r}{E_i}\right)\frac{\sqrt{\varepsilon_2}\cos\theta_2}{\sqrt{\varepsilon_1}\cos\theta_1}$$

or
$$\left[\frac{E_i - E_r}{E_i}\right] = \left[\frac{E_i + E_r}{E_i}\right]\frac{\sqrt{\varepsilon_2}\cos\theta_2}{\sqrt{\varepsilon_1}\cos\theta_1}$$

or
$$(E_i - E_r) = (E_i + E_r)\frac{\sqrt{\varepsilon_2}\cos\theta_2}{\sqrt{\varepsilon_1}\cos\theta_1}$$

or
$$(E_i - E_r)\sqrt{\varepsilon_1}\cos\theta_1 = (E_i + E_r)\sqrt{\varepsilon_2}\cos\theta_2$$

or
$$E_i\left[\sqrt{\varepsilon_1}\cos\theta_1 - \sqrt{\varepsilon_2}\cos\theta_2\right] = E_r\left[\sqrt{\varepsilon_2}\cos\theta_2 + \sqrt{\varepsilon_1}\cos\theta_1\right]$$

$\therefore$
$$\boxed{\frac{E_r}{E_i} = \frac{\sqrt{\varepsilon_1}\cos\theta_1 - \sqrt{\varepsilon_2}\cos\theta_2}{\sqrt{\varepsilon_1}\cos\theta_1 + \sqrt{\varepsilon_2}\cos\theta_2}} \qquad \text{...(11)}$$

The ***reflection coefficient*** for horizontal polarization is given by

$$\Gamma_\perp = \frac{E_r}{E_i} = \frac{\sqrt{\varepsilon_1}\cos\theta_1 - \sqrt{\varepsilon_2}\cos\theta_2}{\sqrt{\varepsilon_1}\cos\theta_1 + \sqrt{\varepsilon_2}\cos\theta_2} \qquad \text{...(12)}$$

Let us consider ***transmission coefficient*** of horizontal polarization. This is defined as

$$\tau_\perp = \frac{E_t}{E_i} = 1 + \frac{E_r}{E_i} \qquad \text{...(13)}$$

Substituting the value of (E_r/E_i) from eq. (12), we get

$$\tau_\perp = 1 + \left[\frac{\sqrt{\varepsilon_1}\cos\theta_1 - \sqrt{\varepsilon_2}\cos\theta_2}{\sqrt{\varepsilon_1}\cos\theta_1 + \sqrt{\varepsilon_2}\cos\theta_2}\right]$$

$$= \frac{\left[\sqrt{\varepsilon_1}\cos\theta_1 + \sqrt{\varepsilon_2}\cos\theta_2\right] + \left[\sqrt{\varepsilon_1}\cos\theta_1 - \sqrt{\varepsilon_2}\cos\theta_2\right]}{\sqrt{\varepsilon_1}\cos\theta_1 + \sqrt{\varepsilon_2}\cos\theta_2}$$

or
$$\tau_\perp = \frac{2\sqrt{\varepsilon_1}\cos\theta_1}{\sqrt{\varepsilon_1}\cos\theta_1 + \sqrt{\varepsilon_2}\cos\theta_2} \quad ...(14)$$

(2) Electric vector E parallel to the plane of incidence (Parallel polarization)

In this case, the electric vector **E** is parallel to the plane of incidence and magnetic vector **H** is perpendicular to the plane of incidence. This is known as parallel polarization because there is no component of **E** in a plane which is perpendicular to plane of incidence. This is shown in Fig. (6).

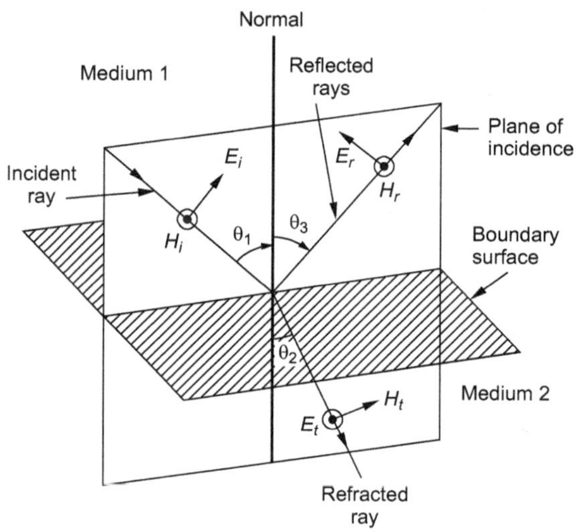

Fig. (6) Showing vertical polarisation

Applying the boundary condition that the tangential component of electric field **E** is continuous across the boundary, *i.e.*,

$$E_i\cos\theta_1 - E_r\cos\theta_1 = E_t\cos\theta_2 \quad (\because \theta_1 = \theta_3)$$

or
$$\frac{E_t}{E_i} = \left(1 - \frac{E_r}{E_i}\right)\frac{\cos\theta_1}{\cos\theta_2}$$

$\therefore$
$$\left(\frac{E_t}{E_i}\right)^2 = \left(1 - \frac{E_r}{E_i}\right)^2 \frac{\cos^2\theta_1}{\cos^2\theta_2} \quad ...(15)$$

Substituting the value of $(E_t / E_i)^2$ from eq. (15) in eq. (9), we get

$$\left(\frac{E_t}{E_i}\right)^2 = 1 - \frac{\sqrt{\varepsilon_2}}{\sqrt{\varepsilon_1}} \cdot \left[\left(1 - \frac{E_r}{E_i}\right)^2 \frac{\cos^2\theta_1}{\cos^2\theta_2}\right] \cdot \frac{\cos\theta_2}{\cos\theta_1}$$

or
$$\left(\frac{E_r}{E_i}\right)^2 = 1 - \frac{\sqrt{\varepsilon_2}}{\sqrt{\varepsilon_1}} \cdot \left(1 - \frac{E_r}{E_i}\right)^2 \cdot \frac{\cos\theta_1}{\cos\theta_2}$$

or $$1-\left(\frac{E_r}{E_i}\right)^2 = \frac{\sqrt{\varepsilon_2}}{\sqrt{\varepsilon_1}}.\left(1-\frac{E_r}{E_i}\right)^2.\frac{\cos\theta_1}{\cos\theta_2}$$

or $$\left(1-\frac{E_r}{E_i}\right)\left(1+\frac{E_r}{E_i}\right) = \frac{\sqrt{\varepsilon_2}}{\sqrt{\varepsilon_1}}.\left(1-\frac{E_r}{E_i}\right)^2\frac{\cos\theta_1}{\cos\theta_2}$$

or $$\left(1+\frac{E_r}{E_i}\right) = \frac{\sqrt{\varepsilon_2}}{\sqrt{\varepsilon_1}}.\left(1-\frac{E_r}{E_i}\right)\frac{\cos\theta_1}{\cos\theta_2}$$

or $$\left(\frac{E_i+E_r}{E_i}\right) = \frac{\sqrt{\varepsilon_2}}{\sqrt{\varepsilon_1}}.\left(\frac{E_i-E_r}{E_i}\right)\frac{\cos\theta_1}{\cos\theta_2}$$

or $$E_i+E_r = (E_i-E_r)\frac{\sqrt{\varepsilon_2}\cos\theta_1}{\sqrt{\varepsilon_1}\cos\theta_2}$$

or $$(E_i+E_r)\sqrt{\varepsilon_1}\cos\theta_2 = (E_i-E_r)\sqrt{\varepsilon_2}\cos\theta_1$$

or $$E_r[\sqrt{\varepsilon_1}\cos\theta_2+\sqrt{\varepsilon_2}\cos\theta_1] = E_i[\sqrt{\varepsilon_2}\cos\theta_1-\sqrt{\varepsilon_1}\cos\theta_2]$$

$\therefore$ $$\frac{E_r}{E_i} = \frac{\sqrt{\varepsilon_2}\cos\theta_1-\sqrt{\varepsilon_1}\cos\theta_2}{\sqrt{\varepsilon_2}\cos\theta_1+\sqrt{\varepsilon_1}\cos\theta_2} \quad ...(16)$$

The reflection coefficient for vertical polarization is given by

$$\boxed{\Gamma_{\parallel} = \frac{E_r}{E_i}\frac{\sqrt{\varepsilon_2}\cos\theta_1-\sqrt{\varepsilon_1}\cos\theta_2}{\sqrt{\varepsilon_2}\cos\theta_1+\sqrt{\varepsilon_1}\cos\theta_2}} \quad ...(17)$$

This equation is also known as *Fresnel reflection for parallel polarization.* The transmission coefficient for vertical polarization is given by

$$\tau_{\parallel} = \frac{E_t}{E_i} = 1+\frac{E_r}{E_i}$$

or $$\tau_{\parallel} = 1+\left[\frac{\sqrt{\varepsilon_2}\cos\theta_1-\sqrt{\varepsilon_1}\cos\theta_2}{\sqrt{\varepsilon_2}\cos\theta_1+\varepsilon_1\cos\theta_2}\right]$$

or $$\boxed{\tau_{\parallel} = \left[\frac{2\sqrt{\varepsilon_2}\cos\theta_1}{\sqrt{\varepsilon_2}\cos\theta_1+\sqrt{\varepsilon_1}\cos\theta_2}\right]} \quad ...(18)$$

This expression gives the transmission coefficient for vertical polarization.

8.4 BREWSTER ANGLE

Brewster observed that in case of parallel polarization, there is a particular angle of incidence for which no reflected wave is observed. The angle of incidence is known as ***Brewster angle.*** This is denoted by θ_B. The law is known as ***Brewster Law. According to Brewster law, the tangent of angle of incidence is equal to the square root of the ratio of the permittivies of two media,*** *i.e.,*

$$\boxed{\tan\theta_B = \sqrt{\left(\frac{\varepsilon_2}{\varepsilon_1}\right)}} \quad ...(1)$$

Now, we shall prove this law in the following way:

The Brewster angle is the angle for which angle of reflection is zero, *i.e.*, $(E_r/E_i) = 0$. From eq. (16) of previous article

$$\frac{E_r}{E_i} = \frac{\sqrt{\varepsilon_2}\cos\theta_1 - \sqrt{\varepsilon_1}\cos\theta_2}{\sqrt{\varepsilon_2}\cos\theta_1 + \sqrt{\varepsilon_1}\cos\theta_1} = 0 \qquad ...(2)$$

$$\therefore \quad \sqrt{\varepsilon_2}\cos\theta_1 - \sqrt{\varepsilon_1}\cos\theta_2 = 0$$

or

$$\sqrt{\varepsilon_2}\cos\theta_1 = \sqrt{\varepsilon_1}\cos\theta_2$$

Squaring on both sides, we get

$$\varepsilon_2\cos^2\theta_1 = \varepsilon_1\cos^2\theta_2$$

$$\varepsilon_2[1-\sin^2\theta_1] = \varepsilon_1[1-\sin^2\theta_2] \qquad ...(3)$$

We know that $\dfrac{\sin\theta_1}{\sin\theta_2} = \sqrt{\left(\dfrac{\varepsilon_2}{\varepsilon_1}\right)}$ or $\sin\theta_2 = \sqrt{\dfrac{\varepsilon_2}{\varepsilon_1}}\sin\theta_1$

$$\therefore \quad \sin^2\theta_2 = \left(\frac{\varepsilon_1}{\varepsilon_2}\right)\sin^2\theta_1 \; \sin^2\theta_1 \qquad ...(4)$$

Substitution the value of $\sin^2\theta_2$ from eq. (4) in eq. (3), we get

$$\varepsilon_2[1-\sin^2\theta_1] = \varepsilon_1\left[1-\left(\frac{\varepsilon_1}{\varepsilon_2}\right)\sin^2\theta_1\right] \qquad ...(5)$$

$$\varepsilon_2 - \varepsilon_2\sin^2\theta_1 = \varepsilon_1 - \frac{\varepsilon_1^{\,2}}{\varepsilon_2}\sin^2\theta_1$$

$$\frac{\varepsilon_1^{\,2}}{\varepsilon_2}\sin^2\theta_1 - \varepsilon_2\sin^2\theta_1 = (\varepsilon_1 - \varepsilon_2)$$

$$\frac{\sin^2\theta_1\left[\varepsilon_1^{\,2} - \varepsilon_2^{\,2}\right]}{\varepsilon_2} = (\varepsilon_1 - \varepsilon_2)$$

or

$$\sin^2\theta_1 = \frac{\sin^2\theta_1[\varepsilon_1^{\,2} - \varepsilon_2^{\,2}]}{\varepsilon^2}$$

$$\therefore \quad \sin^2\theta_1 = \frac{\varepsilon_2}{(\varepsilon_1+\varepsilon_2)} \qquad [\because\ \varepsilon_1^{\,2} - \varepsilon_2^{\,2} = (\varepsilon_1+\varepsilon_2)(\varepsilon_1-\varepsilon_2)]$$

Using the formula $\cos^2\theta_1 = 1 - \sin^2\theta_1$, we get

$$\cos^2\theta_1 = \frac{\varepsilon_2}{(\varepsilon_1+\varepsilon_2)} \qquad ...(6)$$

Dividing eq. (5) by eq (6), we get

$$\tan^2\theta_1 = \frac{\varepsilon_2}{\varepsilon_1} \quad \text{or} \quad \tan\theta_1 = \sqrt{\left(\frac{\varepsilon_2}{\varepsilon_1}\right)}$$

Under this situation $\theta_1 = \theta_B$ (Brewster angle)

$$\therefore \quad \boxed{\tan\theta_B = \sqrt{\left(\frac{\varepsilon_2}{\varepsilon_1}\right)}} \qquad ...(7)$$

EXAMPLE 1. *In case of parallel polarization prove that*

$$\Gamma_{\|} = \frac{\tan(\theta_1 - \theta_2)}{\tan(\theta_1 + \theta_2)}$$

where θ_1 is the angle of incidence and θ_2 is the angle of refraction.

Solution In case of parallel polarization, we have

$$\Gamma_{\|} = \frac{\sqrt{\varepsilon_2}\cos\theta_1 - \sqrt{\varepsilon_1}\cos\theta_2}{\sqrt{\varepsilon_1}\cos\theta_2 + \sqrt{\varepsilon_2}\cos\theta_1} \qquad ...(1)$$

We know that

$$\sqrt{\left(\frac{\varepsilon_1}{\varepsilon_2}\right)} = \frac{\sin\theta_2}{\sin\theta_1} \quad \text{or} \quad \sqrt{\varepsilon_1} = \sqrt{\varepsilon_1}\frac{\sin\theta_2}{\sin\theta_1} \qquad ...(2)$$

Substituting the value of $\sqrt{\varepsilon_1}$ from eq. (2), in eq. (1), we get

$$\Gamma_{\|} = \frac{\sin\theta_1\cos\theta_1 - \sin\theta_2\cos\theta_2}{\sin\theta_2\cos\theta_2 + \sin\theta_1\cos\theta_1}$$

or
$$\Gamma_{\|} = \frac{\sin 2\theta_1 - \sin 2\theta_2}{\sin 2\theta_1 + \sin 2\theta_2} \qquad ...(3)$$

or
$$\Gamma_{\|} = \frac{2\cos(\theta_1 + \theta_2)\sin(\theta_1 - \theta_2)}{2\sin(\theta_1 + \theta_2)\cos(\theta_1 - \theta_2)} \qquad ...(4)$$

or
$$\Gamma_{\|} = \frac{\tan(\theta_1 - \theta_2)}{\tan(\theta_1 + \theta_2)} \qquad ...(5)$$

In eq. (4), we have used the relation

$$\sin A - \sin B = 2\cos\left(\frac{A+B}{2}\right)\sin\left(\frac{A-B}{2}\right)$$

and
$$\sin A - \sin B = 2\sin\left(\frac{A+B}{2}\right)\cos\left(\frac{A-B}{2}\right)$$

EXAMPLE 2. *A uniform plane wave is incident from air onto glass at an angle from the normal of* 30°. *Determine the fraction of the incident power that is reflected for perpendicular polarization. Glass has refractive index* $n_2 = 1.45$.

Solution. For air $\quad n_1 = 1$

For glass, $\quad n_2 = \sin^{-1}\left(\frac{\sin 30°}{1.45}\right) = 20.2° \qquad [\because n_1 \sin\theta_1 = n_2 \sin\theta_2$ (Snell's law)]

Now, for perpendicular polarization

$$\eta_1 = \eta_1 \cos 30° = 377 \times (0.866) = 326\ \Omega$$

and
$$\eta_2 = \eta_2 \cos 20.2° = \frac{377}{1.45} \times (0.938) = 244\ \Omega$$

Now,
$$\Gamma_p = \frac{244 - 326}{244 + 326} = -0.144$$

The fraction of incident power that is reflected is

$$\frac{P_r}{P_{inc}} = |\Gamma_p|^2 = \mathbf{0.021}$$

EXAMPLE 3. *A perpendicular polarized wave propagates from the region* ($\varepsilon_r = 9$, $\mu_{r_1} = 1$, $\sigma_1 = 0$) *to a region of free space with an angle of incidence* 15°. *Given* $E_i = 1\mu$ V/m. *Calculate* (*i*) H_i, (*ii*) E_r, (*iii*) E_t, (*iv*) H_r *and* (*v*) H_t

Solution. Here,

$$\eta_1 = \sqrt{\left(\frac{\mu}{\varepsilon}\right)} = \frac{377}{\sqrt{9}} = 125.6\ \Omega$$

$$\eta_2 = \eta_0 = 377\ \Omega$$

$$\frac{\sin\theta_1}{\sin\theta_2} = \sqrt{\left(\frac{\varepsilon_2}{\varepsilon_1}\right)} \quad \text{or} \quad \frac{\sin 15°}{\sin\theta_2} = \frac{1}{3}$$

$$\therefore \quad \sin\theta_2 = 3\sin 15° \quad \text{or} \quad \theta_2 = 50.90°$$

(*i*) $H_i = \dfrac{E_i}{\eta_1} = \dfrac{1\times 10^{-6}}{125.6} = \mathbf{7.96 \times 10^{-9}\ A/m}$

(*ii*) In case of perpendicular polarization

$$\frac{E_r}{E_i} = \frac{\sqrt{\varepsilon_1}\cos\theta_1 - \sqrt{\varepsilon_2}\cos\theta_2}{\sqrt{\varepsilon_1}\cos\theta_1 + \sqrt{\varepsilon_2}\cos\theta_2} = \frac{\eta_2\cos\theta_1 - \eta_1\cos\theta_2}{\eta_1\cos\theta_2 + \eta_2\cos\theta_1}$$

$$= \frac{364.15 - 79.145}{443.3} = 0.643$$

$$\therefore \quad E_r = 0.643 \times (1 \times 10^{-6}) = 0.643 \times 10^{-6}\ \text{V/m} = \mathbf{0.643\ \mu\ V/m}$$

(*iii*) $\dfrac{E_t}{E_i} = \dfrac{2\,\eta_2\cos\theta_1}{\eta_2\cos\theta_1 + \eta_1\cos\theta_2} = \dfrac{728.3}{443.3} = 1.643$

$$\therefore \quad E_t = 1.643 \times E_i = 1.643 \times (1 \times 10^{-6}) = 1.643 \times 10^{-6}\ \text{V/m}$$

$$= \mathbf{1.643\ \mu V/m}$$

(*iv*) $H_r = \dfrac{E_r}{\eta_1} = \dfrac{0.643\mu\ \text{V/m}}{125.6\Omega} = \mathbf{5.12\ nA/m}$

(*v*) $H_t = \dfrac{E_t}{\eta_2} = \dfrac{1.643\mu\ \text{V/m}}{377\Omega} = \mathbf{4.4\ nA/m}$

EXAMPLE 4. *A parallel polarized wave propagates from air into a dielectric* $\varepsilon_{r_2} = 13.93$. *Find the Brewster angle.*

Solution.

$$\tan\theta_B = \sqrt{\left(\frac{\varepsilon_2}{\varepsilon_1}\right)}$$

$$\frac{\varepsilon_2}{\varepsilon_1} = \frac{\varepsilon_{r_2} \times \varepsilon_0}{1 \times \varepsilon_0} = \varepsilon_{r_2} = 13.93$$

$$\therefore \quad \tan\theta_B = \sqrt{13.93}$$

$$\theta_B = \tan^{-1}\sqrt{(13.93)} = \mathbf{75°}$$

8.5 REFLECTION BY A PERFECT CONDUCTOR NORMAL INCIDENCE

Let a plane wave be incident normally on the surface of the conductor. The wave will be entirely reflected in the same medium and no part of the wave is transmitted. In this case:

(*i*) no energy of the incident wave is transmitted

(*ii*) no energy is absorbed by the conductor

(*iii*) the amplitudes of **E** and **H** in the reflected wave are same as that of incident wave.

(*iv*) the direction of power flow is changed.

Let the electric field of incident wave is represented as

$$\mathbf{E_i} = E_i\, e^{j(\varpi t - \beta x)} \quad ...(1)$$

The expression form reflected wave is

$$\mathbf{E}_r = E_r\, e^{j(\varpi t + \beta x)} \quad ...(2)$$

Consider that the surface of the conductor is taken to be at $x = 0$. Now, ε_r can be determined by boundary condition. The tangential component of **E** at $x = 0$ must be zero.

Thus, $$E_i + E_r = 0 \text{ or } E_i = -E_r \quad ...(3)$$

So, E_r has the same magnitude E_i but with phase reversal.

The resultant electric field at any point at a distance $-x$ from the conductor will be sum of field strengths of incident and reflected waves

$$E_T = E_i\, e^{j(\varpi t + \beta x)} + E_r\, e^{j(\varpi t + \beta x)}$$

$$E_i = [\, e^{j(\varpi t + \beta x)} - e^{j(\varpi t + \beta x)}] \qquad (\because E_i = -E_r)$$

$$= E_i\,[e^{j\varpi t}(e^{-j\beta x} - e^{+j\beta x}]$$

$$= E_i\,[e^{j\varpi t}\{(\cos\beta x - j\sin\beta x) - (\cos\beta x + j\sin\beta x)\}]$$

$$\therefore \quad E_T = -2j\,E_i \sin\beta x\, e^{j\varpi t} \quad ...(4)$$

Let us consider eq. (4) with E_i to be real. We have

$$\therefore \quad E_T = -2j\,E_i \sin\beta x\,(\cos\omega t + j\sin\omega t)$$

$$= 2\,E_i \sin\beta x\,[-j\cos\omega t + \sin\omega t]$$

$$= 2\,E_i \sin\beta x \sin\omega t - 2j\,E_i \sin\beta x \cos\omega t$$

Therefore, with E_i to be real

$$E_T = 2\,E_i \sin\beta x \sin\omega t \quad ...(5)$$

Eq. (5) shows that the incident and reflected waves combine to give a standing wave as shown in Fig. (7a). The wave does not progress.

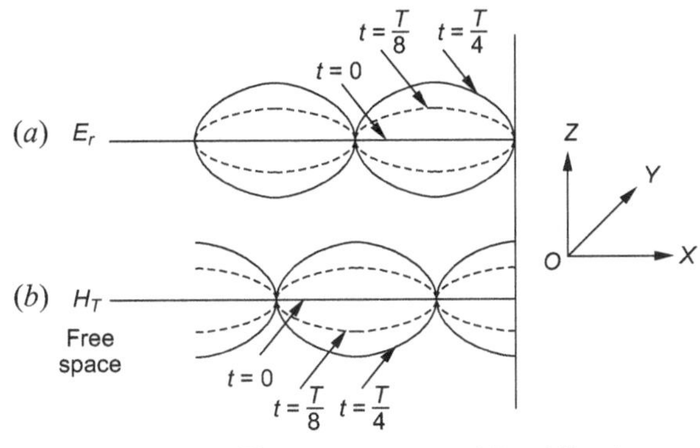

Fig. (7) Standing waves of **E** and **H**

It is observed from Fig. (7a) that the magnitude of electric field of standing wave has a maximum value at distances from the surface that are odd mutliples of quarter wavelength.

The magnetic field is reflected at the surface without phase reversal as shown in Fig. (7 b). The phase of H_r is same as that of E_i at $x = 0$. The resultant field at a distance-x from $x = 0$ is given by

$$\begin{aligned} H_T &= H_i\, e^{j(\omega t - \beta x)} + H_i\, e^{j(\omega t - \beta x)} \\ &= H_i\, e^{j\omega t}(e^{-j\beta x} + e^{j\beta x}) \qquad (\because H_i = H_r) \\ &= 2\, H_i\, e^{j\omega t} \cos \beta x \\ &= 2\, H_i\, [\cos \omega t + j \sin \omega t] \cos \beta x \end{aligned}$$

H_i is real, since, it is in phase with E_i. Therefore,

$$H_T = 2\, H_i \cos \beta x \cos \omega t \qquad \text{...(6)}$$

Eq. (6) also represent a standing wave. H_T has its maximum at the reflecting surface as well as at multiple of half wavelength from the surface. As shown in Fig. (6), E_T and H_T are 90° out of time phase.

8.6 DISPERSION (NORMAL AND ANOMALOUS)

A medium is known as dispersive medium in which the refractive index of the medium varies with frequency (or wavelength). The variation of refractive index with wavelength, *i.e.*, $\left(\frac{dn}{d\lambda}\right)$ constitutes the phenomenon of dispersion. $\left(\frac{dn}{d\lambda}\right)$ is also known as dispersive power.

The dispersion is divided into following two categories:

(1) **Normal dispersion:** When the refractive index decreases with wavelength, then the dispersion is known as normal dispersion.

(2) **Anomalous dispersion:** In small wavelength ranges, there is often increase in refractive index of medium. This is known as anomalous dispersion. The anomalous dispersion is due to an increased absorption of the passing electromagnetic radiation through the medium.

8.7 DISPERSION IN GASES (LORENTZ THEORY)

An adequate theory of dispersion, based on electromagnetic theory, was given by Lorentz. The theory is based on the assumption that the electric field of electromagnetic waves induces a dipole moment in the gas molecules due to relative displacement of electrons and nuclei of the neutral atoms. The theory also assumes that over an atom or a molecule, the electric vector **E** is constant in space, *i.e.*,

$$\mathbf{E} = E_0\, e^{-j(\omega t - k \cdot r)} \approx E_0\, e^{-j\omega t}$$

Let us consider the different forces acting on electron due to electromagnetic radiation. The forces are:

(*i*) Electric force ($e\,\mathbf{E}$) where e is the charge on the electron.

(*ii*) Electric restoring force which is proportional to its displacement $\mathbf{r}$ from equilibrium position. This is denoted by $-k\,\mathbf{r}$ where k is certain positive constant. This depends upon the structure and properties of gas molecules.

(*iii*) A damping frictional force against the motion of electron. This is proportional to its velocity and expressed as $-\alpha\left(\frac{d\mathbf{r}}{dt}\right)$ where α is a positive constant of proportionality.

In general, a force on electron is represented as $m\frac{d^2\mathbf{r}}{dt^2}$ (mass × acceleration) where m is mass of electron. Therefore,

$$m\frac{d^2r}{dt^2} = e\mathbf{E} - \alpha\frac{d\mathbf{r}}{dt} - k\,\mathbf{r}$$

or
$$\frac{d^2\mathbf{r}}{dt^2} + \frac{\alpha}{m}\cdot\frac{d\mathbf{r}}{dt} + \frac{k}{m}\mathbf{r} = \frac{e}{m}\mathbf{E}$$

or
$$\frac{d^2\mathbf{r}}{dt^2} + \gamma\frac{d\mathbf{r}}{dt} + \omega_0^2\,\mathbf{r} = \frac{e}{m}E_0\,e^{-j\omega t} \qquad ...(1)$$

where
$$\gamma = \frac{\alpha}{m} \quad \text{and} \quad \omega_0^2 = \frac{k}{m}$$

Let us solve eq. (1) for **r**. The solution consists of two parts: (*i*) complementary function and (*ii*) particular integral.

(*i*) **Complementary function:** The complementary function is the solution of eq. (1) in the absence of electric field E, *i.e.*,

$$\frac{d^2r}{dt^2} + \gamma\frac{dr}{dt} + \omega_0^2\,r = 0 \qquad ...(2)$$

Let the solution of eq. (2) is given by

$$r = A\,e^{\alpha t}$$

Now,
$$\frac{dr}{dt} = A\,\alpha\,e^{\alpha t} \quad \text{and} \quad \frac{d^2r}{dt^2} = A\,\alpha^2\,e^{\alpha t}$$

Substituting these values in eq. (2), we get

$$A\,\alpha^2\,e^{\alpha t} + \gamma\,A\,\alpha\,e^{\alpha t} + \omega_0^2\,A\,e^{\alpha t} = 0$$

or
$$\alpha^2 + \gamma\,\alpha + \omega_0^2 = 0$$

Solving, we get

$$\alpha = \frac{-\gamma \pm \sqrt{(\gamma^2 - 4\,\omega_0^2)}}{2} = -\frac{\gamma}{2} \pm \sqrt{\left(\frac{\gamma^2}{4} - \omega_0^2\right)}$$

When damping is small, $\frac{\gamma^2}{2} << \omega_0^2$ .Hence,

$$\alpha = -\frac{\gamma}{2} \pm i\sqrt{\left(\omega_0^2 - \frac{\gamma^2}{4}\right)} = -\frac{\gamma}{2} \pm j\beta$$

where
$$\beta = \sqrt{\left(\omega_0^2 - \frac{\gamma^2}{4}\right)}$$

So, the solution of eq. (2), is given as

$$r = A_1 \exp.\left[-\frac{\gamma}{2} + j\beta\right]t_1 + A_2 \exp.\left[-\frac{\gamma}{2} - j\beta\right]t$$

or
$$r = \exp.\left(-\frac{\gamma t}{2}\right)[A_1 \exp.(j\beta t) + A_2 \exp.(-j\beta t)] \qquad ...(a)$$

(*ii*) **Particular integral:** The particular integra is obtained by solving eq. (1), *i.e.*,

$$\frac{d^2r}{dt^2} + \gamma\frac{dr}{dt} + \omega_0^2 r = \frac{e}{m}E_0 e^{-j\omega t} \qquad ...(3)$$

Let the solution of eq. (3) be

$$r = B e^{-j\omega t}$$

where B is constant.

Now,
$$\frac{dr}{dt} = -j\omega B e^{-j\omega t}$$

and
$$\frac{d^2r}{dt^2} = -\omega^2 B e^{-j\omega t}$$

Substituting these values in eq. (3), we get

$$-\omega^2 B e^{-j\omega t} - j\omega\gamma B e^{-j\omega t} + \omega_0^2 B e^{-j\omega t} = \frac{e}{m}E_0 e^{-j\omega t}$$

or
$$-\omega^2 B - j\omega\gamma B + \omega_0^2 B = \frac{e}{m}E_0$$

$\therefore$
$$B = \frac{e E_0}{m(\omega_0^2 - \omega^2 - j\omega\gamma)}$$

So, the particular integral solution is

$$r = \frac{e E_0}{m(\omega_0{}^2 - \omega^2 - j\omega\gamma)} e^{-j\omega t} \qquad ...(b)$$

Therefore, the general solution of eq. (1) is given by

$$r = \exp.\left(-\frac{\gamma t}{2}\right)[A_1 \exp.(j\beta t) + A_2 \exp.(-j\beta t)] + \frac{e E_0}{m(\omega_0{}^2 - \omega^2 - j\omega\gamma)} e^{-j\omega t} \qquad ...(c)$$

In eq. (*c*), the first term represents the free damped vibrations. These vibrations soon die out due to the term $\exp.\left(-\frac{\gamma t}{2}\right)$. So, the solution of eq. (1) becomes,

$$r = \frac{e E_0}{m(\omega_0^2 - \omega^2 - j\omega\gamma)}$$

or
$$\boxed{\mathbf{r} = \frac{e\,\mathbf{E}}{m(\omega_0^2 - \omega^2 - j\omega\gamma)}} \qquad \text{vector form} \quad ...(d)$$

Dispersion Formula

To derive the dispersion formula, we proceed as follow:

(*i*) ***Dipole moment*** **(p):** The dipole moment arises due to the displacement of the electron under consideration. This is given by

$$\mathbf{p} = e\,\mathbf{r}$$

or
$$\mathbf{p} = \frac{e^2\,\mathbf{E}}{m\,(\omega_0^2 - \omega^2 - j\gamma\omega)} \qquad \text{...(4)}$$

(*ii*) ***Polarization vector*** **(*P*):** The polarization is equal to dipole moment per unit volume, So, if there are N dispersion electrons per unit volume, then we may write

$$\mathbf{P} = \mathrm{N}\,\mathbf{p}$$

or
$$\mathbf{P} = \frac{N\,e^2\,\mathbf{E}}{m\,(\omega_0^2 - \omega^2 - j\gamma\omega)} \qquad \text{...(5)}$$

Eq. (5) is derived on the basis that each oscillator has the same natural frequency. To generalise eq. (5), we assume that there are N molecules per unit and each molecule has f_1 oscillators whose constants are ω_1 and γ_1; f_2 oscillators whose constants are ω_2 and γ_2 and so no. Therefore, the generalisation of eq. (5) can be written as

$$\mathbf{P} = N\,e\sum_k f_k\,\mathbf{r}_k$$

or
$$P = \frac{N\,e^2\,\mathbf{E}}{m}\sum_k \frac{f_k}{(\omega_k^2 - \omega^2 - j\gamma_k\omega)} \qquad \text{...(6)}$$

(*iii*) ***Electrical polarisibility*** **(α):** Electrical polarisibility is the ratio of polarization vector (P) and electric vector **E**.

$\therefore$
$$\alpha = \frac{\mathbf{P}}{\mathbf{E}} = \frac{N\,e^2}{m}\sum_k \frac{f_k}{(\omega_k^2 - \omega^2 - j\gamma_k\omega)} \qquad \text{...(7)}$$

(*iv*) ***Dielectric constant*** **(ε_r):** The dielectric constant is given by

$$\varepsilon_r = 1 + \frac{1}{\varepsilon_0}\frac{\mathbf{P}}{\mathbf{E}} \qquad \left(\because\ \varepsilon_r = 1 + \chi = 1 + \frac{1}{\varepsilon_0}\frac{\mathbf{P}}{\mathbf{E}}\right)$$

or
$$\varepsilon_r = 1 + \frac{1}{\varepsilon_0}\cdot\frac{N\,e^2}{m}\sum_k \frac{f_k}{(\omega_k^2 - \omega^2 - j\gamma_k\omega)} \qquad \text{...(8)}$$

Eq. (8) shows that dielectric constant is complex. Thus, we conclude that refractive index is also complex.

(*v*) ***Refractive index*** **(n):** We know that refractive index is square root of dielectric constant, *i.e.*, $n = \sqrt{\varepsilon_r}$ or $n^2 = \varepsilon_r$. Therefore, form eq. (8).

$$\boxed{n^2 = 1 + \frac{1}{4\pi\varepsilon_0}\times\frac{4\pi\,Ne^2}{m}\sum_k \frac{f_k}{(\omega_k^2 - \omega^2 - j\gamma_k\omega)}} \qquad \text{...(9)}$$

This is known as dispersion formula.

Eq. (9) shows that the refractive index is frequency dependent.

(a) Normal Dispersion

In the region remote from the natural frequencies of the oscillator, *i.e.*, the absorption frequencies of the medium, the term $\gamma_k\,\omega$ in the denominator of eq. (9) is so small that it can be neglected in comparison to $(\omega_k^2 - \omega^2)$. So, eq. (9) reduces to

$$\boxed{n^2 = 1 + \frac{1}{4\pi\varepsilon_0} \times \frac{4\pi Ne^2}{m} \sum_k \frac{f_k}{(\omega_k^2 - \omega^2)}} \qquad ...(10)$$

From eq. (10) it is clear that the refractive index is real and increases with the frequency of incident wave.

(b) Anomalous Dispersion

Consider the situation in which the impressed frequencies include one of the so many natural frequencies of the electron. In this case, the normal dispersion is disturbed. The refractive index of the system becomes complex and this is denoted by n^*. For simplicity, we assume that there is only one natural frequency, *i.e.*, $\omega_k = \omega_0$.

Now eq. (9) becomes

$$n^{*2} = 1 + \frac{1}{4\pi\varepsilon_0} \times \frac{4\pi N e^2}{m} \frac{1}{(\omega_0^2 - \omega^2 - j\gamma\omega)}$$

or

$$n^* = \left[1 + \frac{1}{4\pi\varepsilon_0} \times \frac{4\pi N e^2}{m} \frac{1}{(\omega_0^2 - \omega^2 - j\gamma\omega)}\right]^{1/2}$$

or

$$\boxed{n^* = \left[1 + \frac{1}{4\pi\varepsilon_0} \times \frac{2\pi N e^2}{m} \frac{1}{(\omega_0^2 - \omega^2 - j\gamma\omega)}\right]} \qquad ...(11)$$

$$\because \qquad (1+x)^{1+x} = \left(1 + \frac{x}{2}\right)$$

Eq. (11) shows that the refractive index is a complex function of frequency of electromagnetic wave propagation through the gas.

(c) Dispersion Curve

A curve between refractive index plotted as a function of frequency is known as dispersion cure. Fig. (8) shows the real and imaginary part of n^* plotted as a function of frequency ω.

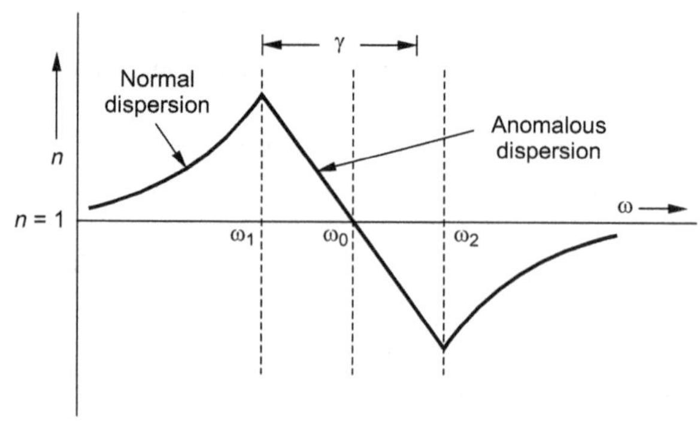

Fig. (8) Showing dispersion curve

It is obvious from Fig. (8) that at low frequencies, n is slightly greater than unity.

First of all, as the frequency (ω) increases, the refractive index increases. It reaches a maximum for ω_1 where $\omega_1 = \left(\omega_0 - \frac{\gamma}{2}\right)$. The curve corresponds to normal dispersion.

After ω_1, the refractive index decreases, becomes 1 for frequency ω_0 and reaches to a minimum value for frequency ω_2. This is known as anomalous dispersion. The refractive index is a complex quantity between frequency region ω_1 to ω_2. This part corresponds to absorption of electromagnetic waves propagating through the gas. After ω_2, the curve again increases and approaches unity asymptotically.

8.8 DISPERSION IN LIQUIDS AND SOLIDS

In liquids and solids, the molecules are sufficiently close to each other. Therefore, the interaction among themselves cannot be neglected. When electromagnetic radiation passes through the material, it is polarized. So, an additional electric field $\mathbf{E}_p$ due to polarization will act on each molecule. Therefore, the actual electric field $\mathbf{E}_{\text{eff.}}$ will be sum of actual field $\mathbf{E}$ plus the electric field $\mathbf{E}_p$ due to polarization. The calculation shows that the contribution of electric field $\mathbf{E}_p$ is equal to $(\mathbf{P}/3\varepsilon_0)$ where $\mathbf{P}$ is polarization vector. Thus,

$$\mathbf{E}_{\text{eff.}} = \mathbf{E} + \frac{\mathbf{P}}{3\,\varepsilon_0} = \mathbf{E} + \frac{N\,e\,\mathbf{r}}{3\,\varepsilon_0} \qquad \text{...(1)}$$

We have derived the equation of motion of electron in the previous article for gas molecules. In case of liquids and gases using eq. (1), the modified form of equation becomes

$$m\frac{d^2\mathbf{r}}{dt^2} = e\left[\mathbf{E} + \frac{N\,e\,\mathbf{r}}{3\,\varepsilon_0}\right] - \alpha\frac{d\mathbf{r}}{dt} - k\,\mathbf{r}$$

or

$$\frac{d^2\mathbf{r}}{dt^2} + \gamma\frac{d\mathbf{r}}{dt} + \left(\omega_0^2 - \frac{N\,e^2}{3\,m\,\varepsilon_0}\right)\mathbf{r} = \frac{e}{m}\mathbf{E} \qquad \text{...(2)}$$

where $\gamma = \frac{\alpha}{m}$ and $\omega_0^2 = \frac{k}{m}$

The solution of eq. (2) is given by

$$\mathbf{r} = \frac{(e/m)\,\mathbf{E}}{\left(\omega_0^2 - \frac{n\,e^2}{3\,m\,\varepsilon_0} - \omega^2 - j\gamma\omega\right)} \qquad \text{...(3)}$$

(1) Dipole Moment

The dipole moment $\mathbf{p}$ is given by

$$\mathbf{p} = e\,\mathbf{r} = \frac{(e^2/m)\,\mathbf{E}}{\left(\omega_0^2 - \frac{N\,e^2}{3\,m\,\varepsilon_0} - \omega^2 - j\gamma\omega\right)} \qquad \text{...(4)}$$

(2) Polarization Vector *P*

The polarization vector **P** is given by

$$\mathbf{P} = N\,\mathbf{p} = \frac{N\,(e^2/m)\,\mathbf{E}}{\left(\omega_0^2 - \dfrac{N e^2}{3\,m\,\varepsilon_0} - \omega^2 - j\gamma\omega\right)} \qquad ...(5)$$

where N = number of electrons per unit volume.

Here, all the electrons are not in identical situations with the molecules. Therefore, we have different pairs of characteristic frequencies and associated damping factors γ. If we define f_k as the probability that an electron has characteristic frequency ω_k and damping coefficient γ_k, the generalisation of eq. (5) yields

$$\mathbf{P} = \frac{N e^2\,\mathbf{E}}{m} = \sum_k \frac{f_k}{\left(\omega_k^2 - \dfrac{N e^2 f_k}{3\,m\,\varepsilon_0} - \omega^2 - j\gamma_k\,\omega\right)} \qquad ...(6)$$

$$\left[\because\ \mathbf{P} = e\sum_k N_k\,\mathbf{r}_k = N\,e\sum_k f_k\,\mathbf{r}_k \text{ as } N_k = N\,f_k\right]$$

(3) Electrical Polarisibility (α)

This is defined as the ratio of polarization vector **P** and electric vector **E**.

$$\alpha = \frac{\mathbf{P}}{\mathbf{E}} = \frac{N e^2}{m}\sum_k \frac{f_k}{\left(\omega_k^2 - \dfrac{N e^2 f_k}{3\,m\,\varepsilon_0} - \omega^2 - j\gamma_k\,\omega\right)} \qquad ...(7)$$

(4) Dielectric Constant (ε_r)

$$\varepsilon_r = 1 + \chi = 1 + \frac{\alpha}{\varepsilon_0}$$

$$\therefore \qquad \varepsilon_r = \frac{N e^2}{m\,\varepsilon_0}\sum_k \frac{f_k}{\left(\omega_k^2 - \dfrac{N e^2 f_k}{3\,m\,\varepsilon_0} - \omega^2 - j\gamma_k\,\omega\right)} \qquad ...(8)$$

We know that $\varepsilon_r = n^2$. Therefore, eq. (8) shows that the refractive index is again complex and depends upon frequency.

Clausius-Mossotti Relation

Eq. (8) can be expressed as

$$(\varepsilon_r - 1)\sum_k\left[\omega_k^2 - \frac{N e^2 f_k}{3\,m\,\varepsilon_0} - \omega^2 - j\gamma_k\,\omega\right] = \frac{N e^2}{m\,\varepsilon_0}\sum_k f_k$$

or $$(\varepsilon_r - 1)\sum_k[\omega_k^2 - \omega^2 - j\gamma_k\,\omega] = \frac{N e^2}{m\,\varepsilon_0}\sum_k f_k + (\varepsilon_r - 1)\frac{N e^2 f_k}{3\,m\,\varepsilon_0}$$

or $$(\varepsilon_r - 1)\sum_k [\omega_k^2 - \omega^2 - j\gamma_k\,\omega] = \frac{N e^2}{m\,\varepsilon_0}\sum_k f_k\left[1 + \frac{(\varepsilon_r - 1)}{3}\right]$$

or $$(\varepsilon_r - 1)\sum_k [\omega_k^2 - \omega^2 - j\gamma_k\,\omega] = \frac{N e^2}{m\,\varepsilon_0}\sum_k f_k\left(\frac{\varepsilon_r + 2}{3}\right)$$

or $$\frac{(\varepsilon_r - 1)}{(\varepsilon_r + 2)} = \frac{N e^2}{3\,m\,\varepsilon_0}\sum_k \frac{f_k}{(\omega_k^2 - \omega^2 - j\gamma_k\,\omega)} \qquad ...(9)$$

If M is the molecular weight of the substance, ρ its density and N_0, Avogadro's number, then

$$N = \frac{N_0\,\rho}{M} \qquad ...(10)$$

Substituting eq. (10) in eq. (9), we get

$$\frac{(\varepsilon_r - 1)}{(\varepsilon_r + 2)} \times \frac{M}{\rho} = \frac{N_0\,\rho\,e^2}{(3\,m\,\varepsilon_0)\,M}\sum_k \frac{f_k}{(\omega_k^2 - \omega^2 - j\gamma_k\,\omega)}$$

or $$\frac{(\varepsilon_r - 1)}{(\varepsilon_r + 2)} \times \frac{M}{\rho} = \frac{N_0\,e^2}{3\,m\,\varepsilon_0}\sum_k \frac{f_k}{(\omega_k^2 - \omega^2 - j\gamma_k\,\omega)} \qquad ...(11)$$

For a given frequency eq. (1), reduces to

$$\boxed{\frac{(\varepsilon_r - 1)}{(\varepsilon_r + 2)} \times \frac{M}{\rho} = \text{constant}} \qquad ...(12)$$

This is known as Clausius-Mossotti relation.

Lorantz-Lorentz Formula

In case of transparent medium $\gamma_k \to 0$, eq. (12) reduces to

$$\left[\frac{(n^2 - 1)}{n^2 + 2}\right]\frac{M}{\rho} = \text{constant} \qquad (n^2 = \varepsilon_r) \;...(13)$$

Eq. (13) shows that how the refractive index depends upon the density of the medium. This is known as Lorentz-Lorentz formula.

Exercises and Problems

EXERCISE

1. A plane electromagnetic is incident normally at the interface of a perfect dielectric media. Discuss the phenomenon of reflection and refraction.
2. Define reflection coefficient and transmission coefficient. Derive their value of **E** vector in case of reflection from perfect dielectric.
3. Discuss the properties of reflection and transmission coefficients. Define standing wave ratio.
4. Establish the laws of reflection and refraction at the boundary of two dielectric media for a plane electromagnetic wave propagating at an oblique incidence.

5. Derive expression for reflection coefficient and transmission coefficient for horizontal polarization
6. Derive expression for reflection coefficient and transmission coefficient for vertical polarization.
7. Derive reflection for uniform plane electromagnetic wave by a perfect dielectric for oblique incidence.
   (*a*) when electric vector **E** is parallel to plane of incidence.
   (*b*) when electric field **E** is perpendicular to plane of incidence.
8. What is Brewster angle? Prove that

$$\tan \theta_B = \sqrt{\left(\frac{\varepsilon_2}{\varepsilon_1}\right)}$$

   where symbols have their usual meanings.
9. Show that when a uniform plane travelling wave is incident normally on a perfect conducting plane, the incident and reflected wave produce standing distribution of **E** and **H,** which does not progress.
10. What do you mean by dispersion of electromagnetic waves? Discuss normal and anomalous dispersions.
11. Discuss the Lorentz theory of dispersion in gases.
12. Derive the dispersion formula in case of dispersion in gases.
13. What are normal and anomalous dispersions? Explain them with the help of dispersion curve.
14. Discuss the phenomenon of dispersion in case of liquids and solids.
15. In connection with the dispersion in case of solids, derive Clausius-Mossotti relation.

PRACTICE PROBLEMS

1. A travelling **E** field in free space of amplitude 100 V/m strikes a perfect dielectric as shown in Fig. (9). Find the electric field strength in medium 3.

   [**Ans.** 59.7 V/m]

   [**Hint.** Here, $\eta_1 = 377\Omega$, $\eta_2 = 377\ \Omega$

$$\therefore \quad \eta_2 = 377\sqrt{\left(\frac{1}{20}\right)} = 84.3\Omega$$

and $$E_i = 100 \text{ V/m}$$

Now, $$\frac{E_r}{E_i} = \frac{2\,\eta_2}{\eta_1 + \eta_2}$$

$$E_i = 36.55 \text{ V/m}$$

$$(E_i)' = E_r e^{-\alpha z} = E_r$$

because $$\alpha = 0$$

Further, $$\frac{E_t}{E_r} = \frac{2\,\eta_3}{\eta_3 + \eta_2} = \frac{2 \times 377}{377 + 84.3}$$

$$E_t = \frac{2 \times 377}{377 + 84.3} \times 36.55 = 59.7 \text{ V/m}]$$

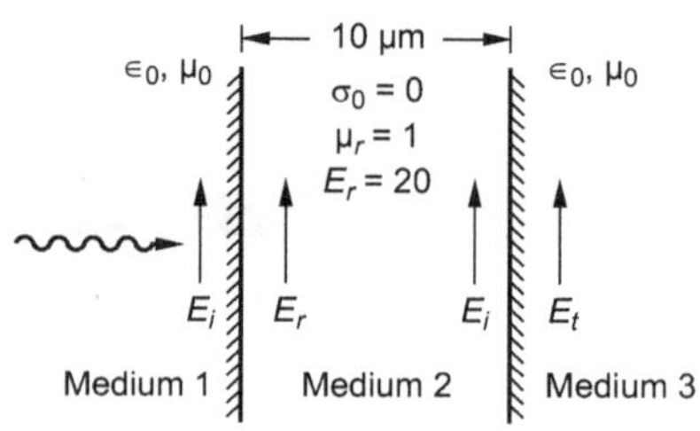

Fig. (9)

2. A travelling electric field in free space has an amplitude of 100 V/m. It strikes a sheet of silver of thickness 5 μm as shown in Fig. (10). The conductivity of silver is $\sigma = 61.7$ ms/m and at frequency $f = 200$ MHz. Find the amplitudes E_2, E_3, E_4.

[**Ans.** 2.68×10^{-3} V/m, 8.88×10^{-4} V/m and 1.78×10^{-3} V/m]

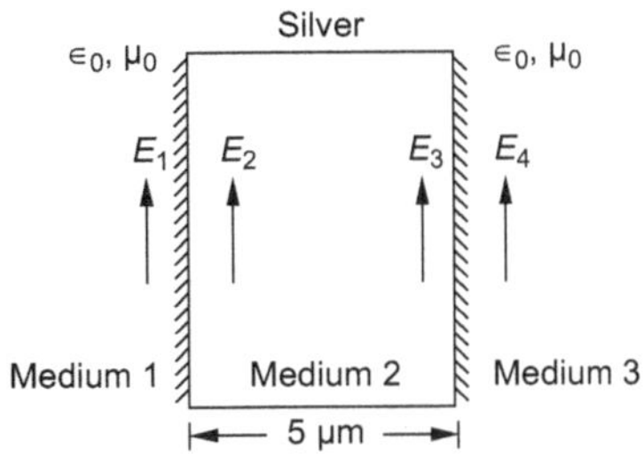

Fig. (10)

[**Hint.** $\eta_1 = 377\ \Omega\ \eta_3 = 377\ \Omega$

and $\eta_2 = 5.06 \times 10^{-3} \angle 45°$

at $f = 200$ MHz $\because\ \eta_2 = \sqrt{\left(\dfrac{\omega\mu}{2\sigma}\right)}$

Now, $$|E_2| = \frac{2\eta_2}{\eta_1 + \eta_2} \times 100 \text{ V/m}$$

$$= 2.68 \times 10^{-3} \text{ V/m}$$

$$E_3 = |E_2| e^{-\alpha z}$$

where $$\alpha = \sqrt{\left(\frac{\omega\mu\sigma}{2}\right)} = \sqrt{(\pi f \mu \sigma)}$$

$$= 2.21 \times 10^5 = 2.68 \times 10^{-3} \exp.[-2.21 \times 10^5 \times (5 \times 10^{-6})]$$

$$= 8.88 \times 10^{-4} \text{ V/m}$$

$$\frac{E_4}{E_3} = \frac{2\eta_3}{\eta_3 + \eta_2} = \frac{2 \times 377}{377 + 5.06 \times 10^{-3} \angle 45°}$$

3. The electric constant of water is 80. A plane wave with perpendicular polarization is incident from air on water surface such that $\theta_1 = 38.62°$ and $\theta_2 = 6.38°$. Find out the reflection coefficient and transmission coefficient for parallel polarization.

[**Ans.** –0.9753 and 0.0247]

4. A parallel polarized wave propagates from air into a dielectric at Brewster angle of 75°. Find ε_r,

[**Ans.** $\varepsilon_r = 13.93$]

[**Hint.** $\tan\theta_B = \sqrt{\left(\dfrac{\varepsilon_2}{\varepsilon_1}\right)}$ or $\tan^2\theta_B = \dfrac{\varepsilon_0 \times \varepsilon_r}{1\,\varepsilon_0}$

$$\varepsilon_r = \tan^2 75°]$$

5. Travelling **E** waves in free space are normally incident on the interface with a perfect dielectric for which $\varepsilon_r = 3.0$. Compute the magnitudes Γ_E and τ_E where Γ_E is reflection coefficient while τ_E is transmission coefficient.

[**Ans.** 0.268 and 0.732]

[**Hint.** $\Gamma_E = \dfrac{\eta_2 - \eta_1}{\eta_1 + \eta_2}$ and $\tau_E = \dfrac{2\eta_2}{\eta_1 + \eta_2}$

where $\eta_1 = 377\ \Omega$ and $\eta_2 = \dfrac{377}{\sqrt{3.0}}\ \Omega$]

UNIT - V

1. Transmission Lines

9

Transmission Lines

INTRODUCTION

In previous chapter, we have studied the wave propagation in space as well as in material media. Generally, the wave propagation is divided into the following two categories:

(1) *Unbounded media propagation*

(2) *Bounded media or guided wave propagation.*

(1) In unbounded media propagation, the propagation is unguided. The uniform plane wave exists throughout all the space. Therefore, the energy spreads over a wide area. This type of propagation is used in radio or T.V. broadcasting.

(2) In bounded media or guided wave propagation, the propagation of plane waves takes place between boundaries. The boundaries are provided by the conductors. This type of propagation is used in telephones, transmission of electrical energy from generating station to load centres, connection between a transmitter to an antenna, connection between computers in network, etc. There are two kinds of wave guiding systems:

(*a*) *Transmission lines*, and (*b*) *Wave guides.*

A transmission line consists of two or more parallel conductors, whereas a waveguide is generally made up of one conductor. The aim of this chapter is to study the basic concepts of transmission lines. Transmission lines are used to transmit electric energy from one point to another point. We can define the transmission lines in the following way:

Transmission line is a device, with minimum energy losses, used to transfer the energy or signal or power, from a source to a load (for example transmitter to receiver).

In this chapter, we shall consider the transverse electromagnetic (TEM) waves guided by transmission lines. In TEM waves, **E** and **H** are both perpendicular to each other and both are transverse to the direction of propagation. The properties of TEM waves are same as those of uniform plane waves.

9.1 TYPES OF TRANSMISSION LINES

Generally the following two types of transmission lines are used:

(1) Two wires transmission line or parallel wire line: Two wires transmission line consists of a pair of parallel conducting wires separated by a uniform distance as shown in Fig. (1*a*). The other type is a shielded pair line. In this type of transmission line, two parallel wires are surrounded by an insulator contained in a cylindrical sheet as shown in Fig. (1*b*)

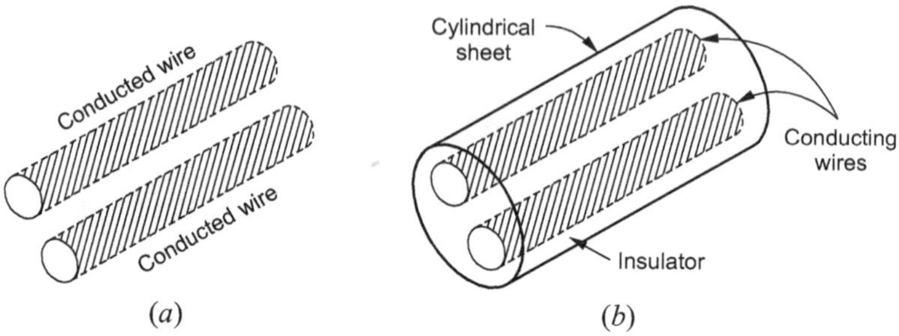

Fig. (1) Two wires transmission line

These are used in power systems or telephone lines. The parallel wire lines are used at frequencies of the order of 100 mega cycles per second. Above 200 mega cycles per second, there is a large loss of energy due to radiation. So, these lines are unsuitable.

(2) **Coaxial transmission line:** The coaxial transmission line is generally used at high frequencies. It consists of two concentric conductors which are separated by an insulation material as shown in Fig. (2).

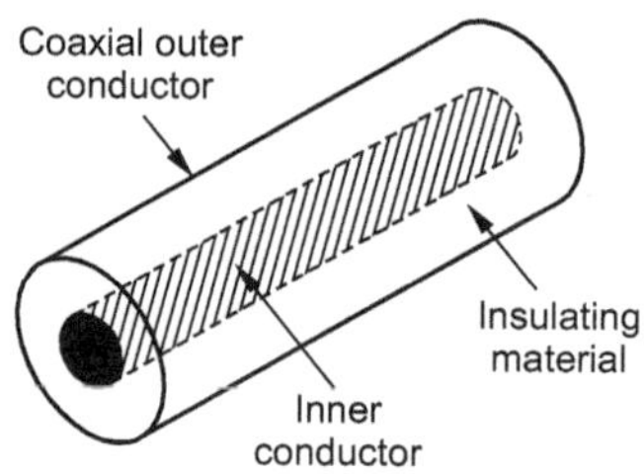

Fig. (2) Coaxial transmission line

In this transmission line, the electric and magnetic fields are only confined to dielectric region. The energy loss due to radiation is quite small. The losses in dielectric medium increase with frequency. Therefore, coaxial lines cannot be used at frequencies above 1 GHz.

9.2 TRANSMISSION LINE EQUATIONS

Consider a section of transmission line of length Δx at a distance x from the sending end as shown in Fig. (3). Let $V(x, t)$ and $I(x, t)$ be the voltage and current at the sending end respectively. Similarly, $V(x + \Delta x, t)$ and $I(x + \Delta x, t)$ be the voltage and current respectively at the receiving end. The transmission line can be represented in terms of four distributed parameters: resistance, inductance and conductance (due to leakage current in insulator). Let R, L, C and G be resistance, inductance, capacitance and conductance per unit length of the conductor respectively. The values of these parameters for length Δx of the transmission line will be $R\,\Delta x$, $L\,\Delta x$, $C\,\Delta x$ and $G\,\Delta x$ respectively. As the wire are close to each other, there would be shunt capacitance between them. Moreover, due to leakage current in the dielectric medium, there would be shunt conductance. These four parameters for length Δx are also shown in Fig. (3).

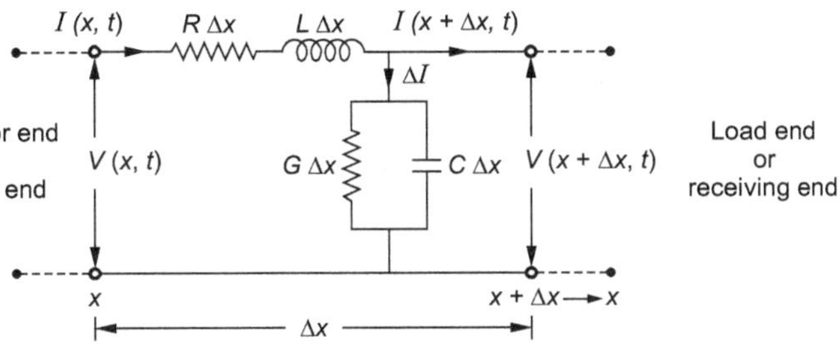

Fig. (3) Showing equivalent circuit of a transmission line for a small length

Voltage Equation

Applying Kirchoff's voltage law to the equivalent circuit of transmission line for a small length Δx, we have

$$V(x,\, t) = (R\,\Delta x)\, I(x,t) + (L\,\Delta x)\frac{\partial I(x,t)}{\partial t} + V(x+\Delta x, t)$$

or $$-V(x+\Delta x,\, t) + V(x,\, t) = \left[R\, I\,(x,t) + L\frac{\partial I(x,t)}{\partial t}\right]\Delta x$$

or $$\frac{-V\,(x+\Delta x, t) + V(x,t)}{\Delta x} = R\, I\,(x,t) + L\frac{\partial I\,(x,t)}{\partial t} \qquad ...(1)$$

as $\Delta x \to 0$, eq. (1) becomes

$$\boxed{-\frac{\partial V}{\partial x} = R\, I + L\frac{\partial I}{\partial t}} \qquad ...(a)$$

Current Equation

The current equation is

$$I(x,\, t) = (G\,\Delta x)\, V(x+\Delta x, t) + (C\,\Delta x)\frac{\partial V\,(x+\Delta x, t)}{\partial t} + I(x+\Delta x, t)$$

or $$-I(x+\Delta x,\, t) + I(x,\, t) = \Delta x\left[G\, V\,(x+\Delta x, t) + C\frac{\partial V\,(x+\Delta x, t)}{\partial t}\right]$$

or $$\frac{-I(x+\Delta x, t) + I(x,t)}{\Delta x} = G\, V\,(x+\Delta x, t) + C\frac{\partial V\,(x+\Delta x, t)}{\partial t} \qquad ...(2)$$

As $\Delta x \to 0$, eq. (2) becomes

$$\boxed{-\frac{\partial I}{\partial x} = G\, V + C\frac{\partial V}{\partial t}} \qquad ...(b)$$

Harmonic Time Dependent Voltage and Current Equation

Considering voltage and current to be harmonic time dependent, the *instantaneous line voltage and current* are given by

$$V(x,\, t) = \text{Re}\,[V_0\,(x)\, e^{j\omega t}] \qquad ...(3)$$

and $$I(x,\, t) = \text{Re}\,[I_0\,(x)\, e^{j\omega t}] \qquad ...(4)$$

where $V_0\,(x)$ and $I_0\,(x)$ are phaser forms of $V(x,\, t)$ and $I(x,\, t)$ respectively. In phaser form $V_0\,(x)$ includes the factor $e^{-(\alpha + j\,\beta)x}$

From eqs. (3) and (4) $\partial V/\partial t = j\,\omega\, V$ and $\partial I/\partial t = j\,\omega\, I$

Substituting these values in eqs. (*a*) and (*b*), we have

$$-\frac{dV}{dx} = RI + L(j\omega I)$$

or $$-\frac{dV}{dx} = (R + j\omega L)I \qquad ...(5)$$

and $$-\frac{dI}{dx} = GV + C(j\omega V)$$

or $$-\frac{dI}{dx} = (G + j\omega C)V \qquad ...(6)$$

Taking the differential of eq. (5), we get

$$-\frac{d^2V}{dx^2} = (R + j\omega L)\frac{dI}{dx}$$

or $$-\frac{d^2V}{dx^2} = (R + j\omega L)[-(G + j\omega C)V]$$

or $$\frac{d^2V}{dx^2} = (R + j\omega L)(G + j\omega C)V \qquad ...(7)$$

Taking the differential of eq. (6), we get

$$-\frac{d^2I}{dx^2} = (G + j\omega C)\frac{dV}{dx}$$

or $$-\frac{d^2I}{dx^2} = -(G + j\omega C)(R + j\omega L)I$$

or $$\frac{d^2I}{dx^2} = (R + j\omega L)(G + j\omega C)I \qquad ...(8)$$

Now, $R + j\omega L = Z$ = Series impedance of the line per unit length (ohm/m)

$G + j\omega C = Y$ = Shunt admittance of line per unit length (mho/m)

Further $$\gamma^2 = (R + j\omega L)(G + j\omega C) \qquad ...(9)$$

or $$\gamma = \sqrt{(R + j\omega L)(G + j\omega C)} = (\alpha + j\beta) \qquad ...(10)$$

Here γ = Propagation constant α = attenuation constant and β = phase constant.

Using these symbols, the transmission line equations for infinite transmission line can be expressed as [using eqs. (5), (6), (7) and (8)]

$$\boxed{-\frac{dV}{dx} = ZI} \qquad ...(11)$$

$$\boxed{-\frac{dI}{dx} = ZV} \qquad ...(12)$$

$$\boxed{\frac{d^2V}{dx^2} = \gamma^2 V} \qquad ...(13)$$

$$\boxed{\frac{d^2I}{dx^2} = \gamma^2 I} \qquad ...(14)$$

These are transmission line equations.

9.3 SOLUTION OF TRANSMISSION LINE EQUATIONS

The solution of eqs. (13) and (14) of previous article (9.2) can be written as

$$V(x) = V^+(x) + V^-(x)$$

or

$$V(x) = V_0^+ e^{-\gamma x} + V_0^- e^{+\gamma x} \quad ...(1)$$

where V_0^+ is the complex amplitude in positive x-direction and V_0^- is the complex amplitude in $-X$ direction.

Similarly,

$$I(x) = I^+(x) + I^-(x)$$

or

$$I(x) = I_0^+ e^{-\gamma x} + I_0^- e^{+\gamma x} \quad ...(2)$$

where I_0^+ is the complex amplitude in $+X$-direction while I_0^- is the complex amplitude in $-X$-direction.

The instantaneous expression for voltage can be expressed as

$$\begin{aligned} V(x, t) &= \text{Re}\,[V_0(x)\, e^{j\omega t}] \\ &= \text{Re}\,[V_0^+ e^{-\gamma x} e^{-j\omega t} + V_0^{-1} e^{+\gamma x} e^{j\omega t}] \\ &= \text{Re}\,[V_0^+ e^{-(\alpha + j\beta)x} e^{j\omega t} + V_0^- e^{(\alpha + j\beta)x} e^{j\omega t}] \\ &= \text{Re}\,[V_0^+ e^{-\alpha x} e^{j(\omega t - \beta x)} + V_0^- e^{\alpha x} e^{j(\omega t + \beta x)}] \\ &= V_0^+ e^{-\alpha x} \cos(\omega t - \beta x) + V_0^- e^{-\alpha x} \cos(\omega t + \beta x) \end{aligned}$$

$\therefore$

$$\boxed{V(x, t) = I_0^+ e^{-a x} \cos(\omega t - \beta x) + V_0^- e^{-a x} \cos(\omega t - \beta x)} \quad ...(3)$$

Similarly

$$\boxed{I(x, t) = I_0^+ e^{-a x} \cos(\omega t - \beta x) + I_0^- e^{-a x} \cos(\omega t - \beta x)} \quad ...(4)$$

9.4 CHARACTERISTIC IMPEDANCE

A circuit containing series and shunt impedances must have an input impedance. In case of a transmission line, this input impedance will depend on the type of line, its length and termination at the far end. For simplicity, the input impedance under certain standard conditions is taken as reference impedance. This is called as characteristic impedance and is denoted by Z_0.

The characteristic impedance of the line is defined as the ratio of positively travelling voltage wave (V_0^+) to current wave (I_0^+) at any point of the line.

$\therefore$

$$Z_0 = \frac{V_0^+}{I_0^+} = \frac{V_0^-}{I_0^-} \quad ...(1)$$

From eq. (5) of article (9.2), we have

$$-\frac{dV}{dx} = (R + j\omega L)\, I \quad ...(2)$$

Considering positively travelling wave, we have

$$V = V_0^+ e^{-\gamma x} \quad \text{and} \quad I = I_0^+ e^{-\gamma x} \quad ...(3)$$

Substituting these values in eq. (2), we get

$$-[-\gamma\, V_0^+ e^{-\gamma x}] = (R + j\omega L)\,[I_0^+ e^{-\gamma x}]$$

$\therefore$

$$\frac{V_0^+}{I_0^+} = \frac{R + j\omega L}{\gamma} \quad ...(4)$$

Therefore, $$Z_0 = \frac{R + j\,\omega\,L}{\gamma} = \frac{(R + j\,\omega\,L)}{\sqrt{(R + j\,\omega\,L)(G + j\,\omega\,C)}}$$

or $$\boxed{Z_0 = \sqrt{\left[\frac{R + j\,\omega\,L}{G + j\,\omega\,C}\right]}} \quad ...(5)$$

Eq. (5) shows that the characteristic impedance depends on *R, L, G, C* and frequency of operation

The characteristic impedance can also be expressed as

$$Z_0 = R_0 + j\,X_0 \quad ...(6)$$

where R_0 and X_0 are the real and imaginary parts of Z_0.

The transmission line can be classified in the following two categories:

(*i*) Lossless line (*ii*) Distortionless line

We shall derive the value of characteristic impedance for these lines.

(i) Characteristic impedance of lossless line.

A characteristic line is said to be lossless, if at radio frequencies, the inductive reactance ($j\,\omega\,L$) is much larger than series resistance *R* and capacitive reactance ($j\,\omega\,C$) is also much larger than shunt conductance *G*. Thus, in case of lossless line

$$R = 0 \quad \text{and} \quad G = 0$$

Substituting these values in eq. (5), we get

$$Z_0 = \sqrt{\left(\frac{j\,\omega\,L}{j\,\omega\,C}\right)} = R_0 + j\,X_0$$

$$\therefore \quad \boxed{R_0 = \sqrt{\left(\frac{L}{C}\right)}} \quad ...(7)$$

and $$X_0 = 0$$

Thus, characteristic impedance is a real quantity. This shows that a loss-less line is purely resistive.

(ii) Distortionless line.

A distortionless line is one for which attenuation constant α is frequency independent while the phase constant β is linearly dependent on frequency. For a distortionaless line,

$$\frac{\boldsymbol{R}}{\boldsymbol{L}} = \frac{\boldsymbol{G}}{\boldsymbol{C}}$$

Now $$Z_0 = \sqrt{\left(\frac{R + j\,\omega\,L}{G + j\,\omega\,C}\right)}$$

$$= \sqrt{\frac{R\left(1 + \dfrac{j\,\omega\,L}{R}\right)}{G\left(1 + \dfrac{j\,\omega\,C}{C}\right)}} = \sqrt{\frac{R\left(1 + \dfrac{j\,\omega\,C}{R}\right)}{G\left(1 + \dfrac{j\,\omega\,C}{G}\right)}}$$

or $$Z_0 = \sqrt{\left(\frac{R}{G}\right)} = \sqrt{\frac{L}{C}} \qquad ...(8)$$

Further $Z_0 = R_0 + j\,X_0 = \sqrt{\frac{L}{C}}$

$$\therefore \quad \boxed{R_0 = \sqrt{\left(\frac{R}{G}\right)}} \quad \text{and} \quad X_0 = 0 \qquad ...(9)$$

Thus, distortionless line has a real characteristic impedance like lossless line. It is important to mention here that a lossless line is also distortioonless line but a distortionless line is not necessarily lossless. In power transmission, a lossless line is desired while in case of telephone lines, a distortionless line is required.

9.5 PROPAGATION CONSTANT AND PHASE VELOCITY

In case of transmission line, the quantity $\sqrt{(R + j\,\omega\,L)\,(G + \omega\,C)}$ is defined as propagation constant and is represented by γ. The real and imaginary parts, α and β are attenuation constant and phase velocity of the respectively. We shall calculate the value of γ and β in case of lossless line and distortionless line.

(i) Lossless line.

In case of lossless line, $R = 0$ and $G = 0$

$$\therefore \quad \gamma = \sqrt{(0 + j\,\omega\,L)\,(0 + j\,\omega\,C)} = j\,\omega\sqrt{LC}$$

$$\gamma = \alpha + j\,\beta = j\,\omega\sqrt{LC} \qquad ...(10)$$

$$\therefore \quad \boxed{\alpha = 0}$$

$$\text{and} \quad \boxed{\beta = \omega\sqrt{(LC)}} \qquad ...(11)$$

Phase velocity, $$v_p = \frac{\omega}{\beta} = \frac{\omega}{\omega\sqrt{LC}} = \frac{1}{\sqrt{LC}}$$

$$\therefore \quad \boxed{v_p = \frac{1}{\sqrt{LC}}} \qquad ...(12)$$

(ii) Distortionless Line

The conduction for distortionless line is

$$\frac{R}{L} = \frac{G}{C}$$

Now $$\gamma = \alpha + j\,\beta = \sqrt{(R + j\,\omega\,L)\left(\frac{RC}{L} + j\,\omega\,C\right)}$$

or $$\gamma = \sqrt{\frac{C}{L}}\,(R + j\,\omega\,L) \qquad ...(13)$$

$$\therefore \qquad \boxed{\alpha = R\sqrt{\left(\frac{C}{L}\right)}}$$

and $$\boxed{\beta = \omega\sqrt{LC}} \qquad ...(14)$$

Thus, α does not depend only frequency but β is a linear function of frequency.

Phase velocity, $$v_p = \frac{\omega}{\beta} = \frac{\omega}{\omega\sqrt{LC}} = \frac{1}{\sqrt{LC}}$$

$$\therefore \qquad \boxed{v_p = \frac{1}{\sqrt{LC}}} \qquad ...(15)$$

EXAMPLE 1. *In a lossless transmission line, the velocity of propagation is* 2.5×10^8 m/s *Capacitance of the line is* 30 pF/m. *Find*

(*i*) *Inductance of the line* (*ii*) *Phase constant at* 100 MHz

(*iii*) *Characteristic impedance of line.*

Solution.

(*i*) The inductance can be calculate as follow:

Phase velocity, $$v_p = \frac{1}{\sqrt{LC}} \text{ or } v_p^2 = \frac{1}{LC}$$

$$\therefore \qquad L = \frac{1}{C\,v_p^2} = \frac{1}{(30\times10^{-12})\,(2.5\times10^8)^2}$$

$$= \frac{1}{187.5\times10^4} = \mathbf{5.33 \times 10^{-7}\ henry/m}$$

(*ii*) Phase constant $$\beta = \frac{\omega}{v_p} \qquad \left(\because v_p = \frac{\omega}{\beta}\right)$$

$$\therefore \qquad \beta = \frac{2\times3.14\times(100\times10^6)}{2.5\times10^8} = \mathbf{2.51\ rad/m}$$

(*iii*) Characteristic impedance $= \sqrt{\dfrac{L}{C}}$

$$\therefore \qquad Z_0 = \sqrt{\frac{5.33\times10^{-7}}{30\times10^{-12}}} = \mathbf{133.28\ \Omega}$$

EXAMPLE 2. *A lossless transmission line has characteristic impedance of* 75 Ω *and phase constant* 3 rad/m *at* 100 Mhz. *Calculate the capacitance and inductance of the line per metre.*

Solution. In case of lossline, $R = 0$, $G = 0$ and $\alpha = 0$.

The characteristic impedance, $Z_0 = \sqrt{(L/C)}$

and phase constant, $\beta = \omega\sqrt{(LC)}$

$$\therefore \qquad \frac{Z_0}{\beta} = \frac{\sqrt{(L/C)}}{\omega\sqrt{LC}} = \frac{1}{\omega C}$$

or
$$C = \frac{\beta}{\beta Z_0} = \frac{\beta}{2\pi f Z_0}$$

Substituting the values, we get

$$C = \frac{3}{2\times 3.14\times(100\times 10^6)\,75}$$

$$= 63.69 \times 10^{-12}\text{ F/m} = \mathbf{63.69\ pF/m}$$

and
$$L = Z_0^2\, C = (75)^2 \times (63.69 \times 10^{-12}) = \mathbf{358.1\ nH/m}$$

EXAMPLE 3. *Calculate the series impedance and shunt admittance of a transmission line at* $\omega = 5.5 \times 10^3$ rad/sec. *The primary parameters of line are* $R = 10.4$ ohm, $L = 3.64$ mH, $C = 0.00825\ \mu$F *and* $G = 0.08\ \mu$ mho.

Solution. Series impedance $= R + j\,\omega\, L$

$$= 10.4 + j(5.5 \times 10^3)(3.64 \times 10^{-3})$$

$$= 10.4 + j(20.02)$$

$$= \sqrt{[R^2 + (\omega L)^2]}\ \underline{|\tan^{-1}(\omega L/R)}$$

$$= \sqrt{[(10.4)^2 + (20.02)^2]}\ \underline{|\tan^{-1}(20.02/10.4)}$$

$$= 22.56\ \underline{|62.55^\circ}$$

Shunt admittance $= G + j\,\omega\, C$

$$= (0.08 \times 10^{-6}) + j\,(5.5 \times 10^3)(0.00825 \times 10^{-6})$$

$$= 0.08 \times 10^{-6} + j(45.37 \times 10^{-6})$$

$$= \sqrt{[(0.08\times10^{-6})^2 + (45.37\times10^{-6})^2]}\ \underline{|\tan^{-1}}\,(45.37\times10^{6}/0.08\times10^{-6})$$

$$= 45.37 \times 10^{-6}\ \underline{|89.9^\circ}$$

EXAMPLE 4. *An open wire transmission line has* $R = 84\ \Omega$/km, $G = 10^{-6}$ mho/km, $L = 0.01$ H/km, $C = 0.06\ \mu$F/km and $f = 10^3$ Hz. *Find the characteristic impedance, propagation constant and velocity of propagation.*

Solution. The characteristic impedance is given by

$$Z_0 = \sqrt{\left(\frac{R + j\,\omega\, L}{G + j\,\omega\, C}\right)}$$

Here, $\quad R + j\,\omega\, L = 84 + j \times (2 \times 3.14 \times 10^3 \times 0.01) = 84 + j\,62.83$

or $\quad (R + j\,\omega\, L) = \sqrt{(84)^2 + (62.83)^2}\ \underline{|\tan^{-1}(62.83/84)} = 104.9\ \underline{|36.8^\circ}\ \Omega$/km

and $\quad G + j\,\omega\, C = 10^{-6} + j \times 2 \times 3.14 \times 10^3 \times 0.061 \times 10^{-6}$

$$= 10^{-6} + j\,383.27 \times 10^{-6} = 383.27 \times 10^{-6}\ \underline{|36.8^\circ}\ \text{mho/km}$$

Substituting these values, we get

$$Z_0 = \sqrt{\left[\frac{104.9\angle 36.8^\circ}{383.27\times10^{-6}\angle 89.85^\circ}\right]} = 523.16\angle 26.53^\circ\Omega$$

Propagation constant γ is given by

$$\gamma = \sqrt{(R+j\,\omega L)(G+j\,\omega C)}$$

$$= \sqrt{[(104.9\angle 36.8^\circ)(383.27\times10^{-6}\angle 89.85^\circ)]}$$

$$= 0.201\ \angle 63.33^\circ = \alpha + j\,\beta = 0.09 + j\,0.18 \text{ per km}$$

$$\therefore \qquad \beta = 0.18 \text{ rad/km}$$

Velocity of propagation $v_p = \dfrac{\omega}{\beta} = \dfrac{2\pi f}{\beta} = \dfrac{2\pi\times10^3}{0.18\times10^{-3}} = \mathbf{34.9\times10^6\ m/s}$

9.6 INPUT IMPEDANCE OF TRANSMISSION LINE

Consider a transmission line of length l which extends from $x=0$ to $x=l$. The line is characterised by propagation constant γ and characteristic impedance Z_0. To the input side, a source (or generator) with voltage V_g and impedance Z_g is connected as shown in Fig. (4). To the output side a load Z_L is connected. The aim of this article is to determine the input impedance.

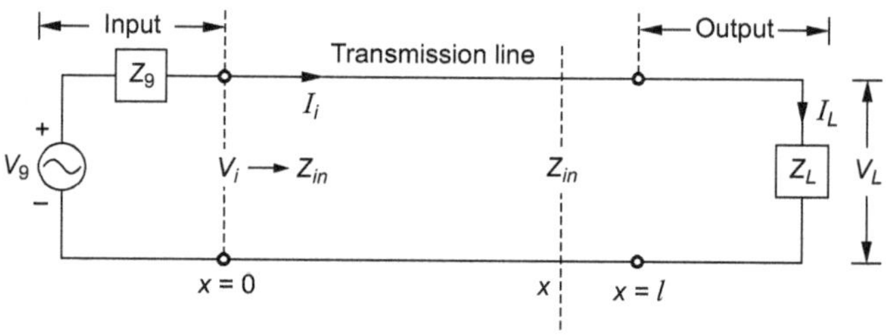

Fig. (4) Finite transmission line terminated with load impedance Z_L

In general, the line impedance of the transmission line is the complex ratio of voltage (phasor) at any point (say x) to the current (phasor) at that point, *i.e.*,

$$Z = \frac{V(x)}{I(x)} \qquad ...(1)$$

The general transmission line equations for voltage and current are

$$V(x) = V_0^+ e^{-\gamma x} + V_0^- e^{\gamma x}.$$

and

$$I(x) = L_0^+ e^{-\gamma x} - I_0^- e^{\gamma x} \qquad ...(2)$$

or

$$I(x) = \frac{V_0^+}{Z_0} e^{-\gamma x} - \frac{V_0^-}{Z_0} e^{\gamma x} \qquad ...(3)$$

Voltage and Current at Input ($x = 0$)

Substituting $x = 0$ in eq. (2) and eq. (3), we get

$$V_i = V_0^+ + V_0^- \qquad (\because V(x) = V_i) \quad ...(4)$$

and

$$I_i = \frac{I}{Z_0}[V_0^+ - V_0^-] \qquad ...(5)$$

Voltage and Current Output Side ($x = L$)

Substituting $x = L$ in eqs. (2) and (3), we get

$$V_L = V_0^+ e^{-\gamma l} + V_0^- e^{\gamma l} \quad ...(6)$$

and

$$I_L = \frac{1}{Z_0}[V_0^+ e^{-\gamma l} - V_0^- e^{\gamma l}] \quad ...(7)$$

Solving eqs. (6) and (7), we get

$$V_0^+ = \frac{1}{2}[V_L + Z_0 I_L]e^{\gamma l} \quad ...(8)$$

$$V_0^- = \frac{1}{2}[V_L - Z_0 I_L]e^{-\gamma l} \quad ...(9)$$

Input Impedance

At generator end, input impedance is given by

$$Z_{in} = \frac{V_i}{I_i} = \frac{V_0^+ + V_0^-}{\frac{1}{Z_0}[V_0^+ - V_0^-]} = \frac{Z_0 [V_0^+ + V_0^-]}{[V_0^+ - V_0^-]} \quad ...(10)$$

Substituting the values of V_0^+ and V_0^- from eqs. (8) and (9) in eq. (10), we get

$$Z_{in} = Z_0 \left[\frac{(V_L + Z_0 I_L) e^{\gamma l} + (V_L + Z_0 I_L)e^{-\gamma l}}{(V_L + Z_0 I_L)e^{\gamma l} - I_L (V_L + Z_0 I_L)e^{-\gamma l}}\right]$$

$$= Z_0 \left[\frac{I_L (Z_L + Z_0) e^{\gamma l} + I_L(Z_L + Z_0)e^{-\gamma l}}{I_L (Z_L + Z_0) e^{\gamma l} - I_L(Z_L + Z_0)e^{-\gamma l}}\right]$$

$$= Z_0 \left[\frac{(Z_L + Z_0) e^{\gamma l} + (Z_L + Z_0)e^{-\gamma l}}{(Z_L + Z_0) e^{\gamma l} - (Z_L + Z_0)e^{-\gamma l}}\right]$$

$$= Z_0 \left[\frac{Z_L(e^{\gamma l} + e^{-\gamma l}) + Z_0(e^{\gamma l} - e^{-\gamma l})}{Z_L(e^{\gamma l} - e^{-\gamma l}) - Z_0(e^{\gamma l} - e^{-\gamma l})}\right]$$

$$= Z_0 \left[\frac{Z_L \cosh(\gamma l) + Z_0 \sinh(\gamma l)}{Z_L \sinh(\gamma l) + Z_0 \cosh(\gamma l)}\right]$$

$\therefore$

$$\boxed{Z_{in} = Z_0 \left[\frac{Z_L + Z_0 \tanh(\gamma l)}{Z_0 + Z_L \tanh(\gamma l)}\right]} \quad ...(11)$$

where we have used $\dfrac{e^{\gamma l} + e^{\gamma l}}{2} = \cosh \gamma l$ and $\dfrac{e^{\gamma l} - e^{-\gamma l}}{2} = \sinh(\gamma l)$

Special Cases

(1) ***For a loss line.*** $\gamma = j\beta$ and $\tanh(\gamma l) = j \tan \beta l$

$\therefore$

$$Z_{in} = Z_0 \left[\frac{Z_L + j Z_0 \tan \beta l}{Z_0 + j Z_L \tan \beta l}\right] \quad ...(12)$$

where βl is electrical length of the line in degree or radian

(2) ***For a short circuited line.*** $Z_L = 0$

Substituting $Z_L = 0$, in eq. (12), we get

$$[Z_{in}]_{SC} = Z_0 \left[\frac{j\,Z_0 \tan \beta\, l}{Z_0} \right] = j\,Z_0 \tan \beta\, l \quad \text{...(13)}$$

(3) ***For open circuited line.*** $Z_L = \infty$

From eq. (12)

$$[Z_{in}]_{OC} = Z_0 \left[\frac{Z_L \left(1 + j \frac{Z_0}{Z_L} \tan \beta\, l\right)}{Z_L \left(\frac{Z_0}{Z_L} + j \tan \beta\, l\right)} \right] = Z_0 \left[\frac{1 + j \frac{Z_0}{Z_L} \tan \beta\, l}{\frac{Z_0}{Z_L} + j \tan \beta\, l} \right]$$

$$= Z_0 \left[\frac{1+0}{0 + j \tan \beta\, l} \right] = -j\,Z_0 \cot \beta\, l \quad \text{...(14)}$$

EXAMPLE 1. *An ideal lossless $\lambda/4$ extension line of $Z_0 = 60\ \Omega$ is terminated with Z_L. Find Z_{in} of the extension when (i) $Z_L = 0$, (ii) $Z_L = \infty$ and (iii) $Z_L = 60\ \Omega$.*

Solution. Given, $Z_{in} = Z_0$ and $\beta\, l = \frac{2\pi}{\lambda} \times \frac{\lambda}{4} = \pi/2$

(*i*) In this case, $Z_{in} = j\,Z_0 \tan \beta\, l = j\,60 \tan (\pi/2) = \infty$

(*ii*) For open circuited line, $Z_{in} = -j\,Z_0 \cot \beta\, l = -j\,60 \cot \pi/2 = 0$

(*iii*) When $Z_L = 60\ \Omega$, the input impedance is given by

$$Z_{in} = 60 \left[\frac{60 + j\,60 \tan \pi/2}{60 + j\,60 \tan \pi/2} \right] = 60 \left[\frac{60 + \infty}{60 + \infty} \right] = \mathbf{60\ \Omega}$$

EXAMPLE 2. *Find the input impedance for a short cirucited transmission line of length $\lambda/8$ and having $Z_0 = 200\ \Omega$.*

Solution. In case of short circuited line

$$Z_{in} = j\,Z_0 \tan \beta\, l$$

$$= j\,200 \tan \left(\frac{2\pi}{\lambda} \times \frac{\lambda}{8} \right) = j\,200 \tan 45°$$

$$= j\,200\,(1) = j\,\mathbf{200\ \Omega}$$

9.7 TRANSMISSION LINE AS CIRCUIT ELEMENTS

Under normal frequencies, the transmission line acts as a communication channel between transmitter and receiver. At very high frequency [300 mHz to 3 GHz], it also acts as circuit elements like inductor and capacitor. Let us study their behaviour as circuit elements.

Considering a lossless line and substituting $Z_0 = R_0$, the input impedance is given by

$$Z_{in} = R_0 \left[\frac{Z_L + j\,R_0 \tan \beta\, l}{R0 + j\,Z_L \tan \beta\, l} \right] \quad \text{...(1)}$$

Now, we consider the following special cases:

(1) Open circuited line ($Z_L = \infty$)

In case of open circuited line, $Z_L \to \infty$ The input impedance is given by

$$(Z_{in})_{OC} = -j\,R_0 \cot \beta\, l \qquad \text{[eq. (14), article 9.6]} \quad ...(2)$$

This equation shows that input impedance of an open circuited lossless line is ***purely reactive.*** It can be capacitive or inductive depending upon the value of cos $\beta\, l$ when it is positive or negative respectively. A plot of input reactance $X_{in} = -R_0 \cot \beta\, l$ against l is shown in Fig. (5).

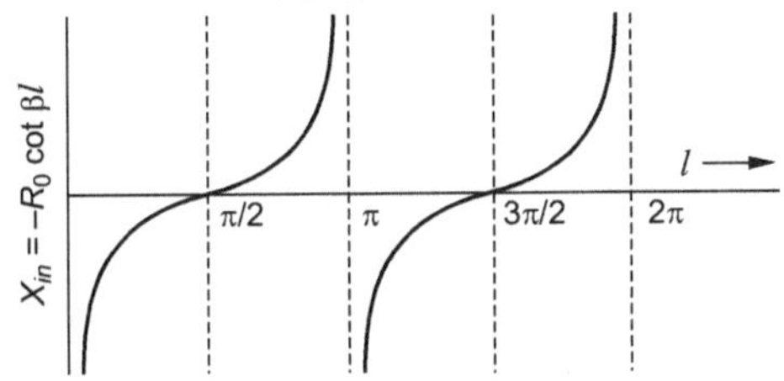

Fig. (5) Input reactance of open circuited transmission line

For very short length of line $\beta\, l << l$, *i.e.*, $\tan \beta\, l \approx \beta\, l$

$$\therefore \qquad Z_{in} = \frac{-j\,R_0}{\tan \beta\, l} \approx -j\frac{R_0}{\beta\, l} = \frac{-j\sqrt{(L/C)}}{l\,\omega\sqrt{LC}}$$

$$Z_{in} = \frac{1}{l\,j\,\omega\,C}$$

This is impedance of a ***Capacitor***.

(2) Short circuited line ($Z_L = 0$)

In case short circuited line, $Z_L \approx 0$. The input impedance is given by

$$Z_{in} = j\,R_0 \tan \beta\, l = j\,X_{in} \qquad \text{[ed. (13), article 9.6]} \quad ...(3)$$

Here $\beta\, l$ can vary from $-\infty$ to $+\infty$. Therefore, input impedance may be purely inductive or purely capacitive depending upon the value of $\beta\, l$. A plot of $X_{in} = R_0 \tan \beta\, l$ as a function of l is shown in Fig. (6).

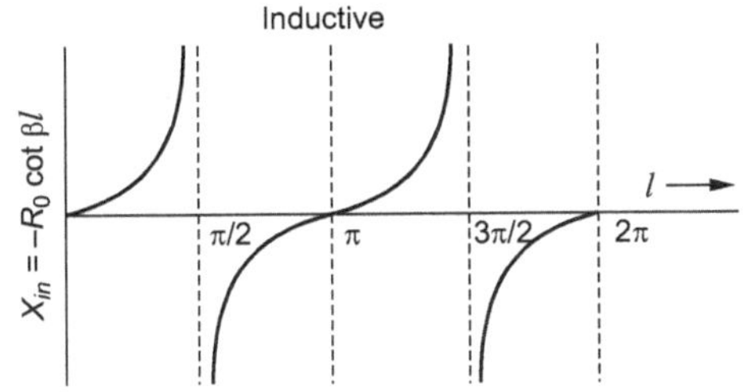

Fig. (6) Input reactance of a short circuited transmission line

For very short line *i.e.*, $\tan \beta l \approx \beta l$

$$(Z_{in})_{SC} = j R_0 \tan \beta l \approx j R_0 \beta l$$

$$\approx j R_0 \sqrt{\left(\frac{L}{C}\right)} \times \omega \sqrt{LC}\; l \approx l(j \omega L) \qquad ...(4)$$

Eq. (4) shows that input impedance is ***inductive reactance***.

(3) Half-wave transmission line ($l = \lambda/2$)

When the length of the line is an integer multiple of $\lambda/2$ *i.e.*, $l = n (\lambda/2)$ where $n = 1, 2, 3, ...$, then

$$\beta l = \frac{2\pi}{\lambda} \times \left(\frac{n\lambda}{2}\right) = n\pi$$

Now $\tan \beta l = 0$

Therefore, the input impedance

$$(Z_i)_{\lambda/2} = Z_L$$

Thus, a half wave lossless line transforms the load impedance to input impedance without any change.

(4) Quarter-wave transmission line ($l = \lambda/4$)

When the length of line is an odd multiple of $\lambda/4$ *i.e.*, $l = (2n - 1)\lambda/4$, where $n = 1, 2, 3, ...$, then

$$\beta l = \frac{2\pi}{\lambda}(2n-1)\frac{\lambda}{4} = (2n-1)\frac{\pi}{2}$$

Now $$\tan \beta l = \tan\left[(2n-1)\frac{\pi}{2}\right] \to \pm\infty$$

Therefore, $$(Z_l)_{\lambda/4} = \frac{R_0^2}{Z_L}$$

So, a *quarter wave line transmission the load impedance to the input terminals as its inverse multiplied by the square of characteristic resistance.* This is also known as ***quarter wave transformer***.

9.8 REFLECTION CO-EFFICIENT AND STANDING WAVE RATIO

Reflection Coefficient

Consider the case that power is applied by the generator at the input end of a transmission line. There are two possibilities:

(1) The transmission line is of infinite length or terminated by its characteristic impedance. Under this situation. all power is absorbed by the load at far end. There is *no reflected wave.*

(2) The transmission line is of finite length or terminated by the impedance other than its characteristic impedance. In this case, some power is absorbed at the load end and the remaining power is reflected back to the generator. Therefore, reflected waves are set up. There is a loss of power due to reflection.

The incident and reflected waves combine to produce standing waves of current and voltage along the transmission line. The resultant voltage (or current) at any point along the line will be a phasor sum of incident and reflected wave components. At certain points, the two waves interfere in such a way that the resultant voltage (or current) becomes maximum. On the other hand, at certain points, the two waves interfere in such a way that resultant voltage (or current) becomes minimum. A plot of resultant voltage or current along the length of line is termed as standing waves of voltage and current as shown in Fig. (7).

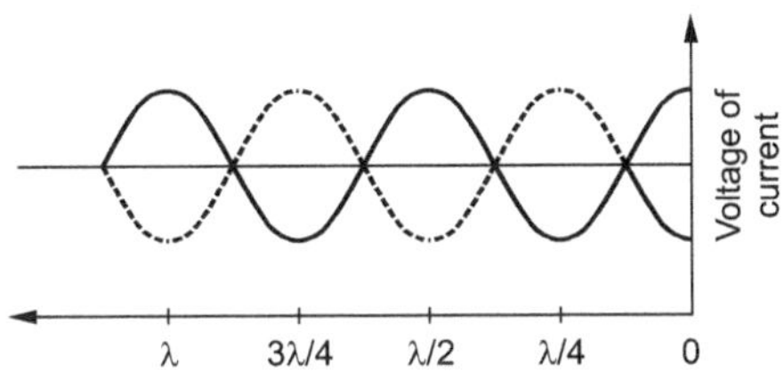

Fig. (7) Showing standing wave

Consider a transmission line of length l terminated in a load impedance different from characteristic impedance Z_0. The line voltage, current and input impedance are expressed as

$$V_L = V_0^+ e^{-\gamma l} + V_0^- e^{\gamma l} \qquad \text{...(1)}$$

$$I_L = \frac{1}{Z_0}[V_0^+ e^{-\gamma l} - V_0^- e^{\gamma l}] \qquad \text{...(2)}$$

$$Z_L = \frac{V_L}{I_L} = Z_0 \left[\frac{V_0^+ e^{-\gamma l} + V_0^- e^{\gamma l}}{V_0^+ e^{-\gamma l} + V_0^- e^{\gamma l}}\right] \qquad \text{...(3)}$$

The voltage reflection coefficient Γ_L is defined as the ratio of voltage of reflected wave to the voltage of incident wave at the load, *i.e.*,

$$\Gamma_L = \frac{V_0^- e^{\gamma l}}{V_0^+ e^{-\gamma l}} \qquad \text{...(4)}$$

To obtain the value of Γ_L. we proceed as follow:

From eq. (3),
$$Z_L = Z_0 \left(\frac{V_0^+ e^{-\gamma l}}{V_0^+ e^{-\gamma l}}\right) \left[\frac{1 + (V_0^- e^{\gamma l} / V_0^+ e^{-\gamma l})}{1 - (V_0^- e^{\gamma l} / V_0^+ e^{-\gamma l})}\right]$$

or
$$Z_L = Z_0 \left[\frac{1+\Gamma_L}{1-\Gamma_L}\right]$$

or
$$Z_L (1 - \Gamma_L) = Z_0 (1 + \Gamma_L)$$
$$Z_L - Z_L \Gamma_L = Z_0 + Z_0 \Gamma_L$$
$$Z_L - Z_0 = \Gamma_L (Z_L + Z_0)$$

∴
$$\boxed{\Gamma_L = \left[\frac{Z_L - Z_0}{Z_L + Z_0}\right]} \qquad \text{...(5)}$$

Eq. (5) gives the voltage reflection coefficient at the load end. Similarly, the current reflection coefficient can be obtained.

At another point x where $x' = l - x$, the voltage reflection coefficient is given by

$$\Gamma_{x'} = \frac{V_0^- e^{\gamma x}}{V_0^- e^{-\gamma x}} = \frac{V_0^- e^{\gamma l} \cdot e^{-\gamma x'}}{V_0^+ e^{-\gamma l} \cdot e^{+\gamma x'}} = \Gamma_L e^{-2\gamma x'}$$

$$\therefore \qquad \boxed{\Gamma_{x'} = \Gamma_L e^{-2\gamma x'}} \qquad ...(6)$$

Special Cases: Using eq. (5), we consider the following special cases

(*i*) If the line is terminated at characteristic impedance, $Z_L = Z_0$, then

$$\Gamma_L = \frac{Z_0 - Z_0}{Z_0 + Z_0} = 0$$

There is no reflected wave.

(*ii*) If the line is short circuited at the receiving end, $Z_L = 0$, then

$$\Gamma_L = \frac{0 - Z_0}{0 + Z_0} = -1$$

This shows that reflection coefficient is negative, *i.e.*, reflected wave is reversed.

This shows that reflected wave is identical to incident wave but travelling in opposite direction.

(*iii*) If the line is open circuited at the receiving end, $Z_L = \infty$, then

$$\Gamma_L = \frac{1 - (Z_0/Z_L)}{1 + (Z_0/Z_L)} = \frac{1-0}{1+0} = 1$$

This shows that reflected wave is identical to incident wave and sign.

(*iv*) If the load impedance is greater or less than characteristic impedance Z_0, then reflection coefficient is always less than one.

For example, if $Z_L = 2\,Z_0$, then

$$\Gamma_L = \frac{2\,Z_0 - Z_0}{2\,Z_0 + Z_0} = \frac{Z_0}{3Z_0} = 0.33$$

If $Z_1 = Z_0/3$, then $\quad \Gamma_L = \dfrac{(Z_0/3 - Z_0)}{(Z_0/3 + Z_0)} = -0.5$

It is obvious from the above cases that maximum reflection occurs when either the line is short circuited or open circuited.

Standing Wave Ratio (SWR)

We have already mentioned that if a line is terminated by load other than characteristic impedance, a reflected wave exists on the line. The superposition of incident and reflected wave given rise a *standing wave* or non-progressive wave as shown in Fig. (8)

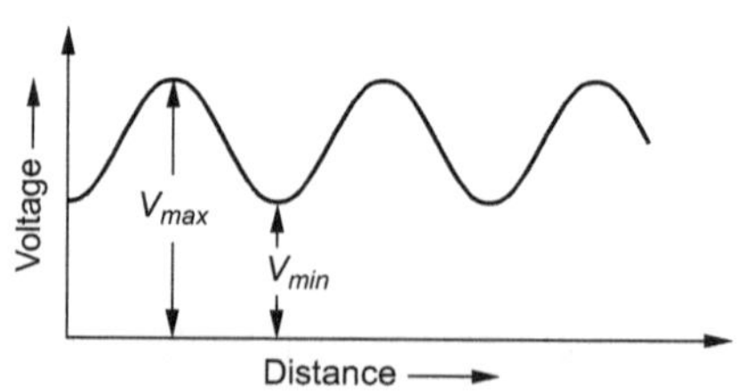

Fig. (8) Showing a standing wave

The ratio of maximum to minimum magnitude of voltage on the line is called as *voltage standing wave ratio, i.e.,*

$$S = \frac{\text{Maximum magnitude of voltage}}{\text{Minimum magnitude of voltage}} = \frac{|V_{max}|}{|V_{min}|} \quad ...(7)$$

Similarly, the *current standing wave ratio* is defined as

$$I = \frac{\text{Maximum magnitude of current}}{\text{Minimum magnitude of current}} = \frac{|I_{max}|}{|I_{min}|} \quad ...(8)$$

The significance of SWR is that it gives a measure of mismatch between load and transmission line. For a matched line, SWR is unity. For a mismatched line SWR is greater than unity. Therefore, higher is the value of SWR, the greater is mismatch between load and line. It should be noted that a high standing wave ratio on the line is undesirable because it causes a larger power loss and noise.

We know that for a corrected transmission line, the voltage maxima occurs at those points where the incident and reflected waves are in phase. The two waves are added up at these points and gives the maximum voltage. On the other hand, at those points where the two waves are out of phase they give minimum voltage. Let V_i and V_r be the r.m.s values of incident and reflected voltages. Then

$$|V_{max}| = |V_i| + |V_r| \quad \text{and} \quad |V_{min}| = |V_i| - |V_r|$$

The voltage standing wave ratio is given by

$$S = \frac{|V_{max}|}{|V_{min}|} = \frac{|V_i| + |V_r|}{|V_i| - |V_r|}$$

$$\boxed{S = \frac{1 + |V_r|/|V_i|}{1 - |V_r|/|V_i|} = \frac{1 + |\Gamma_L|}{1 - |\Gamma_L|}} \quad ...(9)$$

where Γ is reflection coefficient.

From eq. (9),

$$S(1 - |\Gamma_L|) = 1 + |\Gamma_L| \quad \text{or} \quad S - S|\Gamma_L| = 1 + |\Gamma_L|$$

or $$(S - 1) = |\Gamma_L| (S + 1)$$

or $$\boxed{|\Gamma_L| = \frac{(S-1)}{(S+1)}} \quad ...(10)$$

Eq. (10) gives the relation between voltage standing wave ratio VSWR and reflection coefficient Γ_L.

In case of real (resistive load), we can also express SWR in terms of load impedance Z_L and characteristic impedance Z_0 in the following way:

$$SWR = \frac{1 + |\Gamma L|}{1 - |\Gamma_L|} = \frac{1 + \dfrac{(Z_L - Z_0)}{(Z_L + Z_0)}}{1 - \dfrac{(Z_L - Z_0)}{(Z_L + Z_0)}} = \frac{Z_L + Z_0 + Z_L - Z_0}{Z_L + Z_0 - Z_L + Z_0} = \frac{2Z_L}{2Z_0} = \frac{Z_L}{Z_0}$$

$\therefore$ $$\boxed{SWR = \frac{Z_L}{Z_0}} \quad ....(11)$$

EXAMPLE 1. *An open wire transmission line with $Z_0 = 300\ \Omega$ is terminated by $Z_L = 600\ \Omega$. Find the reflection coefficient and standing wave ratio.*

Solution. The reflection coefficient is given by

$$\Gamma_L = \frac{Z_L - Z_0}{Z_L + Z_0} = \frac{600-300}{600+300} = \frac{300}{900} = \frac{1}{3} = \mathbf{0.33}$$

$$SWR = \frac{1+|\Gamma_L|}{1-|\Gamma_L|} = \frac{1+0.33}{1-0.33} = \frac{1.33}{0.67} = \mathbf{1.9}$$

EXAMPLE 2. *An ideal lossless transmission line of $Z_0 = 100\ \Omega$ is connected to unknown Z_L. If S.W.R. = 5, find the reflection coefficient.*

Solution. $$S = \frac{Z_L}{Z_0} \quad \text{or} \quad Z_L = S \times Z_0$$

$\therefore$ $$Z_L = 5 \times 100 = 500\ \Omega$$

Now, Reflection coefficient $\Gamma_L = \dfrac{Z_L - Z_0}{Z_L + Z_0}$

$\therefore$ $$\Gamma_L = \frac{500-100}{500+100} = \frac{400}{600} = \frac{2}{3} = \mathbf{0.66}$$

EXAMPLE 3. *A transmission line with a characteristic impedance of 400 Ω is terminated in a purely resistive load. While making standing wave ratio measurement, the metre reads a maximum voltage of 8.0 μV and a minimum of 5 μV. What should be the load resistance?*

Solution. $$SWR = \frac{|V_{max}|}{|V_{min}|} = \frac{Z_L}{Z_0}$$

or $$Z_L = Z_0 \times \frac{|V_{max}|}{|V_{min}|}$$

$\therefore$ $$Z_L = \frac{400 \times 8}{5} = 640\ \Omega$$

9.9 IMPEDANCE MATCHING

Matched and Unmatched Transmission Line

A transmission line is said to be matched when the load impedance is equal to the characteristic impedance. In this situation maximum power is transmitted to the load with minimum losses. Therefore, for a matched line

$$Z_{in} = Z_0$$

and reflection coefficient $\Gamma_L = 0$, SWR (S) = 1.

If the above condition is not satisfied in case of a transmission line, *the line is said to be mismatched.* Mismatched loads and junctions will result in echoes and will distort the information carrying signal. Therefore, it is necessary to match the transmission line properly. For this purpose, various methods are used. Few methods are discussed below:

1. Impedance matching by quarter-wave transformer.

The mismatching between load Z_L and characteristic impedance Z_0 can be removed by inserting a line of length λ/4 with characteristic impedance Z_0' between original transmission line and load as shown in Fig. (9).

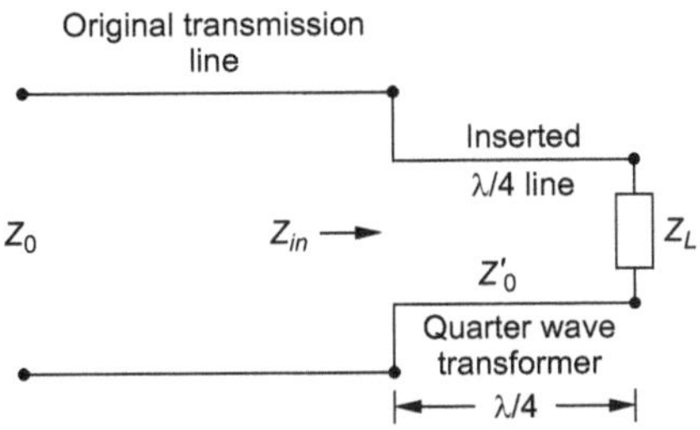

Fig (9) Load matching using λ/4 transformer

The λ/4 transmission line is called as quarter wave transformer because the line acts as a transformer. Let us calculate the characteristic impedance Z_0' of the line.

For a lossless line, the input impedance as observed by looking into quarter wave transformer is

$$Z_{in} = Z_0' \left[\frac{Z_L + j Z_0' \tan \beta l}{Z_0' + j Z_L \tan \beta l} \right]$$

For a λ/4 line, $l = \lambda/4, \quad i.e., \beta l = \frac{2\pi}{\lambda} \times \frac{\pi}{4} = \frac{\pi}{2}$

So $\tan \beta l = \tan \frac{\pi}{2} = \infty$

Now $$Z_{in} = Z_0' \left[\frac{(Z_L / j \tan \beta l) + Z_0'}{Z_0' / j \tan \beta l + Z_L} \right] = Z_0' \left[\frac{0 + Z_0'}{0 + Z_L} \right] = \frac{(Z_0')^2}{Z_L}$$

∴ $(Z_0')^2 = Z_L Z_{in}$

or $Z_0' = \sqrt{(Z_{in} Z_L)} = \sqrt{(Z_0 Z_L)}$ $[\because Z_{in} = Z_0]$

Z_0' is quarter wave transformer impedance. Here Z_0', Z_0 and Z_L are all real.

A complex load impedance cannot be matched by quarter wave transformer.

Single Stub Matching

The impedance matching can also be achieved by connecting sections of open or short circuited lines called stub or tuning stub in shunt with the main line at some point or points. This method of achieving matching is called as *stub matching*. There are basically two types of stub matching, *i.e.*, (*i*) single stub matching and (*ii*) double stub matching. Here we shall discuss single stub matching using short circuited stub.

In this method, a short circuited stub of length l is connected in parallel with the main line at some distance d from the load as shown in Fig. (10). The short circuited stubs are preferred as compared to open circuited stub because open circuited stub radiates some energy at high frequencies.

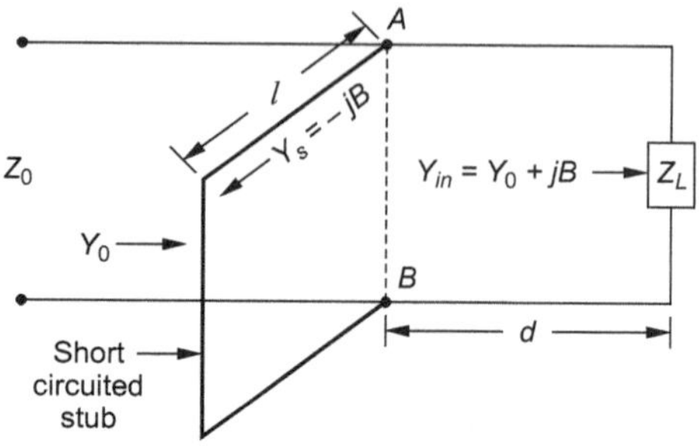

Fig. (10) Matching with a single stub

Let, Line characteristic admittance = $Y_0 = 1/Z_0$

where Z_0 = characteristic impedance

Admittance of the stub = $Y_S = -j\,B$

where B = susceptance of the stub

Line input admittance at $AB = Y_{in} = Y_0 + j\,B$

where Y_0 is real part while $j\,B$ is imaginary part.

The principle of this technique is to connect a short circuited stub of appropriate length l and at a certain distance d from load end in parallel such that *its input admittance is equal to the negative of imaginary part of line input admittance to the right of stub*. Now, the line admittance towards the left of the stub

$$= (Y_0 + j\,B) - j\,B = Y_0$$

Therefore, match is achieved.

Theoretically, we can add inductors or capacitors in parallel with the transmission line but these elements are usually too lossy at the desired frequency. So, their use is avoided.

Double Stub Matching

A single stub cannot be placed physically in an ideal location. So, a double stub technique is employed. In this method, two short circuited parallel stubs are attached to the transmission line at fixed positions and the lengths are adjusted in such a way to provide perfect matching. The spacing between two stubs is usually (λ/8, 3λ/8, 5λ/8, etc). The separation of 3λ/8 is an optimum. The double stub matching is shown in Fig. (11).

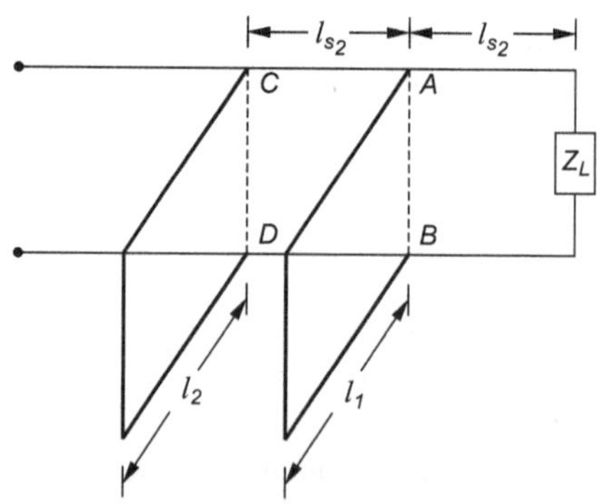

Fig. (11) Showing double stub matching

9.10 SMITH CHART

Smith chart is a graphical method used to solve the problems of transmission line. Smith chart shows, in a very simple way, the standing wave ratio (*SWR*) corresponding to a given load impedance and also line impedance at any desired points) as one moves along the line. It consists of a group of concentric circles (constant resistance and conductance circles). On this chart, the normalized values of impedance and admittance are used. The normalized impedance z_l is the ratio of load impedance Z_L to characteristic impedance Z_0 (*i.e.*, $z_l = Z_L/Z_0$) and normalized admittance $Y_l = 1/Z_L$.

Construction

The Smith chart consists of two sets of perpendicular circles. One set indicates the ratio (Z_L/Z_0) or (R_L/Z_0) where R_L is the resistive component of line impedance and Z_0 is the characteristic impedance which is also resistive for loss-less line. The second set of circles gives the ratio jX/Z_0, where X is the reactive component of the line impedance. First of all, we shall determine the equation of these circles.

The reflection coefficient of a transmission line is given by

$$\Gamma_L = \frac{Z_L - Z_0}{Z_L + Z_0} = (\Gamma_L)_r + j(\Gamma_L)_i \qquad \text{...(1)}$$

where $(\Gamma_L)_r$ and $(\Gamma_L)_i$ are real and imaginary parts of complex reflection coefficient.

In simple form they are expressed as $(\Gamma)_r$ and $(\Gamma)_i$ respectively.

The normalized impedance z_l is given by

$$z_l = \frac{Z_L}{Z_0} = r + jx \qquad \text{...(2)}$$

From eq. (1),

$$\Gamma_L = \frac{(Z_L/Z_0) - 1}{(Z_L/Z_0) + 1} = \frac{z_l - 1}{z_l + 1}$$

or

$$(\Gamma)_r + j(\Gamma)_i = \frac{(r + jx) - 1}{(r + jx) + 1}$$

Solving, we get

$$r + jx = \frac{(1 + \Gamma_r) + j\Gamma_i}{(1 - \Gamma_r) - j\Gamma_i} \qquad \text{...(3)}$$

Multiplying and dividing by $(1 - \Gamma_r) + j\Gamma_i$ on upper and lower right side, we get

$$r + jx = \frac{(1 + \Gamma_r) + j\Gamma_i}{(1 - \Gamma_r) - j\Gamma_i} \times \frac{(1 - \Gamma_r) + j\Gamma_i}{(1 - \Gamma_r) + j\Gamma_i}$$

or

$$r + jx = \frac{1 - \Gamma_r^2 - \Gamma_i^2 + 2j\Gamma_i}{(1 + \Gamma_r)^2 + \Gamma_i^2}$$

or

$$r + jx = \frac{1 - \Gamma_r^2 - \Gamma_i^2}{(1 + \Gamma_r)^2 + \Gamma_i^2} + j\frac{2\Gamma_i}{(1 - \Gamma_r)^2 + \Gamma_i^2} \qquad \text{...(4)}$$

Comparing the real and imaginary parts on both sides, we get

$$r = \frac{1-\Gamma_r^2-\Gamma_i^2}{(1+\Gamma_r)^2+\Gamma_i^2}$$

and

$$x = \frac{2\Gamma_i}{(1-\Gamma_r)^2+\Gamma_r^2} \qquad ...(5)$$

Rearranging the equation of r, we get

$$\left[\Gamma_r - \frac{r}{(1+r)}\right]^2 + \Gamma_i^2 = \left(\frac{1}{1+r}\right)^2 \qquad ...(6)$$

Eq. (6) is identical to the equation of a circle of the forms

$$(x-g)^2+(y-h)^2 = a^2 \qquad ...(7)$$

with radius and centre (g, h).

Comparing eqs. (6) and (7), we can say that the centre of the constant resistance circles is $\left(\frac{r}{1+r}, 0\right)$ with radius $\frac{1}{(1+r)}$. The constant r circle shown in Fig. (12).

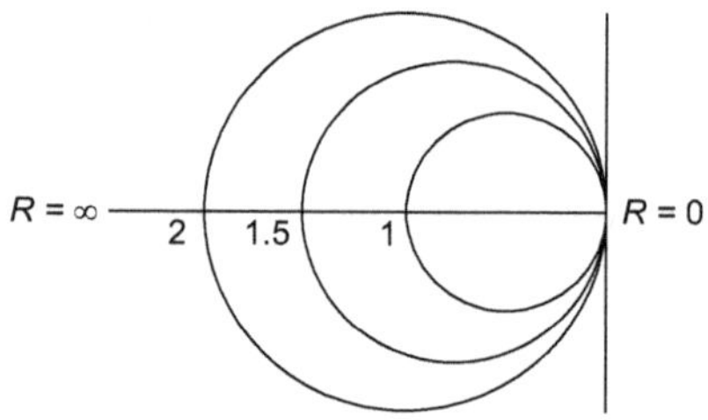

Fig. (12) Constant *r* circles

Similarly, arranging the equation of x, we get

$$(\Gamma_r - 1)^2 + \left(\Gamma_i - \frac{1}{x}\right)^2 = \left(\frac{1}{x^2}\right) \qquad ...(8)$$

Eq. (8) represents the constant reactance circle with centre $\left(1, \frac{1}{2}\right)$ and radius $\frac{1}{x}$. The constant x circles are shown in Fig. (13).

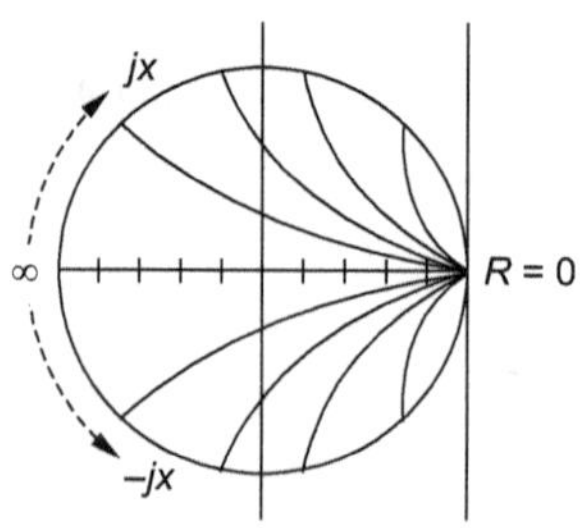

Fig. (13) Constant ***x*** circles

Figs. (12) and (13) shows the Smith charts.

Applications of Smith Chart

Smith chart has wide applications in the study of transmission line. Few of them are:

1. To find the location of maximum voltage and minimum voltage from the load.
2. To find the standing wave ratio (SWR).
3. To find unknown load impedance from reflection coefficient.
4. To find the position and length of the stub used in impedance matching

Exercises and Problems

EXERCISE

1. What are transmission lines? Explain different types of transmission lines.
2. Derive transmission line equations.
3. Given the solution of transmission line equations.
4. Define characteristic impedance. Derive the expressions of characteristic impedance for lossless and distortionless lines.
5. Differentiate between characteristic impedance and line impedance.
6. Define propagation constant and phase velocity.
7. Derive expressions for propagation constant and phase velocity in case of lossless and distortionless lines.
8. On what factors does the input impedance of a transmission line depend?
9. Derive an expression for input impedance at ant point along a transmission line.
10. Discuss how the transmission line acts as circuit elements.
11. Write short notes on the following:
    (*a*) Reflection Coefficient (*b*) Standing wave ratio
12. Define voltage reflection coefficient. Is it same as current reflection coefficient? Explain.
13. Define standing wave ratio. How it is related to voltage and current reflection coefficients?
14. Why it is desirable to achieve an impedance match in a transmission line? Discuss how $\lambda/4$ line acts as an impedance transmission?
15. Explain the principle of impedance matching through stub.
16. Explain the different methods of impedance matching.
17. What is a Smith chart? What are its applications?

PRACTICE PROBLEMS

1. In a lossless transmission line, velocity of propagation is 2.5×10^8 m/s. Capacitance of the line is 30 pF/m, find
   (*i*) Phase constant at 200 mHz
   (*ii*) Inductance of the line
   [**Ans.** (*i*) 5.02 rad/m (*ii*) 2.1×10^{-7} henry/m]

   [**Hint.** (*i*) $v_p = \dfrac{\omega}{\beta}$ or $\beta = \dfrac{\omega}{v_p}$

   $$\therefore\ \beta = \frac{2\pi \times 200 \times 10^6}{2.5 \times 10^8} \text{ rad/m.}$$

$$(ii)\ L = \frac{1}{C v_p^2} = \frac{1}{(30\times10^{-12})\times(5.10)^2\times10^{16}}\Bigg]$$

2. An open wire transmission line has
$R = 5\ \Omega$/m, $L = 5.2 \times 10^8$ h/m, $G = 6.2 \times 10^{-3}$ mho/m, $C = 2.13 \times 10^{-13}$ F/m, frequency = 4 GHz.
Find Z_0, γ and v_p.
[**Ans.** $12.64\underline{|24.49^\circ}, 103.37\underline{|65.23^\circ}$ and 3.0×10^8 m/s]

[**Hint.**
$$v_p = \frac{1}{\sqrt{LC}} = \frac{1}{\sqrt{(5.2\times10^{-8})\times(2.13\times10^{-10})}}$$
$$\omega = 2\,\pi \times 4 \times 10^9 = 8\,\pi \times 10^9 = 2.512 \times 10^{16}\ \text{rad.}$$
$$Z_0 = \sqrt{\left(\frac{R + j\,\omega\,L}{G + j\,\omega\,C}\right)}$$
$$(R + j\,\omega\,L) = 5 + j \times 2.512 \times 10^{16} \times 5.2 \times 10^{-8} = 5 + j\,1306.24$$
$$= 1306.25\underline{|89.78^\circ}$$
$$(G + j\,\omega\,C) = 6.2 \times 10^{-3} + j\,5.35 = 8.18\underline{|24.49}$$
$$\gamma = \sqrt{(R + j\,\omega\,L)(G + j\,\omega\,C)}\ ]$$

3. An air line has characteristic impedance of 75 Ω and a phase constant 3 rad/m at 100 MHz. Calculate the capacitance and inductance of the line per metre.
[**Ans.** 63.66 pF/m, 358.1 nH/m.]
[**Hint.** Air is line treated as lossless line, *i.e.*, $R = 0$, G = 0 and $\alpha = 0$]

Now $Z_0 + \sqrt{\left(\frac{L}{C}\right)}$ and $\beta = \omega\sqrt{(LC)}$

$$\therefore \qquad \frac{Z_0}{\beta} = \frac{1}{\omega\,C}$$

Given $Z_0 = 75\ \Omega$, $\beta = 3$ rad/m and $\omega = 2\pi \times 100 \times 10^6$ rad/s]

4. An RF extension has $Z_0 = 600\ \Omega$ is terminated with Z_L $(400 + j\,200)\ \Omega$. Find standing wave ratio.
[**Ans.** 1.766]

[**Hint.**
$$\Gamma_L = \frac{Z_L - Z_0}{Z_L + Z_0} = \frac{400 + j\,200 - 600}{100 + j\,200} = \frac{282.8\underline{|-45^\circ}}{1019.8\underline{|11.31^\circ}} = 0.277\ \underline{|-56.31^\circ}$$
$$SWR = \frac{1 + |\Gamma_L|}{1 - |\Gamma_L|} = \frac{1 + 0.277}{1 - 0.277}\Bigg]$$

5. Calculate the characteristic impedance of quarter wave transformer if a 120 Ω load is to be matched to a 75 Ω line.
[**Ans.** 95 Ω]

[**Hint.** $Z_0' = \underline{|Z_0.Z_L} = \sqrt{120 \times 75}$]

Index

T

U

V

W

Kolibri 360 GmbH
Ferdinand-Jühlke-Straße 7
99095 Erfurt, Deutschland
produktsicherheit@kolibri360.de